Chemistry
A Self-Teaching Guide

Wiley Self-Teaching Guides teach practical skills in mathematics and science. Look for them at your local bookstore.

Other Science and Math Wiley Self-Teaching Guides:

Science

Basic Physics: A Self-Teaching Guide, Third Edition by Karl F. Kuhn

Biology: A Self-Teaching Guide, Third Edition by Steven D. Garber

Math

All the Math You'll Ever Need: A Self-Teaching Guide by Steve Slavin

Practical Algebra: A Self-Teaching Guide, Second Edition by Peter H. Selby and Steve Slavin

Quick Algebra Review: A Self-Teaching Guide by Peter H. Selby and Steve Slavin

Quick Business Math: A Self-Teaching Guide by Steve Slavin

Quick Calculus: A Self-Teaching Guide, Second Edition by Daniel Kleppner and Norman Ramsey

Chemistry
A Self-Teaching Guide
Third Edition

Richard Post, M.A.
Assistant Dean Emeritus, Ohio University

Chad A. Snyder, Ph.D.
Professor of Chemistry, Liberty University

Clifford C. Houk, Ph.D.
Professor of Health Science Emeritus, Ohio University

 JOSSEY-BASS™
A Wiley Brand

Published by Jossey-Bass
A Wiley Brand
111 River St, Hoboken NJ 07030
www.josseybass.com

Jossey-Bass books and products are available through most bookstores. To contact Jossey-Bass directly call our Customer Care Department within the U.S. at 800-956-7739, outside the U.S. at 317-572-3986, or fax 317-572-4002.

Wiley also publishes its books in a variety of electronic formats and by print-on-demand. Some material included with standard print versions of this book may not be included in e-books or in print-on-demand. For more information about Wiley products, visit www.wiley.com.

Library of Congress Cataloging-in-Publication Data

ISBNs: 978-1-119-63256-6 (paperback), 978-1-119-63265-8 (ePDF), 978-1-119-63262-7 (ePub)

Cover Design: Wiley
Cover Image: © science photo/Shutterstock

Printed in the United States of America

THIRD EDITION
SKY10020309_080620

Contents

v

Appendix

Useful Tables in This Book

Preface

In our years of teaching the fundamental concepts of chemistry to students with widely divergent backgrounds, levels of preparation, career goals, and motivation, the most frequently asked question by those students has been, "Do you have something that I can study on my own?" followed immediately by, "I need some other review material" or "This is the first time I have encountered this stuff, so I need to start from scratch" or "It has been 3 years since my high school chemistry course. I need something to refresh my memory."

This book has been written to meet such needs. It can stand alone as a "first look" at chemistry or may be used as a supplement to any of the many excellent textbooks or methods of instruction currently in use. The material presumes no previous exposure to chemistry and requires only simple algebra.

There are no secrets. Each chapter includes an introductory statement, a list of objectives, and the main teaching section, which consists of frames of tutorial material with constant practice exercises. Each chapter closes with a self-test. You can use this self-test to assess whether you have mastered the chapter well enough to continue and to identify weaknesses that require additional study. Finally, some chapters include an end-of-chapter or transitional story connecting the material to a relevant topic in chemistry.

The topics presented are usually covered early in a general introductory course. The third edition also contains a new chapter on organic chemistry consistent with the material found in general chemistry textbooks. We have minimized "heavy" theoretical discussions, while emphasizing descriptive and practical concepts. There is enough theoretical explanation to provide a basis for understanding the material but not so much that you will get bogged down trying to work through the book.

Introduction

Chemistry: A Self Teaching Guide is unlike the ordinary textbook. It is designed to be completely self-instructional, requiring no chemistry background. The previous editions have been thoroughly tested as a successful means for self-instruction in chemistry for thousands of students. The book can also be used as a supplementary text for any general chemistry course. Each chapter is divided into objectives, an interactive tutorial study section, a self-test, and test answers.

- *Objectives*. By examining the chapter objectives, you can determine what information is contained in each chapter. If you already know the material, take the self-test at the end of the chapter. Review those questions you missed by checking the frame references given with the answer to each question.

- *Tutorial study frames*. The body of each chapter is divided into numbered frames. Each frame contains new information, a problem, or an example of a concept with one or more questions for you to answer. Answers for the questions in each frame are given immediately below the questions. Years of educational research have proven this method of immediate reinforcement with the correct answer to be the most effective and efficient means of learning for self-instruction. While having the answer just below each question assures immediate feedback to reinforce learning, it also requires discipline on your part to think about the answer before viewing. We suggest using a bookmark such as an index card as you go down each page, covering the answer below while you critically think about the answer or solve a problem. Each question and answer frame is crafted as a small step, like each step on a staircase, designed to lead to a fuller understanding of a concept in chemistry.

- *Self-test*. The self-test at the end of each chapter will help you to determine whether you have mastered the chapter material. After completing the chapter, take the test. Refer back to the chapter only if you need formulas or tables to answer specific questions. Compare your answers with those given immediately following the test. If your answers do not agree with the printed ones, review the appropriate frames cited after each answer.

The authors assume no prerequisites except simple high school algebra. However, each chapter builds upon the information provided in previous chapters, so we recommend that the chapters be covered in sequence.

Although we have tried to make this book as useful as possible to the student, any suggestions for improving future revisions would be appreciated. Please address your comments to:

Editor, Self-Teaching Guides
John Wiley & Sons, Inc.
111 River Street
Hoboken
New Jersey 07030
USA

Acknowledgments

We wish to thank all those students who used the first two editions and took the time to write to us and to John Wiley & Sons, Inc. to express their gratitude for helping them understand chemistry and successfully complete a required chemistry course. They were high school, nursing school, community college, technical school, and university students of all ages. It is because of them we have written this third edition.

We also wish to thank those faculty who thought enough of the book that they adopted it for classroom use or recommended it as a self-paced, "second opinion" study guide.

We wish to thank our respective universities, all of our editors for all three editions—especially the late Judy V. Wilson, who had the vision for both the entire *Self-Teaching Guide* series as well as this book—and our publisher, John Wiley & Sons. We also thank the Wiley editorial and production staff for their very thorough editorial comments and enthusiastic encouragement during the preparation of this manuscript.

Chad Snyder would like to thank his wife and children for their love and support through this process. Authors Post and Houk likewise wish to thank their families for their encouragement, patience, and support in the development of this book in its current and previous editions.

APPRECIATING THE CONNECTIONS

The history of science and technology is often based upon a series of individual discoveries and historical events which at first seem unconnected, but in hindsight represent a chain of events that building upon one another result in a new discovery or idea. Scientists often describe this as, "We stood upon the shoulders of giants," acknowledging appreciation for their colleagues and forebears who set the groundwork for their discoveries.

This chain of events is also true of much of history. This book grew out of a need to develop self-instruction for a few concepts in chemistry. The initial authors had no plans for a book. They were directly supported and encouraged in developing their self-instructional materials and related research efforts by their university, where they would continue to spend the major part of their academic careers. Thus without the support of that university, this book would not exist.

By historical connection, that university would itself not exist without an idea presented in 1787, the land grant. As the first university in what was then known as the Northwest Territory, the land west of the Ohio River, Ohio University directly stems from the Northwest Ordinance of 1787, one of the primary documents of American history. The ordinance of Congress called for a public university as part of the settlement and eventual statehood of the Northwest Territory stipulating, "Religion, morality and knowledge being necessary to good government and the happiness of mankind, schools and the means of education shall forever be encouraged." That ordinance and that wording form the basis for a large historical landmark displayed at Ohio University's class gateway.

The authors' self-instructional material eventually came to the attention of Judy V. Wilson, who developed the *Self-Teaching Guide* series for publishers John Wiley & Sons. With the book now in its third edition, as you learn about the interesting and fundamental science of chemistry, you may find yourself becoming part of the chain of events. Let the authors and publisher know how this book contributed to your career through your study of chemistry. Wishing you success.

RP, CS, CCH

How to Use This Book

It is important to note that this book should be used as a tutorial. The content is designed to be interactive. Each separate block of information, called a "frame," ends in a question needing an answer or problem to solve. Each question and answer frame is crafted as a small step, like each step on a staircase, designed to lead to a fuller understanding of a concept in chemistry. Although the answer can be found directly below that block of information, that question or problem is for you to answer before going on to read more. Answering the question or solving the problem will require some critical thinking and application of the material just learned. The answer just below the frame will then provide immediate feedback.

That tutorial technique with immediate feedback has been proven to be a very effective means of learning backed by a great deal of research on instruction. Just cover the answer with an index card or bookmark, think about what you have just read, and provide your own answer. Then uncover the printed answer and compare your answer with that of the book. If your answer and the book answer agree, then go on to the next frame of information. If your answer does not agree, reread the frame and try to determine why.

The information presented has been carefully sequenced for step-by-step learning but requires the discipline for you to answer before checking and moving on to each following step or frame. Each frame is built upon preceding frames. Therefore, if the material is new to you, go through the frames in sequence. Skipping ahead will cause you to miss important information or practice.

Thousands of students have successfully learned the principles of chemistry through the proper use of this book. The authors wish you success as you join their ranks.

1 Atomic Structure, Periodic Table, Electronic Structure

There is a *smallest* unit of substance. This smallest unit may be only a single atom or a group of atoms chemically joined together.

This chapter deals with the structure of the atom, which is the very backbone of chemistry. In this chapter we introduce the three basic subatomic particles in an atom, their arrangement in the atom, and the similarities of this arrangement revealed by the position of the elements in the periodic table. A clear understanding of this chapter will give you a sound basis for learning chemistry.

OBJECTIVES

After completing this chapter, you will be able to

- define, describe, or illustrate: proton, neutron, electron, atom, nucleus, atomic number, shell, orbital, subshell, alkali metal, noble gas, halogen, alkaline earth, period, group, family, oxide, ductile, malleable, metal, nonmetal, metalloid, and Bohr model of an atom;

- determine the numbers of protons, neutrons, and electrons in a neutral atom when given its mass number and atomic number;

- compare and contrast the three fundamental particles in an atom according to mass and charge;

- determine the maximum number of electrons any given shell can hold;

- determine the maximum number of orbitals in any given shell;

- write the electron configuration for any element;

- determine what element is represented when given its electron configuration;

- use the periodic table to locate different families of elements and determine whether an element is a metal, nonmetal, or metalloid.

1 An **atom**, the smallest unit of an element, is composed primarily of three fundamental particles: **electrons**, **protons**, and **neutrons**. The combination of these particles in an atom is distinct for each element. An atom of the element radon is composed primarily of a specific combination of what three basic particles?_____

Answer: electrons, protons, neutrons (any order)

2 Let's forget about neutrons for the moment and consider just electrons and protons. Each atom of the same element has the same combination of protons and electrons. An atom of the element hydrogen in outer space has (the same, a different) _____ combination of electrons and protons as that of an atom of hydrogen on earth.

Answer: the same

3 Each element has a unique combination of protons and electrons in its atoms. The combination of electrons and protons in an atom of one element is different from that in an atom of any other element. Since each element has a known unique number of protons and electrons in its atoms, would it be possible to identify an element if you know the number of protons and electrons in its atoms? _____

Answer: yes (if you could compare the number of electrons and protons in your unknown atom with a list of the electrons and protons in atoms of each known element)

4 Protons are particles with a positive (plus) charge. Electrons are particles with a negative (minus) charge. Unless otherwise stated, an atom is assumed to be **neutral**, with the positive and negative charges being equal. In any neutral atom, the number of electrons (having a negative charge) is always equal to the number of protons (having a positive charge).

An oxygen atom contains eight protons. We assume the atom to be neutral. How many electrons must it have? _____

Answer: eight

5 An atom contains 10 electrons. How many protons does it contain? _____

Answer: 10

6 Each element has a unique number of electrons and protons in its atoms. Since the number of electrons in a neutral atom is equal to the number of protons, do you think we can identify an element if we know just the number of protons in its atoms? _____

Answer: yes (if we could compare the number of protons in an atom of the unknown element with a list or table of the number of protons in atoms of every known element)

7 The **periodic table** is a very useful table describing the atoms of every known element. A complete periodic table is included in Appendix (see page 399) of this book. Each box in the periodic table represents an element. The one- or two-letter symbol in each box is a shorthand notation used to represent a neutral atom of an element. The symbol "C" represents a neutral atom of the element carbon. The symbol "He" represents a neutral atom of the element helium.

The number of protons in an atom is listed above each symbol. (Ignore the number underneath the symbol, called the "atomic weight," for the time being as you will get this information from the periodic table. More on that to come.)

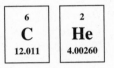

An atom of carbon has six protons. How many protons does an atom of helium have? _____

Answer: two

Note: The table of atomic weights, located in the Appendix along with the periodic table, lists all the elements alphabetically and gives the symbol for each. (Ignore the atomic weights for now.) You'll be using the periodic table and the table of atomic weights throughout this book.

8 The number of protons in an atom of an element is called its **atomic number**. What is the atomic number of the element helium (He)? _____

Answer: 2

9 The element iron (Fe) has an atomic number of 26. How many protons does an atom of iron contain? _____

Answer: 26

10 A neutral atom of iron contains how many electrons? _____

Answer: 26 (the same as the number of protons)

11   Using the periodic table, determine the number of electrons in a neutral atom of zinc (Zn). _____

Answer: 30 (the same as the number of protons)

BOHR ATOMIC MODEL

 12   A Danish physicist, Niels Bohr, came up with a model that pictured the atom with a **nucleus** of protons in the center and electrons spinning in an orbit around it (similar to the movement of the planets around the sun). The following Bohr model contains one orbiting electron and a nucleus of one proton.

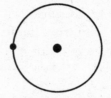

What is the atomic number of the element represented? _____

What element is represented? _____

Answer: 1 (The atomic number equals the number of protons.); hydrogen (H)

13   An electron always carries a negative charge. A proton carries a charge exactly opposite that of the electron. A proton must therefore have a (negative, positive, neutral) _____ charge.

Answer: positive

14   An electron has very little mass when compared to a proton. It takes about 1836 electrons to equal the weight of just one proton. In a hydrogen atom consisting of just one proton and one electron, the greatest proportion by weight is accounted for by the (electron, proton) _____.

Answer: proton (The proton accounts for about 99.95% of the weight of a hydrogen atom and the electron 0.05%.)

15   The element helium (He), represented by the Bohr model below, has an atomic number of _____.

Answer: 2

16   The neutralatom of He contains how many protons? _____
How many electrons? _____

Answer: two; two

17   The weight of an atom of helium is not totally accounted for by the protons and electrons. A third subatomic particle, the **neutron**, is responsible for the additional weight. The neutral atoms of all elements except the most common form of the element hydrogen have one or more neutrons in the nucleus of their atoms. The diagram below shows the neutrons in the corrected Bohr model of helium.

Since a neutral atom contains equal numbers of negatively charged electrons and positively charged protons, what type of electrical charge do you think is possessed by a neutron? _____ (negative, positive, no charge)

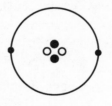

Answer: no charge (The name **neutron** means a neutral particle.)

18   A neutron is slightly heavier than a proton. Of the primary fundamental particles in an atom:

(a) which is the lightest in weight? _____

(b) which is the heaviest? _____

(c) which is between the other two in weight? _____

Answer: (a) the electron; (b) the neutron; (c) the proton

19 In the Bohr model of a lithium atom shown below, which subatomic particle(s) is (are) represented by the circular orbits shown by the larger circles? _____
Which particle(s) make(s) up the nucleus or center of the atom? _____

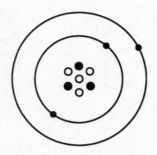

Answer: electrons; protons and neutrons

20 If the negative charge of an electron is represented by −1, the charge on the proton would be (−1, +1, neutral) _____ and the charge on the neutron would be (−1, +1, neutral) _____.

Answer: +1; neutral

21 Neutrons can be found in all atoms of all elements except the most common form of the simplest element. Identify that element. _____ (Hint: If you don't remember, reread frame 17.)

Answer: hydrogen

You have just learned the names, charges, and relative sizes of the fundamental particles that constitute an atom. You have also been shown one model representing the arrangement of these particles in an atom.

We have referred you to the periodic table and hinted that atoms with certain numbers of protons and electrons are located in a specific place in that table. You learned from your introduction to the periodic table that each atom is identified by a symbol.

We continue this chapter by looking more closely at the periodic table. You will be introduced to specific groups of elements and their physical and chemical properties as they relate to their location on the periodic table. We expand upon the use of symbols and the numbers of each particle in an atom as we prepare to study a second model of an atom.

PERIODIC TABLE

 Look at the periodic table. An atom of each element is represented by a one- or two-letter symbol, such as "C" for carbon and "Al" for aluminum. These symbols serve as shorthand notation for the elements. The shorthand symbol in each case indicates a neutral atom. The symbol "Ca" represents a neutral atom of the element calcium. Remembering the definition of a neutral atom, you know that Ca contains 20 protons and how many electrons? _____

Answer: 20 (A neutral atom contains an equal number of protons and electrons.)

 The periodic table of the elements is made up of several rows and some columns. The rows are called **periods** and the columns are called **groups**. The groups are labeled IA, IIA, IIIB, and so on. The elements Be, Mg, Ca, Sr, Ba, and Ra are included in which group? _____

Answer: Group IIA

 The elements Li, Be, B, C, N, O, F, and Ne are all members of a (group, period) _____.

Answer: period

25 Groups are often called **families** because the elements that make up the groups or families have similar chemical properties. Argon (Ar) is part of Group VIIIA. It is a rather unreactive gas. Since families or groups of elements have similar properties, would you expect krypton (Kr) to be a highly reactive gas? _____

Answer: no (All of the elements in Group VIIIA are rather unreactive.)

26 Because all Group VIIIA elements are rather unreactive and are gaseous at room temperature, they have been named the **noble gas** family. An element in Group VIIIA may be generalized by its family name as a(n) (noble gas, alkaline earth, alkali metal) _____.

Answer: noble gas

27 Group IA on the left side of the chart is often called by the family name of **alkali metals** (with the exception of hydrogen). These elements can react vigorously with water to form strong alkaline solutions. If a friend told you that aluminum (Al) was an alkali metal, would he be right or wrong? _____

Answer: wrong (Aluminum is located in Group IIIA and the alkali metals are all located in Group IA.)

28 Group IIA elements are known as the **alkaline earth** metals because the oxides of these metals (chemical compounds of the metals and oxygen) form alkaline solutions in water. The element potassium (K) can be classified as a(n) (noble gas, alkaline earth, alkali metal) _____.

Answer: alkali metal (Group IA)

29 The element Ba (barium) can be classified as a(n) (alkali metal, alkaline earth, or noble gas) _____.

Answer: alkaline earth (Group IIA)

30 An unknown element is placed in water. A vigorous reaction takes place, and the result is an alkaline solution. Of which family is the element probably a member: alkaline earth, alkali metal, or noble gas?_____

Answer: alkali metal (Alkali metals react *directly* with water to form alkaline solutions. The *oxides* of alkaline earth elements react with water to form alkaline solutions.)

31 The elements in Group VIIA are known as the **halogens**, which means "salt formers." Elements from the halogen family combine with metals to form compounds known as salts. Common table salt (NaCl) is made up of sodium (Na) and chlorine (Cl). These two elements (Na and Cl) are members of what families or groups?

Na: _____
Cl: _____

Answer: Group IA, the alkali metals (either answer is acceptable); Group VIIA, the halogens (either answer is acceptable).

32 Strontium (Sr) is an element in the _____ family. Iodine (I) is an element in the _____ family.

Answer: alkaline earth; halogen

METALS, NONMETALS, AND METALLOIDS

33 The periodic table can also be divided into just three classes of elements: the metals, the nonmetals, and the metalloids. In the periodic table, you may have noticed a steplike line. Elements to the left of this line can be classified as **metals** (with the exception of hydrogen). A friend informed you that the element Cu (copper) is a metal. Is your friend correct? _____

Answer: Yes, copper can be classified as a metal.

34 Certain properties are characteristic of metals. Metals are usually **malleable** (can be beaten into fine sheets) and **ductile** (can be drawn into wires). Gold leaf is a very thin sheet of gold. In making gold leaf, we are using what common property of metals? _____

Answer: the property of malleability

35 Besides being malleable and ductile, metals are also good conductors of heat and electricity. Copper is useful in making electrical wiring. What two metallic properties would be useful in electrical wiring? _____

Answer: The metal is a conductor of electricity and it is ductile (can be drawn into fine wires).

36 Metals have a lustrous or shiny surface and are solid at ordinary room temperature (with the exception of mercury, which is liquid at room temperature). Metal cooking utensils take advantage of what two properties of metal? _____ (conducts electricity, conducts heat, ductile, solid)

Answer: Metal conducts heat and is solid.

37 **Nonmetals** are located on the right side of the steplike line in the periodic table. Which of the following families of elements are classified as nonmetals? _____ (halogens, alkaline earths, noble gases)

Answer: halogens and noble gases

38 Nonmetals have properties almost opposite those of metals. Nonmetals are usually very brittle and do not conduct electricity or heat well. Most nonmetals are gases at ordinary temperatures, although some are liquids or solids. An unknown element exists as a gas at room temperature. How would you classify the unknown element, as a metal or as a nonmetal? _____

Answer: nonmetal (With the exception of mercury, which is liquid at room temperature, all metals are solid at ordinary room temperature.)

 An unknown element is a solid but does not conduct electricity. The element is probably a (metal, nonmetal) _____.

Answer: nonmetal (Some nonmetals are solids, although most are gases at room temperature. Nonmetals do not conduct electricity well, but metals usually do.)

40 A third category of elements is classified as **metalloids** because they don't clearly fall into either the metal or nonmetal categories. Metalloids border the steplike line on the periodic table and include elements such as silicon (Si), germanium (Ge), arsenic (As), antimony (Sb), boron (B), tellurium (Te), polonium (Po), and astatine (At). Metalloids could be expected to have some of the properties of metals and some of the properties of _____.

Answer: nonmetals

41 Which of the following elements is (are) classified as metalloid: silicon (Si), phosphorus (P), and sulfur (S)? _____

Answer: silicon

MASS AND MASS NUMBER

42 The following box represents the element sulfur (S) on the periodic table.

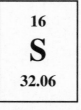

The number 16 represents the atomic number of the element sulfur and also represents the number of protons in an atom of sulfur.

Locate the element phosphorus (P) on the periodic table. The number of protons in an atom of phosphorus is _____.

Answer: 15 (same as the atomic number)

 By convention, the atomic number is often written as a subscript preceding an element's symbol. The symbol and number $_7N$ indicates nitrogen with atomic number of 7. Thus, $_{30}$Zn indicates the element zinc with an atomic number of _____.

Answer: 30

44 Almost all of the mass of an atom (more than 99.9%) is attributed to the nucleus. The nucleus is made up largely of which two fundamental particles? _____ (protons, electrons, neutrons)

Answer: protons and neutrons

This is the first time we have referred to mass in this book. **Mass** is a measure of the amount of matter. The mass of an object determines its **weight**. Weight is the effect of gravity on mass. An astronaut may weigh 180 pounds on earth and 30 pounds on the moon and be weightless in space. That person's mass, however, does not change. In the remaining chapters, we will follow the common practice of most chemistry texts and refer to the masses of objects as their weights to prevent confusion between the two terms. In this chapter and Chapter 2 we will use the term **mass**.

45 Adding together the number of protons and neutrons in the nucleus of an atom results in what is known as the mass number of the atom. The **mass number** is simply the number of protons added to the number of neutrons in an atom. Suppose an atom has a mass number of 15. The atom contains eight protons. How many neutrons does it have? _____

Answer: seven (The total number of neutrons and protons is 15, the mass number. If eight protons are present, there must be seven neutrons, since $15 - 8 = 7$.)

46 The element $_{18}$Ar has a mass number of 40. How many neutrons does its atom contain? _____

Answer: 22 ($_{18}$Ar indicates an atomic number of 18 for element Ar. An atomic number of 18 indicates 18 protons. Mass number is equal to protons plus neutrons, $40 = 18 + 22$.)

47 By convention, the mass number is often written as a superscript in front of the element symbol. $_{18}^{40}$Ar indicates the element argon with a mass number of 40 and atomic number of 18.

$_{80}^{200}$Hg indicates the element mercury with a mass number of _____ and an atomic number of _____.

Answer: 200; 80

48 $^{52}_{24}$Cr indicates chromium with _____ protons and _____ neutrons.

Answer: 24; 28

49 The element scandium (Sc) contains 21 protons and 24 neutrons in its atom. Write the atomic number and mass number to complete the following symbolic expression.

_____Sc

Answer: $^{45}_{21}$Sc

50 The letter A represents an unknown element. Use the periodic table to identify the element.

$^{23}_{11}$A_____

Answer: sodium, Na (The subscript 11 represents the atomic number. Sodium is the element with an atomic number of 11.)

51 Unknown element X has a mass number of 55 and contains 30 neutrons in its atoms. Identify element X. _____

Answer: Mn (manganese) (A mass number of 55 indicates 55 protons and neutrons. Subtract 30 neutrons from 55; this leaves 25 protons. The number of protons is equal to the atomic number, 25. Manganese has an atomic number of 25.)

52 Be (beryllium) contains five neutrons in its atom. Complete the following symbolic expression.

_____Be

Answer: $^{9}_{4}$Be (Find the atomic number of Be on the periodic table. The atomic number indicates the number of protons. Add the neutrons, five, to the protons, four, to find the mass number.)

53 Fill in the required information for the following element: F.

(a) Atomic number: _____

(b) Mass number: _____

(c) Number of protons: _____

(d) Number of neutrons: _____

Answer: (a) 9; (b) 19; (c) 9; (d) 10

You have just learned some properties of metals and nonmetals, the common names of a few families of elements, and how to determine the numbers of protons, neutrons, and electrons in an atom.

Now we are going to look at a model of the atom that helps chemists explain many properties and reactions. Essentially, we will try to develop in your mind a picture of the arrangement of the electrons in an atom and how this arrangement relates to the location of the atom in the periodic table. Later, we will use this arrangement in discussing chemical bonding, chemical reactions, and chemical properties.

QUANTUM ATOMIC MODEL

The model we discuss has evolved from the study of quantum mechanics (a theoretical mathematical approach to the study of atomic and molecular structure). We do not attempt an in-depth presentation here. Instead, we present some of the basic concepts so you may use them later in this book or build upon them in other chemistry courses.

Keep in mind that we are studying the basic model of a *very complex* theory. A good way to help you remember the model is to compare it to an apartment building. An apartment building has different floors, different apartments on each floor, and different rooms within each apartment.

We can look upon the electrons of an atom as rather peculiar apartment dwellers. Electrons prefer the floor closest to the ground and the smallest apartments. Electrons also prefer to live one to a room until each room in an apartment has one occupant. The electrons will then pair up until each room has two. Each room in the apartment can hold only two electrons.

Apartment buildings may have several floors. The model we discuss has several floors, but only the first seven floors will be occupied. All the electrons of the elements known today will fit within seven floors of the building. Additional floors are available but will be occupied only in special cases.

54 The floors in the apartment building are called **shells** in the electron model and are numbered 1 through 7. According to what you have just read, what shell will be occupied first by electrons? _____

Answer: shell 1 (the first floor)

55 Each shell (or floor) in the model has one or more apartments, which are called **subshells**. These subshells are apartments of four sizes: *s, p, d,* and *f*. An *s* subshell (apartment) has only a single room. A *p* subshell has three rooms. A *d* subshell has five rooms, while an *f* subshell has seven rooms. An *s* subshell then will hold a maximum of two electrons according to the model.

(a) A *p* subshell will hold a maximum of how many electrons? _____

(b) How many will a *d* subshell hold? _____

(c) How many will an *f* subshell hold? _____

Answer: (a) six (three rooms × two electrons/room); (b) 10 (five rooms × two electrons/room); (c) 14 (seven rooms × two electrons/room).

56 Each room in a subshell is called an **orbital**. From frame 55 we know, then, that an *s* subshell will consist of one orbital with a capacity (occupancy) of two electrons.

(a) A *p* subshell will consist of three orbitals with a total subshell capacity of _____ electrons.

(b) A *d* subshell will consist of _____ orbitals with a total subshell capacity of _____ electrons.

(c) An *f* subshell will consist of _____ orbitals and hold _____ electrons.

Answer: (a) six (three orbitals × two electrons, or two electrons/orbital); (b) five, 10; (c) seven, 14.

57 The first shell (floor) has only one subshell (apartment), which is an *s* subshell. Because of its location on the first shell, it is called a 1*s* subshell.

(a) How many orbitals (rooms) are there in this 1*s* subshell? _____

(b) How many electrons will the subshell hold? _____

Answer: (a) one (*s* subshells have only one orbital.); (b) two (Each orbital holds only two electrons.)

58 The second shell (floor) only has an *s* subshell (apartment) and a *p* subshell.

(a) If the *s* subshell is called 2*s*, what do you suppose the *p* subshell is called? _____

(b) How many orbitals (rooms) are in that *p* subshell? _____

(c) How many subshells are in the second shell? _____

(d) How many orbitals are there in the second shell? _____

(e) How many electrons can occupy the second shell? _____

Answer: (a) 2*p*; (b) three *(p* subshells have three orbitals.); (c) two (*s* and *p*); (d) four (one *s* orbital and three *p* orbitals); (e) eight (4 orbitals × 2 electrons/orbital).

59   The third shell has three subshells: *s*, *p*, and *d*.

(a) What are they called? _____

(b) How many subshells are in the third shell? _____

(c) How many orbitals are in the third shell? _____

(d) How many electrons can be in the third shell? _____

Answer: (a) 3*s*, 3*p*, 3*d*; (b) three; (c) nine (one *s* orbital, three *p* orbitals, and five *d* orbitals); (d) 18 (nine orbitals × two electrons/orbital).

60   Shells 4 through 7 each have four subshells: *s*, *p*, *d*, and *f*.

(a) What would you call the subshells in the fourth shell? _____

(b) What would you call the subshells in the sixth shell? _____

Answer: (a) 4*s*, 4*p*, 4*d*, 4*f*; (b) 6*s*, 6*p*, 6*d*, 6*f*

61   How many subshells are there in the fifth shell? _____
         How many subshells are there in the seventh shell? _____

Answer: four; four

62   How many orbitals are there in the fourth shell? _____
         How many electrons will that shell hold? _____

Answer: 16 (one *s* orbital, three *p* orbitals, five *d* orbitals, and seven *f* orbitals); 32 (16 orbitals × two electrons/orbital).

63   Let's review what we have just learned. Assume that we have only seven floors in our "building."

(a) A shell may have as many as _____ subshells or as few as _____ subshell(s).

(b) A subshell may have as many as _____ orbitals or as few as _____ orbital(s).

(c) A subshell may hold as many as _____ electrons or as few as _____ electron(s), assuming full occupancy.

(d) A shell may hold as many as _____ electrons or as few as _____ electron(s), assuming full occupancy.

Answer: (a) four, one; (b) seven, one; (c) 14, two; (d) 32, two

As we have mentioned previously, electrons prefer the lower shells (floors) and the smaller subshells (apartments). Electrons prefer the smaller subshells to such a degree that they will sometimes occupy a smaller subshell on the next higher shell rather than the larger subshell on the lower shell.

By experimentation, it has been determined that electrons will fill the 1*s* subshell (apartment) first. They will then fill the 2*s* subshell and then the 2*p* subshell. Next, they will fill the 3*s* subshell and then the 3*p* subshell. However, before going into the large five-orbital 3*d* subshell, electrons will first fill the 4*s* subshell. After filling the 4*s* subshell, electrons will then proceed to fill the 3*d* subshell. The 4*p* subshell is filled next. The electrons prefer to fill the small 5*s* subshell before filling the larger 4*d* subshell. The 4*d* is filled after 5*s*. Next, the electrons fill the 5*p* subshell. Then the small 6*s* subshell is filled. The very large 4*f* subshell is occupied only after 6*s* is filled. After 4*f* comes 5*d*. Next is 6*p*, then 7*s*, and then 5*f*.

A diagram to help you remember the order of filling the subshells appears on page 17.

Note that as we fill consecutive subshells, the energy of the electrons increases. Electrons in the 2*s* subshell have a higher energy than electrons in the 1*s* subshell; 2*p* electrons have a higher energy than 2*s* electrons, and so on.

64 Using the diagram, which subshell is filled first? _____

Answer: 1*s*

65 Is the 4*s* subshell filled before or after the 3*d* subshell? _____

Answer: before

66 Neon has 10 electrons. The order of filling its subshells is first 1*s*, then 2*s*, and finally 2*p*. What is the order of filling the subshells in an atom of magnesium (Mg)? (Use the periodic table to determine the number of electrons in an atom of magnesium.) _____

Answer: Since there are 12 electrons in an atom of magnesium, the order of filling of the subshells is 1*s* 2*s* 2*p* 3*s*.

67 The notation shown on page 17 is used to indicate the number of electrons in each subshell of an atom. For example, neon has 10 electrons; therefore its subshells are written as $1s^2\, 2s^2\, 2p^6$. The numbers to the upper right of each subshell indicate the number of electrons in each subshell. If we add these numbers $(2 + 2 + 6 = 10)$, we get the number of electrons in a neon atom.

How would you use this notation for the magnesium (Mg) atom?

Answer: 1s² 2s² 2p⁶ 3s² (2 + 2 + 6 + 2 = 12)

Order of filling of subshells and approximate energy ranking

ELECTRON CONFIGURATION

 You have just learned the notation a chemist uses to indicate the arrangement of electrons in an atom. This arrangement is called its **electron configuration**. Use the diagram on page 17 to determine the electron configuration of argon, $_{18}$Ar. _____

Answer: $1s^2\ 2s^2\ 2p^6\ 3s^2\ 3p^6$

69 Chlorine ($_{17}$Cl) is an example of an atom in which the last subshell is not completely filled. Its electron configuration is $1s^2\ 2s^2\ 2p^6\ 3s^2\ 3p^5$. Note that the $3p$ subshell has only five electrons and all other subshells are filled.

    Oxygen is another example of an atom in which the last subshell is unfilled. What is its electron configuration? _____

Answer: $1s^2\ 2s^2\ 2p^4$ ($2 + 2 + 4 = 8$ electrons)

70 What are the electron configurations of the following elements?

(a) Potassium (K) _____

(b) Arsenic (As) _____

Answer: (a) $1s^2\ 2s^2\ 2p^6\ 3s^2\ 3p^6\ 4s^1$; (b) $1s^2\ 2s^2\ 2p^6\ 3s^2\ 3p^6\ 4s^2\ 3d^{10}\ 4p^3$

71 We can also identify an atom if we are given its electron configuration. For example, the configuration $1s^2\ 2s^2\ 2p^6\ 3s^2\ 3p^1$ has 13 electrons. Only the aluminum atom has 13 electrons; therefore this configuration must be that of an aluminum atom.

    What atom has the electron configuration $1s^2\ 2s^2\ 2p^6\ 3s^2\ 3p^4$? _____

Answer: sulfur, S ($2 + 2 + 6 + 2 + 4 = 16$)

    Another way to represent the arrangement of electrons around an atom is to use arrows as electrons and boxes to represent orbitals (see frame 72). The boxes become occupied by electrons as we "build up" the atoms of each element in the periodic table. Remember, only one electron will occupy an orbital in a given subshell until all the orbitals in that subshell have one electron in them. Then and only then will a second electron occupy each orbital.

72 Using this method, the electron arrangement for $_{12}Mg$ follows.

$$1s\; \boxed{\uparrow\downarrow} \quad 2s\; \boxed{\uparrow\downarrow} \quad 2p\; \boxed{\uparrow\downarrow}\boxed{\uparrow\downarrow}\boxed{\uparrow\downarrow} \quad 3s\; \boxed{\uparrow\downarrow}$$

The arrow notation for $_7N$ is

$$1s\; \boxed{\uparrow\downarrow} \quad 2s\; \boxed{\uparrow\downarrow} \quad 2p\; \boxed{\uparrow}\boxed{\uparrow}\boxed{\uparrow}$$

Note the unpaired or single electrons in the partially filled $2p$ subshell. The electrons occupy as many orbitals as possible in the same subshell before pairing with another electron. This is known as the **Principle of Maximum Multiplicity**.

Using the Principle of Maximum Multiplicity and the arrow notation, indicate the arrangement of electrons for the following:

(a) $_{14}Si$ _____

(b) $_{16}S$ _____

(c) $_{23}V$ _____

(d) $_{26}Fe$ _____

Answer:

(a) $_{14}Si$ $1s\,\boxed{\uparrow\downarrow}$ $2s\,\boxed{\uparrow\downarrow}$ $2p\,\boxed{\uparrow\downarrow}\boxed{\uparrow\downarrow}\boxed{\uparrow\downarrow}$ $3s\,\boxed{\uparrow\downarrow}$ $3p\,\boxed{\uparrow}\boxed{\uparrow}$

(b) $_{16}S$ $1s\,\boxed{\uparrow\downarrow}$ $2s\,\boxed{\uparrow\downarrow}$ $2p\,\boxed{\uparrow\downarrow}\boxed{\uparrow\downarrow}\boxed{\uparrow\downarrow}$ $3s\,\boxed{\uparrow\downarrow}$ $3p\,\boxed{\uparrow\downarrow}\boxed{\uparrow}\boxed{\uparrow}$

(c) $_{23}V$ $1s\,\boxed{\uparrow\downarrow}$ $2s\,\boxed{\uparrow\downarrow}$ $2p\,\boxed{\uparrow\downarrow}\boxed{\uparrow\downarrow}\boxed{\uparrow\downarrow}$ $3s\,\boxed{\uparrow\downarrow}$ $3p\,\boxed{\uparrow\downarrow}\boxed{\uparrow\downarrow}\boxed{\uparrow\downarrow}$ $4s\,\boxed{\uparrow\downarrow}$ $3d\,\boxed{\uparrow}\boxed{\uparrow}\boxed{\uparrow}$

(d) $_{26}Fe$ $1s\,\boxed{\uparrow\downarrow}$ $2s\,\boxed{\uparrow\downarrow}$ $2p\,\boxed{\uparrow\downarrow}\boxed{\uparrow\downarrow}\boxed{\uparrow\downarrow}$ $3s\,\boxed{\uparrow\downarrow}$ $3p\,\boxed{\uparrow\downarrow}\boxed{\uparrow\downarrow}\boxed{\uparrow\downarrow}$ $4s\,\boxed{\uparrow\downarrow}$ $3d\,\boxed{\uparrow\downarrow}\boxed{\uparrow}\boxed{\uparrow}\boxed{\uparrow}\boxed{\uparrow}$

73 The electron configurations of the naturally occurring noble gases are given below.

$_2He$	$1s^2$
$_{10}Ne$	$1s^2\; 2s^2\; 2p^6$
$_{18}Ar$	$1s^2\; 2s^2\; 2p^6\; 3s^2\; 3p^6$
$_{36}Kr$	$1s^2\; 2s^2\; 2p^6\; 3s^2\; 3p^6\; 4s^2\; 3d^{10}\; 4p^6$
$_{54}Xe$	$1s^2\; 2s^2\; 2p^6\; 3s^2\; 3p^6\; 4s^2\; 3d^{10}\; 4p^6\; 5s^2\; 4d^{10}\; 5p^6$
$_{86}Rn$	$1s^2\; 2s^2\; 2p^6\; 3s^2\; 3p^6\; 4s^2\; 3d^{10}\; 4p^6\; 5s^2\; 4d^{10}\; 5p^6\; 6s^2\; 4f^{14}\; 5d^{10}\; 6p^6$

With the exception of $_2He$, the subshell of greatest energy (last subshell) in each noble gas consists of six electrons occupying a(n) (s, p, d, f) subshell.

_____

Answer: p

74 With the exception of $_2He$, the similar properties of the noble gases are due to their similar electron configuration.

Noble gas	Subshells of the outermost shell
$_{10}$Ne	$2s^2\ 2p^6$
$_{18}$Ar	$3s^2\ 3p^6$
$_{36}$Kr	$4s^2\ 4p^6$
$_{84}$Xe	$5s^2\ 5p^6$
$_{86}$Rn	$6s^2\ 6p^6$

The subshell of greatest energy of each noble gas (mark the correct answer):

_____ (a) is completely filled with electrons.
_____ (b) is half–filled with electrons.
_____ (c) can take one more electron each.

Answer: (a) (Only six electrons can occupy the orbitals in a *p* subshell.)

75 The electron configurations of the naturally occurring halogens are as follows.

$_{9}$F	$1s^2\ 2s^2\ 2p^5$
$_{17}$Cl	$1s^2\ 2s^2\ 2p^6\ 3s^2\ 3p^5$
$_{35}$Br	$1s^2\ 2s^2\ 2p^6\ 3s^2\ 3p^6 4s^2\ 3d^{10}\ 4p^5$
$_{53}$I	$1s^2\ 2s^2\ 2p^6\ 3s^2\ 3p^6\ 4s^2\ 3d^{10}\ 4p^6\ 5s^2\ 4d^{10}\ 5p^5$
$_{85}$At	$1s^2\ 2s^2\ 2p^6\ 3s^2\ 3p^6\ 4s^2\ 3d^{10}\ 4p^6\ 5s^2\ 4d^{10}\ 5p^6\ 6s^2\ 4f^{14}\ 5d^{10}\ 6p^5$

The incomplete subshell in each halogen is made up of how many electrons? _____ In what subshell? _____

Answer: five; *p*

76 The electron configurations of the alkaline earth metals are as follows.

$_{4}$Be	$1s^2\ 2s^2$
$_{12}$Mg	$1s^2\ 2s^2\ 2p^6\ 3s^2$
$_{20}$Ca	$1s^2\ 2s^2\ 2p^6\ 3s^2\ 3p^6\ 4s^2$
$_{38}$Sr	$1s^2\ 2s^2\ 2p^6\ 3s^2\ 3p^6\ 4s^2\ 3d^{10}\ 4p^6\ 5s^2$
$_{56}$Ba	$1s^2\ 2s^2\ 2p^6\ 3s^2\ 3p^6\ 4s^2\ 3d^{10}\ 4p^6\ 5s^2\ 4d^{10}\ 5p^6\ 6s^2$
$_{88}$Ra	$1s^2\ 2s^2\ 2p^6\ 3s^2\ 3p^6\ 4s^2\ 3d^{10}\ 4p^6\ 5s^2\ 4d^{10}\ 5p^6\ 6s^2\ 4f^{14}\ 5d^{10}\ 6p^6\ 7s^2$

The subshell of the outermost shell in each alkaline earth is made up of _____ electrons in a(n) _____ subshell.

Answer: two; *s*

77 Each group of elements in the periodic table has similar subshells with similar numbers of electrons in the **outermost shell**. The outermost shell consists of the subshells that are filled last. This situation serves to explain the (similar, greatly different) _____ chemical properties of elements within the same groups.

Answer: similar

The knowledge of what constitutes an atom is important to the discussion of atomic weights and molecular weights. The arrangement of the electrons around the atom is important to the discussion of chemical bonding, chemical formulas, and chemical properties—all topics of later chapters. What you have learned so far will be the springboard to a greater understanding of chemistry as you continue your study.

Self-Test

This self-test is designed to show how well you have mastered this chapter's objectives. Correct answers and review instructions follow the test.

1. Write the number of the item on the right that *best* describes each item on the left. You may use the periodic table if you wish.

_____	(a) proton	(1) an alkaline earth
_____	(b) Sr	(2) a halogen
_____	(c) Li	(3) a noble gas
_____	(d) Br	(4) an alkali metal
_____	(e) electron	(5) responsible for nuclear charge
		(6) occupies subshells

2. How many electrons, protons, and neutrons does a neutral K atom have?

_____ protons

_____ neutrons

_____ electrons

3. What is the electron configuration of the element in question 2?

_____

4. How many electrons, protons, and neutrons does a neutral Mg atom have?

_____ protons

_____ neutrons

_____ electrons

5. What is the electron configuration of the element in question 4? _____

6. What is the outermost subshell electron configuration common to all of the halogens? _____

7. What is the outermost subshell electron configuration common to all of the alkali metals? _____

8. A substance that shines and conducts heat and electricity is a (metal, non-metal, metalloid) _____.

9. A substance that is usually very brittle and does not heat well is a (metal, nonmetal, metalloid) _____.

10. Silicon and antimony belong to the class of elements known as the (metals, nonmetals, metalloids) _____.

11. Iodine and xenon belong to the class of elements known as the (metals, nonmetals, metalloids) _____.

12. What element has the electron configuration $1s^2 \ 2s^2 \ 2p^6 \ 3s^2 \ 3p^3$? _____ In what group would it be found in the periodic table? _____

13. What element has the electron configuration $1s^2 \ 2s^2 \ 2p^5$? _____ In what group would it be found in the periodic table? _____

14. How would you write the box and arrow notation for $_{25}Mn$? _____

15. How would you write the box and arrow notation for $_{15}P$? _____

Answers

Compare your answers to the self-test with those given below. If you answer all questions correctly, you are ready to proceed to the next chapter. If you miss any, review the frames indicated in parentheses after the answers. If you miss several questions, you should probably reread the chapter carefully.

1.

(a) 5 (frames 12, 13)

(b) 1 (frames 28, 76)

(c) 4 (frames 27, 30)

(d) 2 (frames 31, 75)

(e) 6 (frames 54–63)

2. 19; 20; 19 (frames 42–53)

3. $1s^2\ 2s^2\ 2p^6\ 3s^2\ 3p^6\ 4s^1$ (frames 64–76)

4. 6; 6; 6 (frames 42–53)

5. $1s^2\ 2s^2\ 2p^2$ (frames 64–76)

6. p^5 (frame 75)

7. s^2 (frame 76)

8. metal (frame 35)

9. nonmetal (frame 38)

10. metalloid (frame 40)

11. nonmetals (frame 37)

12. phosphorus, Group VA (frames 71–77)

13. fluorine, Group VIIA (frames 71–77)

14.

$1s$ [↑↓] $2s$ [↑↓] $2p$ [↑↓|↑↓|↑↓] $3s$ [↑↓] $3p$ [↑↓|↑↓|↑↓] $4s$ [↑↓] $3d$ [↑|↑|↑|↑|↑]

(frames 72–76)

15.

$1s$ [↑↓] $2s$ [↑↓] $2p$ [↑↓|↑↓|↑↓] $3s$ [↑↓] $3p$ [↑|↑|↑]

(frames 72–76)

ELECTRONS IN FORENSIC CHEMISTRY

The electronic structure of an atom can provide us a wide range of information. In this chapter you learned about protons, neutrons, and electrons as well as where these subatomic particles are located within an atom. Not only that, you also learned specifically about electrons and their unique configurations within an atom's orbitals. However, did you know that scientific research, as well as forensic science, makes use of these electrons in performing investigative analyses?

For instance, residue is deposited on the hands and clothing of a person who discharges a firearm. This is known as gunshot residue (GSR). A person's clothing and skin can be analyzed to see whether the person has discharged a firearm. When a gun is fired, burned and unburned particles from the primer are blown back onto the person who pulled the trigger. This residue usually consists of lead, antimony, and barium or at least antimony and barium. The suspect's hand is carefully swabbed, and the residue is collected for analysis by cyclic voltammetry or a potentiometer.

A variety of techniques exists for analyzing GSR, but one older method focuses on exciting the electrons found within lead, antimony, and barium. Once a GSR sample is obtained, it can be analyzed using atomic absorption spectrophotometry (AAS).

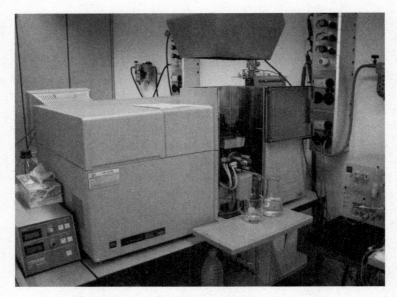

Figure 1 An atomic absorption spectrometer

In AAS, the GSR sample is atomized (sprayed as a fine droplet mist) into the AAS's flame or graphite furnace to make gaseous atoms. The atoms within the spray are exposed to ultraviolet or visible light in the hot furnace. As this process occurs, the free atoms of lead, antimony, and barium absorb ultraviolet (UV) or visible light and become excited. Other atoms within the GSR sample will become excited as well.

Before we go any further, let's look at lead's electronic configuration. The lead atom has a total of 82 electrons with the following electronic configuration within its normal, or ground, state:

$1s^2\ 2s^2\ 2p^6\ 3s^2\ 3p^6\ 4s^2\ 3d^{10}\ 4p^6\ 5s^2\ 4d^{10}\ 5p^6\ 6s^2\ 4f^{14}\ 5d^{10}\ 6p^2$

(Ground state electronic configuration is needed for reference to the excited state. The ground state of an atom is its "resting" state.)

An excited lead atom can have the following electronic configuration:

$1s^2\ 2s^2\ 2p^6\ 3s^2\ 3p^6\ 4s^2\ 3d^{10}\ 4p^6\ 5s^2\ 4d^{10}\ 5p^6\ 6s^2\ 4f^{14}\ 5d^9\ 6p^3$

Notice that an electron from the outer $5d$ orbital was excited to the outer $6p^2$ orbital, resulting in $6p^3$. This excitation of an electron is the result of the lead atom absorbing radiation. In order for the atom to return to its normal state, its promoted electron must return back to the $5d$ orbital.

When this transition occurs, UV or visible light is released by the atom and picked up by the detector. Each atom has its own unique wavelengths of emitted UV or visible light. Based on known standards, chemists can confirm the presence or absence of a particular element in a sample. In this case, lead can be confirmed along with antimony and barium.

Developments in instrumentation continue, and AAS often has been replaced by more advanced techniques. However, AAS utilizes electrons from excited atoms to confirm the presence and quantity of such elements. This technique will always have a home in instrumental analysis as well as in forensic chemistry.

2 Atomic Weights

Now that you have some idea of what is in an atom, let's look at the weight of atoms. Each atom has a definite and characteristic weight. This weight provides a very convenient way to state the amount of substance required for a chemical reaction.

In this chapter we discuss how the weights of atoms were determined experimentally. You will encounter for the first time a very formidable number — 602,200,000,000,000,000,000,000 (6.022×10^{23}) — called Avogadro's number. The number becomes very important in later chapters, so be sure you understand its significance!

OBJECTIVES

After completing this chapter, you will be able to

- recognize and apply or illustrate: isotope, atomic weight, atomic mass unit (amu), mass spectrograph, gram atomic weight, and Avogadro's number;

- explain the fractional atomic weights listed in the periodic table;

- calculate the number of atoms in a given weight of an element and vice versa;

- calculate the approximate atomic weight of an element when given the abundance and approximate mass of its isotopes;

- distinguish between atomic weight expressed in amu and gram atomic weight.

1 Let's review a bit. The notation $^{35}_{17}\text{Cl}$ indicates a neutral atom of chlorine.

(a) What is its atomic number? _____

(b) What is its mass number? _____

(c) How many protons does it have? _____

(d) How many electrons? _____

(e) How many neutrons? _____

Answer: (a) 17; (b) 35; (c) 17; (d) 17; (e) 18

2   Different atoms of the same element can have different numbers of neutrons and, therefore, different mass numbers. Here is another neutral chlorine atom: $^{37}_{17}Cl$.

(a) What is its mass number? _____

(b) How many protons does it have? _____

(c) How many neutrons? _____

Answer: (a) 37; (b) 17; (c) 20

3   Since neutrons and protons combine to make up the mass number, two atoms of the same element can have different mass numbers. Chlorine can exist as $^{37}_{17}Cl$ and as $^{35}_{17}Cl$. The only difference between these atoms of chlorine is that $^{37}_{17}Cl$ contains two more neutrons than $^{35}_{17}Cl$.

Antimony can exist as $^{121}_{51}Sb$ and $^{123}_{51}Sb$. The only difference between these atoms of antimony is that $^{123}_{51}Sb$ contains two more _____ than $^{121}_{51}Sb$.

Answer: neutrons

4 $^{37}_{17}Cl$ has a greater mass than $^{35}_{17}Cl$ because of the two extra neutrons. Which of the following atoms of antimony has the greater mass, $^{121}_{51}Sb$ or $^{123}_{51}Sb$? _____

Answer: $^{123}_{51}Sb$ (because it has two more neutrons)

5   Atoms of the same element having different masses are called **isotopes**. Elements as found in nature are usually mixtures of two or more isotopes. The atom $^{123}_{51}Sb$ is one isotope of the element antimony; $^{121}_{51}Sb$ is another isotope of antimony. The main difference between two isotopes of the same element is the number of (protons, neutrons, electrons) _____.

Answer: neutrons

6   Isotopes exist for every known element. The isotopes of the element neon were first discovered by two English scientists, J. J. Thomson and F. W. Aston. Thomson and Aston continued in their work to discover other isotopes through inventing the **mass spectrograph** (also called the mass spectrometer).

In the mass spectrograph, atoms of different masses of the same element (mixtures of isotopes) are charged (no longer neutral) and accelerated by an electron beam toward a target, such as a photographic plate. A strong magnetic field bends the paths of the charged atoms. Atoms of greater mass have their paths bent to a lesser degree than atoms of lighter mass.

In the diagram of the mass spectrograph (pictured here) where do the lighter atoms strike, point A or point B? _____

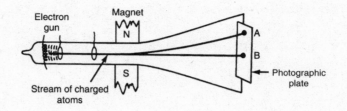

Answer: point A (because the path of the lighter atoms is bent to a greater degree)

7 An analogy to the mass spectrograph would be to roll a bowling ball and a basketball at the same speed at a target while a stiff crosswind is blowing. The bowling ball is considerably heavier than a basketball. Look at the diagram below. Which ball would strike the target at point B? _____

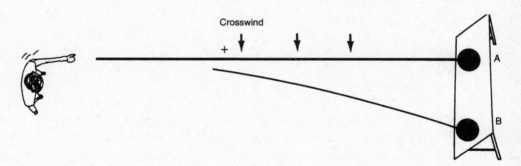

Answer: the basketball (The basketball is lighter; therefore, its path is more readily changed by the crosswind.)

8 In the rolling balls analogy to the spectrograph, the bowling ball and basketball are analogous to isotopes of different mass. The strong crosswind is analogous to the _____.
(Refer to the diagram of the spectrograph, if necessary.)

Answer: magnetic field (or magnet)

9 Thomson and Aston invented an instrument that detects the presence and characteristics of isotopes. What is this instrument called? _____

Answer: mass spectrograph (or mass spectrometer)

10 The **atomic weights** of the elements are listed in the periodic table. The atomic weight of sodium, for example, is listed as 22.990. The atomic weight listed for sodium is actually the atomic weight of a mixture of isotopes, $^{22}_{11}$Na and $^{23}_{11}$Na. The proportion of these isotopes is generally constant wherever sodium is found.

The atomic weight of an element is the average weight of a mixture of two or more _____.

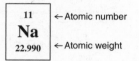

Answer: isotopes

11 The periodic table lists the atomic weight of Al as _____.

Answer: 26.982

12 Atomic weights are based on the carbon 12 scale. That is, carbon 12 or $^{12}_{6}$C, the most abundant isotope of carbon, is used as the standard unit in measuring atomic weights. By international standard, one atom of the $^{12}_{6}$C isotope has an atomic weight of exactly 12 **atomic mass units**, abbreviated **amu**. The atomic weight of the $^{12}_{6}$C isotope is exactly _____ amu.

Answer: 12

13 All atomic weights can be expressed in atomic mass units. By international agreement, 12 amu would equal the mass of a single $^{12}_{6}$C atom. One amu is equal to what fraction of a single $^{12}_{6}$C atom? _____

Answer: $\frac{1}{12}$ the mass of a single $^{12}_{6}$C atom

14 Although the $^{12}_{6}$C isotope weighs exactly 12.000 amu by definition, the atomic weight of C as listed on the periodic table is 12.011 amu. The atomic weight of carbon as listed on the periodic table is greater than that of the $^{12}_{6}$C isotope. Why? _____

Answer: The atomic weight of an element is the average weight of a mixture of two or more isotopes.

15 While the element carbon as found in nature is made up largely of the $^{12}_{6}$C isotope (98.9%), a small quantity of $^{13}_{6}$C isotope (1.1%) is mixed uniformly as part of the element. The $^{12}_{6}$C isotope has an atomic weight of 12.000 amu. The

$^{13}_{6}C$ isotope has an atomic weight of 13.003 amu. The resultant average atomic weight would be slightly (heavier, lighter) _____ than 12.000 amu.

Answer: heavier (In fact, the periodic table lists the atomic weight of carbon as 12.011 amu.)

16 The atomic weights on the periodic table are the average atomic weights of the isotopic mixtures in the element. We can determine the average atomic weight of an element if we know the approximate mass of each isotope and the proportion of each isotope within the element.

Here are the steps for calculating the average atomic weight of the element carbon. Look at the table below as you read through the steps.

Multiply the mass of the $^{12}_{6}C$ isotope by its decimal proportion (12.000×0.989).

Multiply the mass of the $^{13}_{6}C$ isotope by its decimal proportion (13.003×0.011).

Add the results to find the average atomic weight of the element C.

Element	Isotope	Mass of isotope		Proportion in element		Mass × proportion		Sum
C	$^{12}_{6}C$	12.000	×	0.989	=	11.868	+ =	____
	$^{13}_{6}C$	13.003	×	0.011	=	0.143		

Now you do the final calculation step, adding the two results. Fill in the blank in the "Sum" column. Round off this answer and all others to the nearest hundredth (two places to the right of the decimal) unless otherwise indicated.

Answer: The calculated atomic weight of C is 12.011, rounded off to 12.01 amu.

17 Now calculate the atomic weight for fluorine.

Element	Isotope	Mass of isotope		Proportion in element		Mass × proportion		Sum
F	$^{19}_{9}F$	19.000	×	99.7% or 0.997	=	____	+ =	____
	$^{18}_{9}F$	18.000	×	0.3% or 0.003	=	____		

Answer:

Mass × proportion	Sum
18.943	+ = 18.997
0.054	

The calculated atomic weight of F is rounded off to 19.00 amu.

18 Sodium has two isotopes, $^{23}_{11}$Na and $^{22}_{11}$Na. The isotope $^{23}_{11}$Na has an atomic mass of approximately 23.000 amu, and its proportion in the element is 99.2%. The isotope $^{22}_{11}$Na has a mass of approximately 22.000 amu and a proportion within the element of 0.8%. Determine the atomic weight of sodium. (Remember, 0.8% = 0.008 and 99.2% = 0.992.) Use a separate sheet of paper to set up a table as we have done in the past few frames. Here again are the column headings you will need.

Element	Isotope	Mass of isotope	Proportion in element	Mass × proportion	Sum

Answer:

Element	Isotope	Mass of isotope		Proportion in element		Mass × proportion			Sum
Na	$^{23}_{11}$Na	23.000	×	0.992	=	22.816			
	$^{22}_{11}$Na	22.000	×	0.008	=	0.176	+	=	22.992

The calculated atomic weight of Na is 22.99 amu.

19 The element cobalt has an isotope $^{60}_{27}$Co that has an approximate mass of 60.00 and constitutes 48.0% of the element. Another isotope, $^{58}_{27}$Co, has an approximate mass of 58.00 and constitutes 52.0% of the element. Calculate the atomic weight of Co using the given data. Use a separate sheet of paper to set up a table of calculations.

Answer:

Element	Isotope	Mass of isotope		Proportion in element		Mass × proportion			Sum
Co	$^{60}_{27}$Co	60.00	×	0.480	=	28.80			
	$^{58}_{27}$Co	58.00	×	0.520	=	30.16	+	=	58.96

The calculated atomic weight of Co is 58.96 amu.

20 The percentage proportions of all the isotopes within an element must add up to a total of _____ %. The decimal proportions of all the isotopes within an element must add up to _____ (Hint: See above example, 0.480 + 0.520.)

Answer: 100; 1

21 We've been calculating atomic weight given the mass and proportion of isotopes. We can also determine the proportion of each individual isotope within an element if we know the atomic weight of the element. Set up a table like the ones you have been using and fill in the information given below. Use letters such as A and B to represent unknown proportions.

The element Cr, which has an overall atomic weight of 51.996 amu, has two isotopes: $^{52}_{24}Cr$ with atomic mass of 52.000 amu and $^{52}_{24}Cr$ with an atomic mass of 51.000 amu. Just fill in the "givens" and "unknowns" for now; don't try to solve the problem yet.

Answer:

Element	Isotope	Mass of isotope		Proportion in element		Mass × proportion			Sum
Cr	$^{52}_{24}Cr$	52.000	×	A	=	$52.000 \times A$			51.996
	$^{51}_{24}Cr$	51.000	×	B	=	$51.000 \times B$	+	=	

22 To solve this equation with two unknowns, you must form a second equation showing the relationship between A and B. You have already learned the answer to the following question in frame 20. Add the decimal proportions. $A + B =$ _____

Answer: 1

23 Modify the equation in frame 22 so that just B remains on the left side of the equation. $B =$ _____

Answer: $1 - A$

24 Here is the equation you need for calculating the sum of the isotopes' mass × proportion: $(52.000A) + (51.000B) = 51.996$. Substitute the expression $(1 - A)$ for B in the equation _____

Answer: $[52.000A] + [51.000(1 - A)] = 51.996$

25 Solve the equation derived in frame 24 to determine the value of A (the proportion of $^{52}_{24}Cr$ in an average mixture of chromium) to the nearest thousandth. Use a separate sheet of paper for your calculations.

Answer: $[52.000A] + [51.000(1 - A)] = 51.996$

$(52.000A) + (51.000) - (51.000A) = 51.996$

$1.000A + 51.000 = 51.996$

$1.000A = 0.996$

$A = 0.996$

26 Since $B = 1 - A$, what is the value of B (nearest thousandth)? $B = $ _____

Answer: $B = 1 - A$

$B = 1 - 0.996$

$B = 0.004$

27 Here is a similar problem. In the next few frames, you will determine the proportion of the $^{35}_{17}Cl$ isotope in an average mixture of chlorine, which is made up of both $^{35}_{17}Cl$ and $^{37}_{17}Cl$. Here is our table of givens and unknowns.

Element	Isotope	Mass of isotope	Proportion in element		Mass × proportion			Atomic weight
Cl	$^{35}_{17}Cl$	34.97 ×	A	=	$34.97 \times A$			
	$^{37}_{17}Cl$	36.97 ×	B	=	$36.97 \times B$	+	=	35.45

The values of A and B added together must equal _____.

Answer: 1

28 Since $A + B = 1$, then $B = $ _____

Answer: 1 − A

29 The proportion $1 - A$ has been substituted for B in the following table.

Element	Isotope	Mass of isotope	Proportion in element		Mass × proportion			Atomic weight
Cl	$^{35}_{17}Cl$	34.97 ×	A	=	$34.97 \times A$			
	$^{37}_{17}Cl$	36.97 ×	$(1 - A)$	=	$36.97 \times (1 - A)$	+	=	35.45

Using the table above, determine the proportion of $^{35}_{17}Cl$ within the element chlorine (find the value of A). Use a separate sheet of paper for your set of equations.

Answer: $^{35}_{17}Cl$ mass × A + $^{37}_{17}Cl$ mass × $(1 - A) = 35.45$

$34.97\,A + 36.97(1 - A) = 35.45$

$34.97\,A + 36.97 - 36.97\,A = 35.45$

$36.97 - 2.00\,A = 35.45$

$36.97 = 35.45 + 2.00\,A$

$1.52 = 2.00\,A$

$0.76 = A$

30 Determine the proportion of $^{37}_{17}Cl$ in Cl (find B).

Answer: The proportion of $^{37}_{17}Cl$ in Cl has been given the value of B in the table.

$B = 1 - A$

$B = 1 - 0.76$

$B = 0.24$

31 Neon has two isotopes: $^{22}_{10}Ne$ and $^{20}_{10}Ne$. The approximate mass of $^{22}_{10}Ne$ is 22.000 amu and the mass of $^{20}_{10}Ne$ is approximately 20.000 amu. The atomic weight of neon is 20.179 amu. Determine the proportion (to the nearest thousandth) of the $^{22}_{10}Ne$ isotope within the element. (Let the proportion of $^{20}_{10}Ne$ within Ne be equal to B.) Use a separate sheet of paper for your table of calculations and set of equations.

Answer:

Element	Isotope	Mass of isotope		Proportion in element		Mass × proportion			Atomic weight
Ne	$^{22}_{10}Ne$	22.000	×	$A =$	=	$(22.000 \times A)$	+	=	20.179
	$^{20}_{10}Ne$	20.000	×	$B =$	=	$(20.000 \times B)$			

$(22.000A) + (20.000B) = 20.179$

$B = 1 - A$

$22.000A + 20.000(1 - A) = 20.179$

$22.000A + 20.000 - 20.000A = 20.179$

$2.000A + 20.000 = 20.179$

$2.000A = 0.179$

$A = 0.090$ (the proportion of $^{22}_{10}Ne$)

32 What is the proportion (to the nearest thousandth) of the other isotope, $^{20}_{10}Ne$, within the element Ne? _____

Answer: The proportion of $^{20}_{10}Ne$ is equal to the value of B in the table.

$B = 1 - A$

$B = 1 - 0.090$

$B = 0.910$

33 So far, we have considered all atomic weights in terms of atomic mass units (amu).

(a) An atomic weight expressed in amu represents the average weight of how many atom(s) of an element? _____

(b) Carbon has an atomic weight of 12.011 amu, which represents the average weight of how many atom(s) of carbon? _____

Answer: (a) one (An atomic weight expressed in amu represents the average weight of one single atom of an element.); (b) one

GRAM ATOMIC WEIGHT

34 Since it is impossible to measure the weight of one atom with a laboratory balance, another unit for expressing atomic weight must be used. Atomic weight can be expressed in grams as well as amu. An atomic weight expressed in grams (called a **gram atomic weight**) contains 6.022×10^{23} atoms. This number, called **Avogadro's number**, will be encountered often in this book.

(a) If the atomic weight of carbon is expressed as 12.011 amu, it represents the average weight of how many atom(s)? _____

(b) If the atomic weight of carbon is expressed as 12.011 grams, it represents the average weight of how many atom(s)? _____

Answer: (a) one; (b) 6.022×10^{23}

35 Using information in frame 34, answer the following question. One gram is how many times heavier than 1 amu? _____

Answer: 6.022×10^{23} (One gram is equivalent to 6.022×10^{23} amu.)

36 One ton is equivalent to 2000 pounds. One pound represents $\frac{1}{2000}$ of a ton. 6.022×10^{23} amu is equivalent to 1 gram. One amu represents what fraction of a gram? _____

Answer: $\dfrac{1}{6.022 \times 10^{23}}$

37 Avogadro's number (6.022×10^{23}) is written in **exponential notation** (sometimes called **scientific notation**). It actually represents a very large number: 602,200,000,000,000,000,000,000. Exponential notation will be used throughout this book, as in most other chemistry texts. You may already be familiar with exponential notation and with multiplying and dividing numbers with exponents. If so, you may skip to frame 38. If you need a quick refresher of exponential notation, we have summarized the basic rules in the examples below.

$$400 = 4 \times 10 \times 10 = 4 \times 10^2$$
$$5000 = 5 \times 10 \times 10 \times 10 = 5 \times 10^3$$
$$645,000 = 6.45 \times 10 \times 10 \times 10 \times 10 \times 10 = 6.45 \times 10^5$$

The exponent indicates the number of places that the decimal must be moved. The number 645,000 has the decimal moved five places to the left (645,000.). The result is 6.45×10^5.

We can also use exponential notation as in the examples below.

$$0.00004 = 4 \times 10^{-5}$$
$$0.073 = 7.3 \times 10^{-2}$$

In these two cases, we moved the decimal to the right, with the result that the exponent is negative.

0.00004 (Move the decimal five places to the right to arrive at the notation 4×10^{-5}.)

0.073 (Move the decimal two places to the right to arrive at the notation 7.3×10^{-2}.)

For 0.073, we could have moved the decimal three places to the right, with a result of 73×10^{-3}. We could also move the decimal one place to the right, with a result of 0.73×10^{-1}. Normally, in chemistry calculations we write a single digit of value 1 through 9 to the left of the decimal.

To *multiply* two numbers with exponential notation, multiply the decimal portions of the numbers and *add* the *exponents*.

$$(2 \times 10^5) \times (4 \times 10^3) = (2 \times 4) \times 10^{(5+3)} = 8 \times 10^8$$
$$(3 \times 10^4) \times (2 \times 10^{-3}) = (3 \times 2) \times 10^{(4-3)} = 6 \times 10^1$$

To *divide* two numbers with exponential notation, divide the decimal portions of the numbers and *subtract* the *exponent* in the denominator from the exponent in the numerator.

$$(6 \times 10^4)/(2 \times 10^2) = (6/2) \times 10^{(4-2)} = 3 \times 10^2$$
$$(8 \times 10^6)/(2 \times 10^{-4}) = (8/2) \times 10^{(6-(-4))} = 4 \times 10^{10}$$
$$(8 \times 10^{-4})/(4 \times 10^{-2}) = (8/4) \times 10^{(-4-(-2))} = 2 \times 10^{-2}$$

Try these problems.

(a) $42,000 = 4.2 \times$ _____

(b) $0.00465 = 4.65 \times$ _____

(c) $(7.0 \times 10^{-5}) \times (4.0 \times 10^8) =$ _____

(d) $\dfrac{9.0 \times 10^8}{3.0 \times 10^5} =$ _____

(e) $\dfrac{6.0 \times 10^{-4}}{3.0 \times 10^{-6}} =$ _____

Answer:

(a) 4.2×10^4

(b) 4.65×10^{-3}

(c) $(7.0 \times 4.0) \times 10^{(-5+8)} = 28 \times 10^3 = 2.8 \times 10^4$

(d) $\left(\dfrac{9.0}{3.0}\right) \times 10^{(8-5)} = 3.0 \times 10^3$

(e) $\left(\dfrac{6.0}{3.0}\right) \times 10^{(-4-(-6))} = 2.0 \times 10^2$

38 Look at the following examples of fractions that are converted to exponential notation. Then fill in the correct answer for the equivalence of amu to grams by using the same process of eliminating the fraction and expressing the equivalence in proper exponential notation.

1 pound equals $\dfrac{1}{2000}$ ton

1 pound equals $\dfrac{1}{(2 \times 10^3)}$ ton

1 pound equals $\frac{1}{2} \times 10^{-3}$ ton

1 pound equals 0.5×10^{-3} ton

1 pound equals 5.0×10^{-4} ton

1 amu equals $\dfrac{1}{(6.022 \times 10^{23})}$ grams

1 amu equals _____

Answer: $\dfrac{1}{(6.022 \times 10^{-23})} = \left(\dfrac{1}{6.022}\right) \times 10^{23} = 0.166 \times 10^{-23} = 1.66 \times 10^{-24}$ grams

39 You should memorize Avogadro's number (6.022×10^{23}) and its meaning.

(a) How many atoms are in an atomic weight expressed in grams? _____

(b) One gram is equivalent to the weight of how many amu? _____

Answer: (a) 6.022×10^{23} atoms per gram atomic weight; (b) 6.022×10^{23} amu per gram

40   The average atomic weight of neon is 20.180 according to the periodic table (rounded off to the nearest thousandth).

(a) The average weight of one neon atom is how many amu? _____

(b) The average weight of 6.022×10^{23} neon atoms is how many grams? _____

Answer: (a) 20.180; (b) 20.180

41   There are 6.022×10^{23} atoms in 1 gram atomic weight (atomic weight expressed in grams). The element neon has a gram atomic weight of 20.180 g. How many atoms are contained in 10.090 grams of neon? _____ (Hint: 10.090 grams of neon is half of the gram atomic weight of neon.)

Answer: Since 10.090 grams is half of a gram atomic weight, the number of atoms is half of 6.022×10^{23} atoms, or 3.011×10^{23} atoms.

UNIT FACTOR ANALYSIS (FACTOR LABEL ANALYSIS/DIMENSIONAL ANALYSIS)

42   If the problem in frame 41 had been more difficult, we would have used the **unit factor method** (also called **factor label analysis** and **dimensional analysis** in some texts) for solving problems, a mathematical procedure used in most chemistry textbooks for its convenience in calculations. We would have set up the problem in the manner shown below.

$$10.090 \text{ grams of Ne} \times \frac{1 \text{ gram atomic weight}}{20.180 \text{ grams of Ne}} \times \frac{6.022 \times 10^{23} \text{ atoms}}{1 \text{ gram atomic weight}}$$

$$= 3.011 \times 10^{23} \text{ atoms}$$

All of the unit names except "atoms" cancel out during multiplication.

$$10.090 \; \cancel{\text{grams of Ne}} \times \frac{1 \; \cancel{\text{gram atomic weight}}}{20.180 \; \cancel{\text{grams of Ne}}} \times \frac{6.022 \times 10^{23} \text{ atoms}}{1 \; \cancel{\text{gram atomic weight}}}$$

$$= 3.011 \times 10^{23} \text{ atoms}$$

Just the numbers and the name "atoms" are left after cancelation of unit names.

$$10.090 \times \frac{1}{20.180} \times \frac{6.022 \times 10^{23} \text{ atoms}}{1} = 3.011 \times 10^{23} \text{ atoms}$$

The unit factor method involves multiplying the given value by one or more conversion factors. In this problem, we multiplied the 10.0895 grams of Ne by the conversion factors of

$$\frac{1 \text{ gram atomic weight}}{20.180 \text{ grams of Ne}} \text{ and } \frac{6.022 \times 10^{23} \text{ atoms}}{1 \text{ gram atomic weight}}$$

The conversion factors come from definitions. For example, 1 gram atomic weight is equal to 6.022×10^{23} atoms, and 20.180 grams of Ne are equal to 1 gram atomic weight of Ne. The conversion factors are arranged so that the unit names will cancel out. They convert the units of the given values to those of the answer being sought and will give us the correct numerical answer.

Here is another example using the unit factor method. The necessary definitions are: 1 meter = 39.37 inches, and 1 yard = 36 inches.

The first conversion factor can be $\dfrac{(39.37 \text{ inches})}{(1 \text{ meter})}$ or $\dfrac{(1 \text{ meter})}{(39.37 \text{ inches})}$.

The second conversion factor can be $\dfrac{(36 \text{ inches})}{(1 \text{ yard})}$ or $\dfrac{(1 \text{ yard})}{(36 \text{ inches})}$.

Suppose we wish to determine the number of meters in 2.12 yards. We arrange the conversion factors so that the answer will be in meters and all other unit names will cancel.

$$2.12 \text{ yards} \times \frac{1 \text{ meter}}{39.37 \text{ inches}} \times \frac{36 \text{ inches}}{1 \text{ yards}}$$

$$= 1.94 \text{ meters (rounded to the nearest hundredth)}$$

Use these conversion factors to determine the number of yards in 3.55 meters (nearest hundredth).

Answer: $3.55 \text{ meters} \times \dfrac{39.37 \text{ inches}}{1 \text{ meter}} \times \dfrac{1 \text{ yard}}{36 \text{ inches}} = 3.88 \text{ yards}$

43 The gram atomic weight of silicon is 28.09 grams. Using the unit factor method shown in frame 42, calculate the number of atoms in 2.00 grams of silicon (Si).

Answer: $2.00 \text{ grams of Si} \times \dfrac{1 \text{ gram atomic weight}}{28.09 \text{ grams of Si}} \times \dfrac{6.022 \times 10^{23} \text{ atoms}}{1 \text{ gram atomic weight}}$

$$= 4.29 \times 10^{22} \text{ atoms}$$

44 The atomic weight of silicon is 28.09. Determine the average weight (in amu) of one silicon atom. _____ amu

Answer: **28.09 (same as the atomic weight).**

45 There is a big difference between an atomic weight expressed in amu and an atomic weight expressed in grams. You have already determined that one silicon atom weighs 28.09 amu. Now calculate the weight of one silicon atom in grams.

$$\frac{28.09 \text{ grams of Si}}{1 \text{ gram atomic weight}} \times \frac{1 \text{ gram atomic weight}}{6.022 \times 10^{23} \text{ atoms}} = \underline{\qquad} \text{ grams/atom}$$

Answer: **4.66 × 10⁻²³**

46 Determine the mass in grams of one boron (B) atom. (The gram atomic weight of boron is 10.81 grams.)

Answer: $\dfrac{10.81 \text{ grams of B}}{1 \text{ gram atomic weight}} \times \dfrac{1 \text{ gram atomic weight}}{6.022 \times 10^{23} \text{ atoms}}$

$= 1.795 \times 10^{-23} \text{ grams/atom}$

47 If one boron atom weighs 1.795×10^{-23} grams, what would 1,000,000 atoms of boron weigh? $(1,000,000 = 1 \times 10^6)$

Answer: $1 \times 10^6 \text{ atoms of B} \times \dfrac{1.795 \times 10^{-23} \text{ grams}}{1 \text{ atom of B}} = 1.795 \times 10^{-17} \text{grams}$

48 The atomic weight of gold is 196.97. If you were offered an atomic weight of gold for only one dollar, would you buy it? Why or why not?

Answer: We wouldn't. The atomic weight of gold would be expressed as 196.97 amu, so you are buying one atom of gold for a dollar, definitely no bargain. (However, a *gram* atomic weight of gold weighs 196.97 grams, which would definitely be a bargain for one dollar.)

49 How much would a billion billion atoms $(1 \times 10^{18} \text{ atoms})$ of magnesium weigh? The gram atomic weight of magnesium is 24.305 grams. (Hint: Find the weight of one atom of magnesium and multiply the result by 1×10^{18}.)

Answer: Here are two methods for solving this problem. The first method is one long step.

$$1 \times 10^{18} \text{ atoms of Mg} \times \frac{1 \text{ gram atomic weight}}{6.022 \times 10^{23} \text{ atoms}} \times \frac{24.305 \text{ grams}}{1 \text{ gram atomic weight}}$$

$= 4.04 \times 10^{-5} \text{ grams of Mg}$

The second method involves two shorter steps.

$$\frac{24.305 \text{ grams}}{1 \text{ gram atomic weight}} \times \frac{1 \text{ gram atomic weight}}{6.022 \times 10^{23} \text{ atoms}} = 4.04 \times 10^{-23} \text{ grams/atom}$$

$$1 \times 10^{18} \text{ atoms of Mg} \times \frac{4.04 \times 10^{-23} \text{ grams}}{1 \text{ atom}} = 4.04 \times 10^{-5} \text{ grams of Mg}$$

What are the most important concepts in this chapter?

- The weights assigned to atoms are relative weights. That is, all atoms are compared to the weight of a single $^{12}_{6}C$ atom.

- All atoms of the same element do not have the same weight (elements exist as isotopes).

- The isotopes of an element are not present in nature in equal amounts.

- The atomic weight of an element may be determined experimentally.

- Atoms have very small weights. A chemist deals with large numbers of atoms (on the order of 6×10^{23}) in the laboratory.

In Chapter 3 you will see that atomic weight is one of several periodic properties. You will encounter atomic weights throughout the remainder of this book, especially in Chapters 4 and 7, so you should have their meaning and significance clearly in mind before proceeding.

SELF-TEST

This self-test is designed to show how well you have mastered this chapter's objectives. Correct answers and review instructions follow the test. Calculate answers to the nearest hundredth unless otherwise indicated.

1. Isotopes of the element neon were first discovered by two English scientists, _____ and _____.

2. What is a mass spectrometer used for? _____

3. In a mass spectrometer's magnetic field which atoms are affected to a greater degree? _____ (lighter or heavier mass)

4. The atoms $^{35}_{17}Cl$ and $^{37}_{17}Cl$ are called _____ of chlorine.

5. How many protons, neutrons, and electrons are found in each of the chlorine isotopes in question 4?

6. How many neutrons are in the following carbon isotopes?
   carbon-12 _____, carbon-13 _____, carbon-14 _____

7. What are the atomic weights of the following elements?
chromium (Cr) _____, potassium (K) _____, aluminum (Al)
_____

8. What are the atomic weights of the following elements?
osmium (Os) _____, calcium (Ca) _____, gallium (Ga) _____

9. What elements have the following atomic weights?
4.00 _____
183.84 _____
244 _____

10. You are given 0.100 gram atomic weight of gold (Au). The atomic weight of
gold is 196.97 grams to the nearest hundredth.

    (a) How many atoms would you have? _____

    (b) How many grams? _____

11. A sample of iron has a total of 9.77×10^{21} atoms of iron. Calculate the mass,
in grams, of this sample.

12. Calculate the number of platinum atoms in a 15.5-gram sample of platinum.
The atomic weight of platinum is 195.08 amu to the nearest hundredth.

13. The atomic weight of iron is 55.85 amu to the nearest hundredth and it has
isotopes with approximate masses of 55.000 amu and 56.000 amu. What is
the proportion of the $^{56}_{26}$Fe isotope? (The other isotope is $^{55}_{26}$Fe.)

14. The atomic weight of sodium is 22.99 amu to the nearest hundredth. A sample has isotopes with approximate masses of 20.998 and 23.991 amu. What
is the proportion of the ^{24}Na isotope? (The other isotope is ^{21}Na.)

15. If you could cash it in, which would you rather have, 1×10^{40} amu of silver or
150 grams? (Hint: The problem is to determine which value has the greater
number of atoms.)

ANSWERS

Compare your answers to the self-test with those given below. If you answer all questions
correctly, you are ready to proceed to the next chapter. If you miss any, review the frames
indicated in parentheses following the answers. If you miss several questions, you should
probably reread the chapter carefully.

1. J.J. Thomson and F. W. Aston (frame 6)
2. to detect the presence of isotopes of each element (frame 6)
3. *lighter* atoms are affected to a greater degree (frame 6)
4. isotopes (frames 5, 6)
5. Chlorine-35 has 17 protons, 18 neutrons, and 17 electrons. Chlorine-37 has 17 protons,
20 neutrons, and 17 electrons (frames 1–5).

6. Carbon-12 has six neutrons, carbon-13 has seven neutrons, and carbon-14 has eight neutrons (frames 1–5).

7. Cr = 51.996, K = 39.0983, Al = 26.98154 (frame 10)

8. Os = 190.23, Ca = 40.078, Ga = 69.723 (frame 10)

9. He (4.00), W (183.84), Pu (244) (frame 10)

10. (a) $0.100 \text{ gram atomic weight} \times \dfrac{6.022 \times 10^{23} \text{ atoms}}{1 \text{ gram atomic weight}} = 6.022 \times 10^{22} \text{ atoms Au}$

    (b) $0.100 \text{ gram atomic weight} \times \dfrac{196.97 \text{ grams}}{1 \text{ gram atomic weight}} = 19.70 \text{ grams Au}$

    (frames 34, 39–43)

11. 0.906 g Fe (frames 42–49)

12. 4.78×10^{22} atoms Pt

13. Following the examples in the tables in frames 27-31, let A represent the proportion of $^{56}_{26}\text{Fe}$ isotope, and let $1 - A$ represent the proportion $^{55}_{26}\text{Fe}$ isotope.

    $56.000A + 55.000(1 - A) = 55.85$

    $56.000A + 55.000 - 55.000A = 55.85$

    $1.000A = 0.85$

    $A = 0.85$ (proportion of Fe which is the $^{56}_{26}\text{Fe}$ isotope)

    (frames 21–32)

14. Let A represent proportion: ^{24}Na isotope, and let $1 - A$ represent proportion ^{21}Na isotope.

    $23.991A + 20.998(1 - A) = 22.99$

    $23.991A + 20.998 - 20.998A = 22.99$

    $2.993A = 1.99$

    $A = 0.66$ (proportion of ^{24}Na)

    (frames 21–32)

15. Ag = 107.87 amu/atom, so our two equivalences are:

    $$\text{atoms} = 1 \times 10^{40} \text{ amu} \times \frac{1 \text{ atom}}{107.87 \text{ amu}} = 9.27 \times 10^{37} \text{ atoms}$$

    $$\text{atoms} = 150 \text{ grams} \times \frac{1 \text{ gram atomic weight}}{107.87 \text{ grams}} \times \frac{6.022 \times 10^{23} \text{ atoms}}{1 \text{ gram atomic weight}}$$

    $$= 8.38 \times 10^{23} \text{ atoms}$$

    You would probably rather cash in the 1×10^{40} amu.

    (frames 47–49)

KILOGRAM'S CHANGE

Have you ever considered how important it is to have standards of measure? Without them how would we know how much mass an object possesses or how much time it takes for a chemical reaction to occur? There are many standards of measure for all kinds of measurements whether it be for mass, length, volume, time, etc.

Around the world the standard unit for mass is the kilogram. For more than 100 years the kilogram was defined by the mass of a platinum–iridium alloy that was housed at the International Bureau of Weights and Measures in Paris, France. The kilogram mass has served as the base unit of mass in the International System of Units (SI) from 1889 until present day.

The international unit for mass, the kilogram, is used in chemistry and physics for a variety of applications. For example, if a chemist needs to know how many kilograms are in a sample of carbon s/he can obtain that through the atomic mass of carbon from the periodic table and then convert to kilograms.

Also, we can convert kilograms to another unit of measure. For instance, the mass of a 1983 penny is 2.5×10^{-3} kilograms, but we can convert this amount into grams.

$$2.5 \times 10^{-3} \text{ kilograms} \times \frac{1000 \text{ grams}}{1 \text{ kilogram}} = 2.5 \text{ grams}$$

The kilogram is not only involved in converting between moles and kilograms, or kilograms to another unit of measure. The unit of mass used to express atomic and molecular weights is equal to one-twelfth of the mass of an atom of carbon-12, which is approximately 1.66×10^{-27} kilograms.

In addition to the kilogram there are 17 derived units that are defined in relation to the kilogram. In other words the kilogram is part of their unit of measure. Some of these include the Newton (N), Pascal (Pa), Joule (J), volt (V), Ohm (Ω), and Sievert (Sv). If we look at forces and Newton's second law of motion we see that force (F) $= ma$. Force is measured in Newtons, where m is the mass in kilograms and a is the acceleration of the object. One Newton is the equivalent of one kilogram multiplied by a meter per second squared.

$$1 \text{ Newton} = 1 \text{ kilogram} \times 1 \text{ meter/second}^2$$

As you can imagine, how a kilogram is defined affects every derived unit mentioned. As of 20 May 2019 the International Committee for Weights and Measures approved a proposed redefinition of the kilogram. This redefinition, although small in its change, defines the kilogram in terms of the second and the meter. With the change in the kilogram's definition we can obtain even greater accuracy with the results obtained from a variety of experiments.

3 Periodic Properties and Chemical Bonding

In Chapter 1 you learned that the elements in a horizontal row of the periodic table show regular variation in properties from left to right. The elements are arranged in the table in order by increasing atomic number (reading the table left to right, line by line, in the way you are reading this paragraph). The reason this arrangement works so well is that all atoms consist of electrons, protons, and neutrons. The neutrons and protons are in the nucleus with electrons arranged in "shells" around the nucleus.

Why did we consider the electronic arrangement of atoms in such detail? Because the chemical properties of an element depend upon the number of electrons in its outermost shell, the energy levels of its outermost electrons, and the size of the atom. These details of atomic structure determine what kinds and how many chemical bonds can be formed by an atom.

In this chapter we discuss several properties not mentioned in Chapter 1 that depend upon the outermost shell electronic structure. We will review electron configuration and introduce new "dot" symbols. We will then discuss whether atoms gain, lose, or share electrons, and how many, when they combine to form new substances. The major portion of the chapter is devoted to the types of chemical bonds (ionic, covalent, polar covalent) formed between atoms in chemical compounds. Finally, we look briefly at the shapes of compounds and ionization energy.

OBJECTIVES

After completing this chapter, you will be able to

- recognize and apply or illustrate: ion, ionic, covalent, coordinate covalent, molecule, compound, electron dot symbol, formula, electronegativity, polar, ionization energy, metallic character, the octet rule, nonmetallic character, and valence shell electrons;

- write electron dot symbols that represent the number of valence shell electrons for the Group A elements and noble gases;

- write the symbol for the ion formed by any of the Group A elements;

- write equations using dot symbols to show the formation of ionic, covalent, and coordinate covalent compounds;

- predict the type of bond (ionic, covalent, polar covalent) formed when any two elements combine when given the electronegativity of the elements;

- use the octet rule to choose a correct dot structure for a compound;

- predict whether a molecule containing three or more atoms would be polar or nonpolar if given its shape;

- predict whether an element, according to its position in the periodic table: (1) is metallic or nonmetallic, (2) has a high or low electronegativity, (3) has a high or low ionization energy, and (4) will gain or lose electrons and how many.

OUTER SHELL ELECTRONS

1 The following chart represents the first 20 elements of the periodic table. Only the symbols and atomic numbers have been included.

IA							VIIIA
1 **H**	IIA	IIIA	IVA	VA	VIA	VIIA	2 **He**
3 **Li**	4 **Be**	5 **B**	6 **C**	7 **N**	8 **O**	9 **F**	10 **Ne**
11 **Na**	12 **Mg**	13 **Al**	14 **Si**	15 **P**	16 **S**	17 **Cl**	18 **Ar**
19 **K**	20 **Ca**						

The atomic number increases by one each time as the elements are viewed from left to right (11 to 12 to 13 to 14 to 15, and so on). Each time the atomic number increases by one, how many electrons are added?_____

Answer: one (Remember that the atomic number represents the number of protons. The symbols in the periodic table represent atoms of neutral elements. In a neutral atom, the number of protons equals the number of electrons.)

2   Periodic tables can be made to include different kinds of information. The following table includes the electronic structure for the *outermost shell* for each of the first 20 elements.

IA							VIIIA
H $1s^1$	**IIA**	**IIIA** **IVA** **VA** **VIA** **VIIA**					**He** $1s^2$
Li $2s^1$	**Be** $2s^2$	**B** $2s^22p^1$	**C** $2s^22p^2$ **N** $2s^22p^3$ **O** $2s^22p^4$ **F** $2s^22p^5$				**Ne** $2s^22p^6$
Na $3s^1$	**Mg** $3s^2$	**Al** $3s^23p^1$	**Si** $3s^23p^2$ **P** $3s^23p^3$ **S** $3s^23p^4$ **Cl** $3s^23p^5$				**Ar** $3s^23p^6$
K $4s^1$	**Ca** $4s^2$						

There are similarities in the structure of elements within groups. Notice the structure of each outermost shell in Group IA. How many electrons are present in the outer shell of each element in Group IA (H, Li, Na, K, and others)? _____

Answer: one (Each element in Group IA has only one electron in its outermost shell.)

3   Use the periodic table in frame 2 for the following statements.

(a) Group IIA elements each have two electrons in their outermost shell. Group IIIA elements each have _____ electrons in their outermost shell.

(b) Group IVA elements each have _____ electrons in their outermost shell.

(c) Group VA elements each have five, Group VIA elements each have _____, and Group VIIA elements each have _____ electrons in their outermost shell.

(d) Group VIIIA elements (the noble gases) each have eight electrons in their outermost shell *with the exception of helium*, which has only _____ electrons.

Answer: (a) three; (b) four; (c) six, seven; (d) two

4 Nitrogen (N), a Group VA element, has _____ electrons in its outermost shell. Aluminum (Al), a Group IIIA element, has _____ electrons in its outermost shell.

Answer: five; three

5 The outer shell electrons are also known as **valence electrons**. The periodic table on the next page includes all of the symbols for the first 20 elements. The numbers of valence electrons are listed for lithium (Li), carbon (C), and argon (Ar). Fill in the number of valence electrons for each remaining element in this periodic table.

Remember that helium (He) is an exception in Group VIIIA, since it only has a total of two electrons in its entire atom, both of which are valence electrons.

IA							VIIIA
( ) H							( ) He
	IIA	IIIA	IVA	VA	VIA	VIIA	
(1) Li	( ) Be	( ) B	(4) C	( ) N	( ) O	( ) F	( ) Ne
( ) Na	( ) Mg	( ) Al	( ) Si	( ) P	( ) S	( ) Cl	(8) Ar
( ) K	( ) Ca						

Answer:

IA							VIIIA
(1) H							(2) He
	IIA	IIIA	IVA	VA	VIA	VIIA	
(1) Li	(2) Be	(3) B	(4) C	(5) N	(6) O	(7) F	(8) Ne
(1) Na	(2) Mg	(3) Al	(4) Si	(5) P	(6) S	(7) Cl	(8) Ar
(1) K	(2) Ca						

6 The outer shell or valence electrons are especially important because those electrons are involved when atoms unite chemically to form compounds.

When an atom of magnesium unites with another different atom to form a compound, what electrons of the magnesium atom are primarily involved?

Answer: the two valence or outer shell electrons

ELECTRON DOT SYMBOLS (LEWIS SYMBOLS)

7 The outer shell or valence electrons may also be represented by a series of dots. Beryllium, with two valence electrons, can be represented as Be: (each dot represents one valence electron).

:C̈l· represents a chlorine atom with how many valence electrons? _____

Answer: seven

8 The first 20 elements and their electron dot symbols are as follows.

·H							:He
·Li	:Be	:Ḃ	:Ċ·	:N̈·	:Ö·	:F̈·	:N̈e:
·Na	:Mg	:Ȧl	:Ṡi·	:P̈·	:S̈·	:C̈l·	:Är:
·K	:Ca						

Note that some books represent boron as ·Ḃ·, aluminum as ·Ȧl·, carbon as ·Ċ·, and silicon as ·Ṡi·. Such a representation is used to simplify the explanation of how many bonds a given element may have or may form with other elements, but it does not agree with the quantum mechanical concept of the atom (Chapter 1).

We will discuss the *number* of bonds an element has in a compound in Chapter 14, Organic Chemistry. Later in this chapter we discuss a type of bonding that requires agreement with what is called the octet rule and involves pairs of electrons, so we have chosen to use the scheme shown in this table.

Which elements in the table have five valence electrons? _____

Answer: :N̈· and :P̈·

9 Notice in the table that, as we go from left to right, the first two electrons pair up, while all additional electrons remain single until all four sides of a symbol are occupied. The symbol :Mg is the same as M̈g or Mg: or ῼg. The placement of the dots is not critical as long as the proper number of electron pairs and unpaired electrons are represented. The symbol Äl is the same as Äl· or ·Äl or Äl· or any arrangement of one pair and one single electron. Note that ·Äl· is *not correct* since one pair and a single electron are called for, *not* three single electrons.

Nitrogen is pictured in electron dot symbols as :N̈· with one pair and three single electrons. Which of the following are also correct symbols for nitrogen? :N̈· ·N̈· :N̈: ·N̈: _____

Answer: ·N̈· and ·N̈: (since they show three single electrons and one pair)

10 The dot symbols showing the valence electrons for the third period elements of the periodic table are as follows.

Na·   Mg: Äl·   ·S̈i·   ·P̈·   ·S̈:   :C̈l·   :Är:

Remember that the first two electrons pair up while all additional electrons remain single until all four sides of a symbol are occupied. The first pair of electrons represent an *s* subshell while the next six electrons represent a *p* subshell. By the **Principle of Maximum Multiplicity**, electrons in a subshell prefer to remain unpaired until each orbital contains one electron. The dot diagrams reflect this principle.

The element arsenic (As) has five valence electrons in its outermost shell. The outermost shell consists of two electrons in an *s* subshell and three electrons in a *p* subshell. Draw an electron dot symbol for As. _____

Answer: :Äs· (or any arrangement with one pair and three single electrons)

11 The element tellurium (Te) has six valence electrons. Its outermost shell consists of a completed *s* subshell and four electrons in a *p* subshell. Draw an electron dot symbol for Te. _____

Answer: $\cdot\ddot{Te}\colon$ (or any arrangement with two single electrons and two pairs of electrons)

At this point, you may be wondering how these electron dot symbols can be used. You already know that the dots represent valence electrons and that the valence electrons are especially important because they are involved when two or more atoms combine chemically to form compounds. The electron dot symbol arrangement changes when atoms combine to form compounds.

Before you can use the dot symbols to show compounds, however, you must learn to distinguish between two major categories of chemical compounds.

These two categories, ionic and covalent, will be introduced next. The valence electron arrangements for these two kinds of compounds are different, but both arrangements can be shown through the use of dot symbols.

IONS

12 An **ion** is an atom or group of atoms that is no longer neutral. The numbers of electrons and protons in an ion are *not equal*. You have learned that a proton has a positive (+) charge and an electron has a negative (−) charge. A neutral atom, in contrast, has an equal number of + and − charges, with the net result that the charges cancel each other. The number of protons (atomic number) remains unchanged when an atom becomes an ion, but one or more electrons may be gained or lost. This results in an ion with either fewer or more electrons than protons.

(a) Oxygen, with an atomic number of 8, was found to have 10 electrons. Is it an ion or a neutral atom? _____

(b) Aluminum, with an atomic number of 13, also was found to have 10 electrons. Is it an ion or a neutral atom? _____

(c) Magnesium, with an atomic number of 12, was found to have 12 electrons. Is it an ion or a neutral atom? _____

Answer: (a) ion; (b) ion; (c) neutral atom

13 The symbol for an ion is determined as follows, using two of the examples from frame 12.

An oxygen ion with 10 electrons and an atomic number of 8 has gained two extra electrons. Since each electron has a negative (−) charge, the oxygen ion has a 2– charge. The 2– is written to the upper right of the (O) symbol to indicate two extra electrons, O^{2-}. The aluminum ion has 10 electrons and an atomic number of 13. It has *lost* three electrons, and its symbol is Al^{3+}. The 3+ to the upper right of Al indicates that the protons outnumber the electrons by three.

The calcium ion (Ca^{2+}) has 20 protons (same as atomic number) and _____ electrons.

Answer: 18 (It has lost two electrons; therefore, the protons outnumber the electrons by two.)

14 A bromide ion has gained one electron. It is symbolized by Br^-. A sodium ion has lost one electron. It is symbolized by Na^+. (Note that the number 1 is implied and is not written as part of the symbol for an ion.)

A lithium ion is symbolized as Li^+. It has (lost, gained) _____ electrons. How many? _____

Answer: lost; one

15 If copper (Cu) has lost two electrons, write the symbol for its ion. _____
If sulfur (S) has gained two electrons, write the symbol for its ion. _____
If fluorine (F) has gained one electron, write the symbol for its ion. _____

Answer: Cu^{2+}, S^{2-}, F^-

16 An ion is formed when a neutral atom gains or loses one or more electrons. You may be wondering why an atom would gain or lose electrons to form an ion. As a general rule, atoms tend to form sets of eight electrons in their outermost shells (valence electrons) when they combine to form compounds. Atoms with six or seven valence electrons have a great attraction for one or two extra electrons, and atoms with only one, two, or three valence electrons have only a weak hold on those electrons. An atom with six or seven valence electrons will tend to gain electrons from another atom with one, two, or three valence electrons when those atoms combine chemically.

Calcium (Ca) has two valence electrons. Sulfur (S) has six valence electrons. If an atom of calcium and an atom of sulfur are combined chemically, which atom would gain electrons and which atom would lose electrons?

_____

Answer: Sulfur would gain electrons while calcium would lose electrons.

17   By gaining two electrons, the sulfur atom, which had six valence (outer shell) electrons, becomes an ion with a total of eight outer shell electrons. The ion symbol is S^{2-}. By losing two valence electrons, the calcium atom becomes a positive ion symbolized by Ca^{2+}. The general tendency of atoms to form sets of eight electrons in their outermost shells (valence electrons) when they combine to form compounds is commonly called the **octet rule**.

(a) When a calcium ion and a sulfur ion combine to form a compound, the sulfur ion (S^{2-}) has a total of how many electrons in its outermost shell? _____

(b) According to what rule? _____

Answer: (a) eight; (b) the octet rule

18   The noble gases (Group VIIIA in the periodic table) already have eight electrons in their outermost shells (with the exception of helium). When atoms combine to form compounds, the atoms generally tend to form electron configurations that are similar to those of the noble gases.

A neutral bromine (Br) atom has 35 electrons in its electron configuration. Its total electron configuration is $1s^2\ 2s^2\ 2p^6\ 3s^2\ 3p^6\ 4s^2\ 3d^{10}\ 4p^5$. Its outermost shell configuration (the fourth shell) is represented by $4s^2\ 4p^5$, which adds up to $2 + 5 = 7$ outer shell (valence) electrons. When combining chemically with another atom such as sodium, the bromine atom will gain one electron to form the bromide ion (Br^-). The bromine *atom* (Br) has an electron configuration of $1s^2\ 2s^2\ 2p^6\ 3s^2\ 3p^6\ 4s^2\ 3d^{10}\ 4p^5$. The bromide *ion* ($Br^-$) has an electron configuration of $1s^2\ 2s^2\ 2p^6\ 3s^2\ 3p^6\ 4s^2\ 3d^{10}\ 4p^6$.

The bromine *atom* has seven outer shell electrons and a total of 35 electrons. The bromide *ion,* however, has how many outer shell electrons? _____ How many total electrons? _____

Answer: eight; 36

19   Remember that when atoms combine to form compounds, the atoms generally tend to form electron configurations that are similar to those of the noble gases. When a bromine atom becomes an ion, it gains one electron. Instead of having 35 electrons in its total electronic structure, it gains one electron to have a total of 36 electrons. The resulting bromide ion has attained an electron configuration that is similar to that of a noble gas atom.

What neutral noble gas atom has a total of 36 electrons? (Hint: Use the periodic table for your answer.) _____

Answer: Krypton (Kr) has an atomic number of 36 and, therefore, has 36 electrons in its neutral atom.

20 When chlorine (Cl) and sodium (Na) combine to form NaCl, both atoms become ions. Chlorine gains one electron from sodium to become the Cl^- ion. The sodium atom loses its one outer shell electron to chlorine and becomes the Na^+ ion.

The sodium atom originally had 11 electrons in its electron configuration ($1s^2\ 2s^2\ 2p^6\ 3s^1$). In becoming an ion, sodium loses its only outer shell electron ($3s^1$) and becomes the Na^+ ion with a total of 10 electrons. What neutral atom has the same number of electrons as the sodium *ion* (Na^+)? _____ To what family of elements does that neutral atom belong? (Use the periodic table.) _____

Answer: Ne (neon) (has 10 electrons); noble gas family

21 All of the elements in Group VIIA of the periodic table (F, Cl, Br, I, At and Ts) have outer shell structures with seven electrons. Each element in Group VIIA has just one less electron than the corresponding noble gas in each period and each, when forming an ion, tends to gain one electron. You are already familiar with the chloride ion, which is symbolized by Cl^-. The single minus sign to the upper right of the Cl symbol indicates a gain of one electron.

Write the symbol for the ion of each of the following neutral atoms in Group VIIA. Assume that each ion has gained one electron.

Atoms	Ions
F, Cl, Br, I, At	___, Cl^-, ___, ___, ___

Answer: Ions: F^-, Cl^-, Br^-, I^-, At^-

22 An atom of each element in Group VIA has six electrons in its outermost shell. In order to become ions with eight electrons, these atoms must gain two electrons. You are already familiar with the symbol for an oxygen ion (O^{2-}). The symbol represents two electrons gained.

Write the symbols for the ions of each of the following neutral atoms from Group VIA. Each has gained two electrons. (The first has been filled in.)

Atoms	Ions
O, S, Se, Te, Po	O^{2-}, ___, ___, ___, ___

Answer: Ions: O^{2-}, S^{2-}, Se^{2-}, Te^{2-}, Po^{2-}

23 An atom of each element in Group IA has only one electron in its outermost shell. An atom of each element in Group IIA has two electrons in its outermost shell. These atoms readily give up their outer shell electrons to become positive ions. You are already familiar with the sodium ion (Na^+), which has lost one electron, and the calcium ion (Ca^{2+}), which has lost two electrons.

Write the symbols for the ions of each of the following neutral atoms from Group IA and Group IIA. Use the periodic table to determine whether an atom is from Group IA or Group IIA.

Atoms	Ions
Ca, K, Ba, Li, Mg, H	Ca^{2+}, ___, ___, ___, ___, ___

Answer: Ions: Ca^{2+}, K^+, Ba^{2+}, Li^+, Mg^{2+}, H^+

24 Many of the elements in groups other than IA, IIA, VIA, and VIIA can commonly form ions with more than one value. In some cases the octet rule is followed, and in some cases it is not. For example, the element copper (Cu) can form Cu^+ as well as Cu^{2+}. In other words, an atom of copper can lose either one electron or two electrons to form two different kinds of ions. When such ions are encountered in this and later chapters, you will be informed of the number of electrons lost or gained or be given the proper symbol for the ion.

Use the periodic table to write the proper ion symbol for each of the following atoms from Groups IA, IIA, VIA, or VIIA.

Atoms	Ions
Br, S, Be, Rb, Na	___, ___, ___, ___, ___

Answer: Ions: Br^-, S^{2-}, Be^{2+}, Rb^+, Na^+

25 A zinc (Zn) atom loses two electrons to become an ion. An iron (Fe) atom loses three electrons to become an ion. Write the proper symbols for these ions.

_____

Answer: Zn^{2+}, Fe^{3+}

26 Metals generally form positive ions (ions with a + charge) while nonmetals generally form negative ions (ions with a − charge). Na^+, Ca^{2+}, and Fe^{3+} are some examples of metallic ions, and Cl^-, O^{2-}, and Br^- are examples of nonmetallic ions.

Metals, in forming ions, would generally be expected to (lose, gain) _____ electrons. Nonmetals, in forming ions, would generally be expected to (lose, gain) _____ electrons.

Answer: lose; gain

IONIC BONDS

27 Compounds can be formed from the combination of a negative ion and a positive ion. Such compounds are called **ionic compounds**. The negative and positive ions of the compound are held together by an **ionic bond** (sometimes called an **electrostatic bond**). You are already familiar with the ionic compound NaCl, which is made up of two ions with opposite charges. One ion is positive and metallic. The other ion is negative and nonmetallic. Write the appropriate symbols for the ions that make up NaCl. _____

Answer: Na^+ and Cl^-

28 An ionic (electrostatic) bond involves oppositely charged ions held together by the attraction from their opposite electrical charges. (Oppositely charged particles attract.) Circle those pairs of ions that could probably form ionic compounds.

K^+ and F^- Mg^{2+} and O^{2-}
Ca^{2+} and Ba^{2+} Br^- and At^-

Answer: K^+ and F^-; Mg^{2+} and O^{2-} (The other pairs of ions have like charges and will not form ionic bonds, since like charges repel.)

29 In an ionic bond, the two ions are of opposite charge. When a potassium (K) atom and a fluorine (F) atom combine to form a compound, the F atom with only seven valence electrons gains one electron from the K atom, which has only one valence electron to lose. The result is the K^+ ion is united in an ionic bond with the F^- ion. Both ions have achieved an electron configuration similar to that of the noble gases and are held together by their opposite electrical charges.

(a) The metallic ion is (K^+, F^-)_____.

(b) The nonmetallic ion is (K^+, F^-)_____.

(c) The K^+ ion has the same number of electrons as which noble gas atom? (Use the periodic table.) _____

(d) The F^- ion has the same number of electrons as which noble gas atom? (Use the periodic table.) _____

Answer: (a) K^+; (b) F^-; (c) Ar (argon) (with 18 electrons); (d) Ne (neon) (with 10 electrons)

30 When the **formula** (set of symbols representing a compound) for an ionic compound is written, the most metallic element is written first. In the case of NaCl, it should be clear that Na is much more metallic than Cl (since Na is a metal and Cl is a nonmetal). As you may remember, the most metallic elements are located on the left side of the periodic table and the most nonmetallic elements are located on the right. Which of the following ionic compounds are correctly written with the most metallic element first?

MgO, FK, LiBr, OCa _____

Answer: MgO and LiBr (The other two are incorrect. Since K is more metallic than Br, it should be written as KF, and since Ca is more metallic than O, the formula is written as CaO.)

31 Up to this point, you have dealt only with compounds made of ions with opposite but equal charges. Suppose we wish to make an ionic compound with Ca^{2+} and Cl^- ions. In this case, the calcium (Ca) atom has two electrons to lose to form an ion, but the chlorine (Cl) atom only needs to gain one electron to form an ion. We can form the compound by using two Cl^- ions for each Ca^{2+} ion. The result is an ionic compound written as $CaCl_2$. The number 2 written at the lower right side (subscript) of Cl in the formula indicates that there are two Cl^- ions for each Ca^{2+} ion. Write the formula for the compound made up of the ions Mg^{2+} and Br^-. _____

Answer: $MgBr_2$ (This requires two Br^- ions for each Mg^{2+} ion.)

32 The formula Li_2O represents an ionic compound. Based upon what you have just learned, how many Li^+ ions are necessary for each O^{2-} ion? _____

Answer: Two Li^+ ions are necessary for each O^{2-} ion.

33 Write the formulas for ionic compounds that are made up of the following sets of ions.

(a) Ca^{2+}, I^- _____
(b) O^{2-}, K^+ _____
(c) Na^+, S^{2-} _____
(d) Fe^{3+}, Cl^- _____

Answer: (a) CaI_2; (b) K_2O; (c) Na_2S; (d) $FeCl_3$

The next few frames discuss a second type of bonding that occurs when atoms *share* electrons instead of actually exchanging them.

COVALENT BONDS

34 In the formation of an ionic (electrostatic) bond, one or more valence electrons are removed from one atom and taken by another atom, and in the process both atoms become ions. The resulting atoms are bonded together by the attraction of their opposite charges.

A different type of bond is the **covalent bond**. In a covalent bond, one or more valence electrons from each atom are *shared*. The resultant compound is made up of **molecules**, not ions.

Carbon monoxide exists as a molecule composed of carbon and oxygen with shared outer shell electrons. Carbon monoxide uses what kind of bonding, ionic or covalent? _____

Answer: covalent

35 Covalent bonds are usually formed between nonmetals. Ionic bonds are usually formed between a metal and a nonmetal. The bond between nitrogen and oxygen in the compound NO would most likely be (ionic, covalent) _____.

Answer: covalent (because both atoms are nonmetals, on the right side of the periodic table)

36 Carbon tetrachloride is composed of one carbon and four chlorine atoms. Both carbon and chlorine are nonmetals. What type of bonding is probably involved in this compound? _____

Answer: covalent

37 Carbon tetrachloride, a compound with covalent bonding, is made up of (ions, molecules) _____.

Answer: molecules

38 Which type of bonding (ionic or covalent) could more likely be expected in a compound whose atoms:

(a) are all nonmetals? _____

(b) share valence electrons? _____

(c) have become ions? _____

(d) have transferred valence electrons from one atom to another? _____

Answer: (a) covalent; (b) covalent; (c) ionic; (d) ionic

39 Electron dot symbols can be used to represent both ionic and covalent bonding. The dot symbols below represent the ionic bonding of **KI** (potassium iodide). A potassium (K) atom transfers its single valence electrons to an iodine (I) atom, and both become ions. The brackets, [ ], show that all eight electrons in the outer shell now belong to the negative ion. The comma placed between the negative and positive ions is there only to separate the two ionic symbols.

$$K\cdot \ + \ :\ddot{\underset{..}{I}}\cdot \ \rightarrow K^+, [:\ddot{\underset{..}{I}}:]^-$$

The following dot symbols represent covalent bonding. Two fluorine atoms share a pair of electrons to become one fluorine gas molecule.

$$:\ddot{\underset{..}{F}}\cdot \ + \ \cdot\ddot{\underset{..}{F}}: \ \rightarrow :\ddot{\underset{..}{F}}\!\!:\!\!\ddot{\underset{..}{F}}:$$

What type of bonding do the following symbols represent?
$$Ca: \ + \ \cdot\ddot{\underset{..}{O}}: \ \rightarrow Ca^{2+}, [:\ddot{\underset{..}{O}}:]^{2-} \ \underline{\qquad\qquad}$$

Answer: ionic bonding (The two valence electrons from Ca were transferred to the O atom, and both atoms become ions.)

40 Assume ionic bonding for the following reaction. Complete the dot symbols equation. The strontium atom gives up both valence electrons to the sulfur atom, and the atoms become ions.

$$Sr: \ + \ \cdot\ddot{\underset{..}{S}}: \ \rightarrow \ \underline{\quad}, [\underline{\quad}]$$

Answer: $\quad Sr^{2+}, [:\ddot{\underset{..}{S}}:]^{2-}$

41 Ionic bonding for gallium iodide, GaI_3, is represented as follows.

$$\ddot{G}\!a\cdot \ + \ 3:\ddot{\underset{..}{I}}: \ \rightarrow Ga^{3+}, 3[:\ddot{\underset{..}{I}}:]^-$$

Note that three iodine atoms each take one electron from the three available valence electrons of gallium. The number 3 placed in front of the brackets indicates three iodide ions each having a charge of 1−.

Complete the ionic bonding equation for $MgCl_2$ in which two chlorine atoms each take one electron from the two available valence electrons of magnesium.

$$\ddot{M}g + 2:\ddot{C}l: \rightarrow \underline{\quad}, [\underline{\quad}]$$

Answer: $Mg^{2+}, 2\ [:\ddot{C}l:]^-$

 42   Look at the following ionic bonding equations.

$$Ca: + \cdot\ddot{O}: \rightarrow Ca^{2+}, [:\ddot{O}:]^{2-}$$

$$\dot{K} + :\ddot{I}\cdot \rightarrow K^+, [:\ddot{I}:]^-$$

$$Sr: + :\ddot{S}\cdot \rightarrow Sr^{2+}, [:\ddot{S}:]^{2-}$$

$$Mg: + 2:\ddot{C}l: \rightarrow Mg^{2+}, 2[:\ddot{C}l:]^-$$

In these equations, how many total outer shell electrons does each negative ion have? (Count the electrons around the negative ion in brackets.) _____

What rule would lead us to expect this? _____

Answer: eight; the octet rule

43   All noble gas atoms, with the exception of helium, have eight outer shell electrons. In following the octet rule, bonding atoms tend to form an outer shell electron configuration similar to that of the noble gases. Fluorine (F_2) is composed of two fluorine atoms and is covalently bonded, as indicated in the following bonding equation.

$$:\ddot{F}\cdot + \cdot\ddot{F}: \rightarrow :\ddot{F}:\ddot{F}:$$

Including the shared electron pair, how many electrons are in the outer shell of each atom? _____

Answer: eight (Note that one pair is shared by both atoms.)

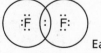 Each circle contains eight electrons.

44   The correct bonding equation for N_2 must be written so that each N atom has eight electrons, following the octet rule. More than one pair of electrons can be shared.

$$\cdot\ddot{N}\cdot + \cdot\ddot{N}\cdot \rightarrow N\ N$$

Knowing that each N atom within the N_2 molecule must have eight electrons, which of the following is the correct outer shell electron structure for

N_2? (You may wish to draw circles around each atom to decide your answer. Be sure to circle *all* of the shared electrons around each atom.)

$$:\overset{\cdot}{N}:\overset{\cdot}{N}: \quad \text{or} \quad :\overset{\cdot}{N}::\overset{\cdot}{N}: \quad \text{or} \quad :N:::N:$$

Answer: :N:::N: (This structure showing three pairs of shared electrons — a triple bond — is the only structure that allows each atom access to eight electrons.)

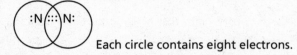

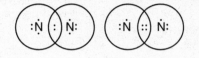 Each circle contains eight electrons.

The others are not correct, as shown by the number of electrons in each circle.

45 Select the proper electron dot structure for the outer shell electrons of the O_2 molecule. Follow the octet rule.

$$:\overset{\cdot\cdot}{\underset{\cdot}{O}}\cdot + :\overset{\cdot\cdot}{\underset{\cdot}{O}}\cdot \rightarrow \underline{\hspace{2cm}} \quad (:\overset{\cdot\cdot}{O}::\overset{\cdot\cdot}{O}: \text{ or } \cdot\overset{\cdot\cdot}{O}::\overset{\cdot\cdot}{O}\cdot \text{ or } :\overset{\cdot\cdot}{O}:::\overset{\cdot\cdot}{O}: \text{ or } :\overset{\cdot\cdot}{O}::\overset{\cdot\cdot}{O}:)$$

Answer: :Ö· or :Ö::Ö: (These structures give each atom eight electrons. Two pairs of electrons must be shared to give each atom eight electrons.)

46 Which of the following is the correct representation of the valence electrons of the carbon dioxide molecule? Four covalent bonds are included in the molecules.

$$\cdot\overset{\cdot\cdot}{C}\cdot + 2\overset{\cdot\cdot}{\underset{\cdot}{O}}\cdot \rightarrow \underline{\hspace{2cm}} \quad (:\overset{\cdot\cdot}{O}: C:\overset{\cdot\cdot}{O}: \text{ or } :\overset{\cdot\cdot}{O}::C::\overset{\cdot\cdot}{O}: \text{ or } :O:::C:::O:)$$

Answer: :Ö:: C ::Ö: (This is the only structure that allows each atom eight electrons, no more or less.)

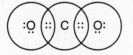

47 The H_2 molecule electron structure shows that each hydrogen atom prefers the helium noble gas structure with two shared electrons.

$$H\cdot + H\cdot \rightarrow H:H$$

Determine the electron outer shell structure of the HCl gas molecule. The hydrogen atom prefers the helium structure and the chlorine atom follows the octet rule.

$$\text{H} \cdot + \; :\!\overset{..}{\underset{..}{\text{Cl}}}\!: \rightarrow \underline{\qquad}$$

Answer: $\text{H} : \overset{..}{\underset{..}{\text{Cl}}} :$ (H has two electrons and Cl has eight electrons.)

48 Not all compounds follow the octet rule, but it serves as a useful rule for showing most electronic bonding structures. The general tendency in bonding is for atoms or ions to attain the noble gas outer shell structure.

With the exception of helium, the noble gases have how many electrons in their outer shells? _____ Helium has how many electrons in its outermost (and only) shell? _____

Answer: eight; two

49 In a regular covalent bond, a shared pair of electrons originates when each of two atoms supplies one electron. A special type of covalent bond is known as the coordinate covalent bond. In a **coordinate covalent bond**, both electrons in a shared pair of electrons come from the same atom. The bonding between the nitrogen atom in ammonia (NH_3) and the boron atom in boron trifluoride (BF3) is an example of a coordinate covalent bond.

$$
\begin{array}{ccccc}
\text{H} & & :\!\overset{..}{\underset{..}{\text{F}}}\!: & & \text{H} : \overset{..}{\underset{..}{\text{F}}} : \\
\text{H} : \overset{..}{\underset{..}{\text{N}}} : & + & \text{B} : \overset{..}{\underset{..}{\text{F}}} : & \rightarrow & \text{H} : \overset{..}{\underset{..}{\text{N}}} : \text{B} : \overset{..}{\underset{..}{\text{F}}} : \\
\text{H} & & :\!\overset{..}{\underset{..}{\text{F}}}\!: & & \text{H} : \overset{..}{\underset{..}{\text{F}}} :
\end{array}
$$

In the above equation, which atom supplied *both* electrons to form the coordinate covalent bond, H, N, B, or F? _____

Answer: The N atom supplied both electrons to form a coordinate covalent bond. (Notice that in NH_3 on the left side of the equation, the N is already shown with eight electrons.)

50 A coordinate covalent bond is like any other covalent bond except for the origin of the shared electron pair. Does the molecule below conform to the octet rule (with the exception of the hydrogen atoms)? _____

$$
\begin{array}{c}
\text{H} : \overset{..}{\underset{..}{\text{F}}} : \\
\text{H} : \overset{..}{\underset{..}{\text{N}}} : \text{B} : \overset{..}{\underset{..}{\text{F}}} : \\
\text{H} : \overset{..}{\underset{..}{\text{F}}} :
\end{array}
$$

Answer: Yes, all the atoms in the molecule (except the hydrogen atoms) have eight outer shell electrons.

51 Review your understanding of ionic, covalent, and coordinate covalent bonding.

(a) A bond in which one or more electrons are removed from one atom and taken by another is called a(n) _____ bond.

(b) A bond in which two atoms share a pair of electrons with one electron coming from each atom is called a(n) _____ bond.

(c) A bond in which two atoms share a pair of electrons with both electrons coming from one atom is called a(n)_____ bond.

Answer: (a) ionic; (b) covalent; (c) coordinate covalent

Up to this point we have assumed that the electrons in a covalent bond are shared equally between the bonding atoms, that is, the electrons are midway between the bonding atoms. In the frames that follow we discuss what happens if the electrons are not equally shared, that is, if the electron pair is closer to one atom than the other.

POLAR BONDS

52 When all the atoms in a covalent molecule are identical, the attraction each atom has for shared electrons is equal. When the atoms in a covalent molecule are different, the attraction for the shared electrons is not equal. In a covalent molecule such as Cl_2, the attraction for the shared electron pair by both atoms within the molecule is equal. This is obvious since the two chlorine atoms are exactly the same. A molecule of HCl gas is also covalently bonded, but the attraction for electrons by the H and Cl atom is *not* the same. The Cl atom has a greater attraction for electrons than the H atom.

Which of the following molecules (all have covalent bonds) should have atoms with *equal* attraction for electrons H_2, F_2, HF, NH_3? _____

Answer: H_2 and F_2 are molecules with an identical pair of atoms that equally attract electrons.

53 The attraction an atom has for the electrons in a covalent bond is called **electronegativity**. In the HCl molecule, the shared pair of electrons spend more time nearer the chlorine atom. In the HF molecule, the shared electrons spend more time nearer the F atom. Based on this information, which of the two atoms would be more electronegative, H or F? _____

Answer: F is more electronegative because it attracts the electron pair more than H.

54 A molecule that has atoms with differing electronegativities has a slight + charge on one side of the molecule and a slight − charge on the other side because the shared electron pair is located nearer the more electronegative atom. When this happens, the molecule has "poles" (like the earth) or regions of + and − charge and is said to be **polar**. Later, in Chapter 11, you will see how polarity affects the properties of a compound. HCl is a covalent molecule with a slight − charge on the chlorine side and a slight + charge on the hydrogen side of the molecule. Is HCl a polar molecule? _____

Answer: yes (It has regions of slight + and − charges in the molecule.)

55 HCl is a polar molecule. The Cl atom has a greater electronegativity than the H atom. As a result, the bonding electron pair is shared unequally and spends more time near the Cl than the H. The HCl molecule therefore has a slight negative charge on the Cl side and a slight positive charge on the H side. The slight negative charge is equal but opposite the slight positive charge. Based upon what you have learned so far, which of the following covalent molecules *cannot* be polar molecules: HF, H2, F2, NH_3? _____

Answer: H_2 and F_2 are definitely *not* polar molecules because the atoms are identical. HF is polar (frame 53), but at this time you have insufficient information to judge NH_3.

56 HBr is a polar covalent molecule. The shared bonding electrons are nearer the Br than the H. Since electrons are negative charges, which of the following correctly describes this polar molecule?

(a) Slightly negative H and slightly positive Br _____
(b) Slightly positive H and slightly negative Br _____

Answer: (b)

57 Iodine is more electronegative than hydrogen, so HI is a polar molecule. Which atom is slightly negative? _____ Which is slightly positive? _____

Answer: I; H

58 The small Greek letter δ (delta) is sometimes used to show a partial positive or negative charge on a polar covalent molecule. For example, the HCl molecule could be symbolized as:

$$^{\delta+}H:\overset{..}{\underset{..}{Cl}}:^{\delta-}$$

The partial positive and negative charges are indicated near the proper atoms.

Fluorine is more electronegative than hydrogen. Which of the following symbols is correct?

(a) $^{\delta+}H:\overset{..}{\underset{..}{F}}:^{\delta-}$ _____

(b) $^{\delta-}H:\overset{..}{\underset{..}{F}}:^{\delta+}$ _____

Answer: (a)

59 Carbon monoxide (CO) is also a polar covalent molecule. The structure for CO is

$$:C:::O:$$

Oxygen is more electronegative than carbon. Which of the following is correct?

(a) $^{\delta+}:C:::O:^{\delta-}$ _____

(b) $^{\delta-}:C:::O:^{\delta+}$ _____

Answer: (a)

60 Note that the partial + and − charges in a polar covalent molecule are much smaller than the unit + or − charges of a proton or electron. The polar charge is always much less than the charges that hold ions together in an ionic compound. Which type of bonding has the greater charge difference holding the atoms together: ionic, polar covalent, or covalent? _____

Answer: ionic

61 For compounds consisting of only two atoms, it is possible to classify the bond(s) as covalent, polar covalent, or ionic based upon the electronegativity difference between the atoms. Electronegativity is abbreviated "en," and the electronegativity difference is shown as Δen. We will use this scheme of classification. The compound is:

> covalent if Δen is less than 0.5 en units;
> polar covalent if Δen is between 0.5 and 1.5 en units;
> ionic if Δen is greater than 1.5 en units.

The purpose of such a classification is to provide a rough rule of thumb for predicting the type of bond that might form when two atoms combine. To be sure, *any* difference in electronegativities will result in a polar bond. However, it is the degree or extent of the difference that will permit you to tell whether one compound is more or less polar than other compounds, or whether a particular bond is more or less polar than another or more ionic than covalent. The only time we have a 100% covalent bond is when Δen = 0, which usually happens only when identical atoms combine. Using the table shown below and the scheme above, CO would be classified as a polar covalent compound.

The electronegativity value is the number below the symbol in the table, C = 2.5 and O = 3.5 units. The difference (Δen) is 3.5–2.5 = 1.0 unit, therefore CO is polar covalent.

1 **H** 2.1																	2 **He** —
3 **Li** 1.0	4 **Be** 1.5			5 **B** 2.0	6 **C** 2.5	7 **N** 3.1	8 **O** 3.5	9 **F** 4.1	10 **Ne** —
11 **Na** 1.0	12 **Mg** 1.3			13 **Al** 1.5	14 **Si** 1.8	15 **P** 2.1	16 **S** 2.4	17 **Cl** 2.9	18 **Ar** —
19 **K** 0.9	20 **Ca** 1.1								

Electronegativities of the First 20 Elements

How should KF be classified: ionic, polar covalent, or covalent?

_____

Answer: Δen = 4.1 (fluorine) – 0.9 (potassium) = 3.2, thus it is an ionic bond

62 Using the table again, how should each of the following be classified?

(a) HCl _____

(b) LiCl _____

(c) F_2 _____

Answer: (a) polar covalent (Δen = 0.8 unit); (b) ionic (Δen = 1.9 units); (c) covalent (Δen = 0.0; they are identical atoms)

SHAPES OF MOLECULES

The classification scheme discussed in frames 61 and 62 works only with compounds composed of two atoms. If we have compounds of three or more atoms, the *shape* of the molecule becomes important in determining its polarity.

A discussion of the shapes of molecules requires an exhaustive look at atomic and molecular orbitals and their shapes and spatial orientation, along with a discussion of the valence bond electron repulsion theory. Such a presentation is beyond the scope of this book. We will limit our discussion to a few simple molecules that are known to be **linear** (all the atoms in the compound can be joined by a single straight line through their nuclei) or **bent** (nuclei cannot be joined by a straight line). Bent molecules may be **planar** (all the atoms fall within the same flat surface, such as this page) or **nonplanar** (one or more atoms fall outside the flat surface — say, above or below the plane of this page).

63 Carbon dioxide (CO) is a linear molecule with the structure shown below.

$$:\ddot{O}::C::\ddot{O}:$$

When a carbon atom and an oxygen atom form a bond or bonds between them, will the electrons be closer to the carbon atom or the oxygen atom?

_____

Answer: oxygen (because it is more electronegative)

64 Carbon dioxide is a linear molecule. Oxygen (1) is pulling on the electrons between it and the carbon to the same extent as oxygen (2) is pulling on its electrons.

$$:\ddot{O}::C::\ddot{O}:$$
$$\;\;(1)\quad\;\;(2)$$

An analogy could be two tug-of-war teams pulling on a rope with neither team moving. Even though each bond *within* the molecule may be polar, the net result, because the molecule is linear, is that the molecule *as a whole* is nonpolar.

Carbon disulfide (CS_2) has a dot structure of:

$$:\ddot{S}::C::\ddot{S}:$$

Would you expect the molecule to be polar or nonpolar? _____

Answer: nonpolar

65 Water (H_2O) is an example of a bent planar molecule. Its line structure is:

$$:\ddot{O}:$$
$$H\diagup\;\;\diagdown H$$

The "tugs" in this case are not along the same straight line so they do not offset each other. Would you expect H_2O to be a polar or nonpolar molecule?

_____

Answer: polar

66 H_2S has a structure similar to water.

(a) What is its dot structure? _____

(b) Would you expect it to be polar or nonpolar? _____
(Note: Sulfur and oxygen are both in Group VIA in the periodic table, so they have the same number of outer shell electrons.)

Answer:

(a)   (Or something similar. The molecule is bent and planar but not linear.)

(b) polar

67 Recall that in frame 55 we could not judge whether ammonia, NH_3, would be polar or nonpolar. Ammonia is a bent molecule and is not planar. It looks like a pyramid with a triangular base. The nitrogen atom serves as the "peak" with the hydrogen atoms located at the corners of the base.

$$\overset{\displaystyle \overset{..}{N}}{\underset{\displaystyle H}{H\diagup\;|\;\diagdown H}}$$

(a) Would you expect the "tug" on the electrons to be equal and offset? _____

(b) Would NH_3 be a polar or nonpolar molecule? _____

Answer: (a) no; (b) polar

68 *Planar* molecules are nonpolar. A planar molecule consists of a molecule with all of its atoms lying in the same plane. SO_3 is both planar and nonlinear.

(a) Which of these structures could represent SO_3?

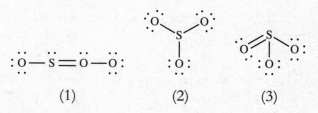

(1)   (2)   (3)

(b) Is SO_3 polar or nonpolar? _____

Answer:

(a) Structure 2 is the only one that is planar and nonlinear.

(b) nonpolar

 69 It is also possible to determine experimentally if compounds are polar covalent, nonpolar covalent, or ionic. A liquefied ionic compound, or an ionic compound in water solution, will conduct electricity and a covalent compound will not. A polar covalent compound will be attracted toward a charged rod. A nonpolar compound will not be attracted.

A compound that will not conduct electricity and will not be attracted by a charged rod should be which kind of compound? _____

Answer: nonpolar covalent

70 A stream of pure water is attracted by a charged rod, but the water does not conduct electricity.

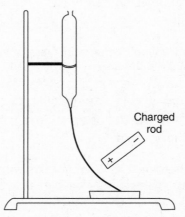

Charged rod

Water must be what kind of compound? _____

Answer: polar covalent

71 A water solution of KCl is a good conductor of electricity. KCl must be what kind of compound? _____

Answer: ionic

PERIODIC PROPERTIES

72 Use the periodic table for the following review questions.

(a) Where are metals located in the periodic table: right side, left side, or both sides of the steplike line? _____

(b) Where are nonmetals located? _____

Answer: (a) left side; (b) right side

73 The alkali metals (in Group IA) are the most metallic family in the periodic table. Metallic characteristics increase with increased atomic number. Given these statements, which element could be expected to be the most metallic? _____

Answer: Fr (francium) would be the most metallic, since it is the element with the largest atomic number in the most metallic group (IA).

74 Of the alkaline earth family (IIA), which element could be expected to be the most metallic? _____

Answer: Ra (radium) is the most metallic element in Group IIA, since it has the largest atomic number in the group.

75 In the same group, nonmetallic properties *decrease* as the atomic number increases. Of the halogens (Group VIIA), which is the most *nonmetallic*? _____

Answer: F (fluorine) (Since nonmetallic properties decrease with increasing atomic number, the most nonmetallic halogen must be the one with the smallest atomic number.)

76 Which of the noble gases (Group VIIIA) could be expected to have the most non-metallic characteristics? _____

Answer: He (helium) (the noble gas with the smallest atomic number; therefore, it is the most nonmetallic of the noble gases)

77 A number of other properties also increase or decrease in relation to position in the periodic table. One such property is electronegativity, which we have just learned is a measure of the tendency of a covalently bonded atom to attract the bonding electron pair. Electronegativity generally increases from left to right across the periodic table and from the bottom to the top. The elements of least electronegativity would be found on the bottom left side of the periodic table. Where would the elements of greatest electronegativity be located?

_____

Answer: top right side of the periodic table (The most electronegative element is fluorine. Note that the electronegativities of the noble gases are unknown.)

78 **Ionization energy** (the energy required to remove electrons from a neutral atom) increases and decreases in the same general way as electronegativity. Elements requiring the *most* ionization energy are found on the top right side of the periodic table. Helium, the element of lowest atomic number in the noble gases, requires the most ionization energy of any element.

Which element within the halogen family (Group VIIA) could be expected to require the *most* ionization energy? _____

Answer: fluorine (Within a group or family, the element of lowest atomic number would require the most ionization energy. The halogen of lowest atomic number is fluorine.)

79 In Group IIA (the alkaline earth metals), which element could be expected to require the greatest ionization energy? _____ Which element in Group IIA has the greatest electronegativity? _____

Answer: Be (beryllium) is the correct answer to both questions. It has the lowest atomic number within the group. Thus, it has the greatest electronegativity and requires the greatest ionization energy within Group IIA.

80 Within the third period (the elements from Na on the left to Ar on the right), which element would be expected to require the greatest ionization energy?

_____

Answer: argon (Ionization energy generally increases from left to right in the periodic table.)

81 Within the fourth period (potassium to krypton), which element would be expected to require the *least* ionization energy? _____

Answer: potassium

You have just learned that the valence shell electron structure of an atom is the controlling factor of several properties of the atom. It determines whether an atom gains or loses electrons and how many. It also determines the type of chemical bond formed when atoms combine and whether atoms have metallic or nonmetallic properties.

In any vertical column of elements in the A groups of the periodic table, as you go from top to bottom:

• the elements become more metallic;

• the number of electron shells increases;

• the number of valence shell electrons remains the same;

• the ionization energy decreases;

• the electronegativity decreases.

In any horizontal row of the elements in the A groups of the periodic table, as you go from left to right:

• the elements become more nonmetallic;

• the number of electron shells remains the same;

• the number of valence shell electrons increases;

• the ionization energy increases;

• the electronegativity increases.

You have been introduced to ions and how they combine to form compounds. You will learn more about molecules, molecular weights, naming of compounds, and writing formulas of compounds in the chapters that follow.

A key point to remember is that even though we focused primarily on the first 20 elements in the periodic table, you can apply what you have learned to any of the elements in the A groups.

SELF-TEST

This self-test is designed to show how well you have mastered this chapter's objectives. Correct answers and review instructions follow the test. Use any of the tables presented in the chapter to answer the questions.

1. A covalent bond is formed when two atoms (share or exchange) _____ two or more electrons.

2. When two atoms form a bond by the exchange of one or more valence electrons, a(n) (covalent or ionic) _____ bond is formed.

3. How many electrons are present in the outermost shell of a neutral S atom? _____

4. How many electrons are present in the outermost shell of a neutral Mg atom? _____

5. Which of the following neutral atoms has three electrons in its outmost shell, silicon (Si), chlorine (Cl), or boron (B)? _____

6. Circle the correct responses in the following questions.

   (a) Which has the lower ionization energy, Li or K?

   (b) Which would be more polar, HF or HBr?

   (c) Which is more nonmetallic, F or I?

   (d) Which is more electronegative, K or Rb?

   (e) Which has more outer shell electrons, Ca or C?

   (f) Which would you expect to be more ionic, LiF or HCl?

7. Circle the correct responses in the following questions.

   (a) Which has the higher ionization energy, Mg or Sr?

   (b) Which would be less polar, HCl or HI?

   (c) Which is more nonmetallic, Cl or Br?

   (d) Which is more electronegative, Cr or Y?

   (e) Which has fewer outer shell electrons, Li or F?

   (f) Which would you expect to be more ionic, CaO or H_2O?

8. Write equations, using electron dot symbols, showing the formation of $MgCl_2$ and H_2O using Mg, Cl, H, and O as starting materials. What type of bond is formed in each case?

9. Write equations, using electron dot symbols, showing the formation of LiF and H_2S using Li, F, H, and S as starting materials. What type of bond is formed in each case?

10. Here are the dot symbols of four fictitious elements: Q· R: :Ẍ· :Ÿ:

    (a) In what groups would elements Q and Y be found in the periodic table?_____

    (b) Would a compound formed between Q and Y more likely be ionic or covalent? _____

    (c) What ion would you expect element X to form? _____

    (d) X and Y combine to form a molecule of XY_2, which is covalent. Which of the following could be the correct dot structure for the compound?

    _____

    :Ÿ::X: Ÿ: :Ẍ: Ÿ: Ÿ: X::Ÿ: :Ẍ::Ÿ:Ÿ:
                      :Ÿ:

         (1)   (2)   (3)   (4)

11. Which noble gas structure does Mg have when it becomes an ion? _____

12. Which noble gas structure does Cl have when it becomes an ion? _____

13. What is the line structure for HBr? _____

14. Nitrogen dioxide (NO_2) is a molecule with a bent structure. The NO_2 molecule is (polar, nonpolar) _____.

15. Phosphorus trichloride (PCl_3) is used in making compounds known as haloalkanes from alcohols. Phosphorus trichloride's phosphorus atom has a lone pair of electrons on its line structure. As a result PCl_3 is (polar, nonpolar) _____.

ANSWERS

Compare your answers to the self-test with those given below. If you answer all questions correctly, you are ready to proceed to the next chapter. If you miss any, review the frames indicated in parentheses following the answers. If you miss several questions, you should probably reread the chapter carefully.

1. share (frame 34)

2. Ionic (frames 27–33)

3. 6 (S is in Group VIA) (frames 2, 5, 8)

4. 2 (Mg is in Group IIA) (frames 2, 5, 8)

5. Boron is in Group IIIA so it will have three electrons in its outmost shell (frames 2, 5, 8)

6. (a) K (frames 78–81 and summary)
   (b) HF (frames 54–61)
   (c) F (frames 73–76)
   (d) K (frames 53, 61)
   (e) C (frames 2–6)
   (f) LiF (frames 54–61)

7. (a) Mg (frames 78–81 and summary)
   (b) HI (frames 54–61)
   (c) Cl (frames 73–76)
   (d) Cr (frames 53, 61)
   (e) Li (frames 2–6)
   (f) CaO (frames 54–61)

8.

$$Mg\!:\, +\, 2\,:\!\overset{\cdot\cdot}{\underset{\cdot\cdot}{Cl}}\!\cdot\, \rightarrow\, Mg^{2+},\, 2\, [:\!\overset{\cdot\cdot}{\underset{\cdot\cdot}{Cl}}\!:]^-,\ \text{ionic}$$

$$2H\!\cdot\, +\, :\!\overset{\cdot\cdot}{\underset{\cdot}{O}}\!\cdot\, \rightarrow\, :\!\overset{\cdot\cdot}{\underset{\cdot\cdot}{O}}\!:H$$
$$\qquad\qquad\qquad H \quad\text{, polar covalent (frames 39–48)}$$

9.

$$Li\!\cdot\, +\, :\!\overset{\cdot\cdot}{\underset{\cdot}{F}}\!\cdot\, \rightarrow\, Li^+,\, \left[:\!\overset{\cdot\cdot}{\underset{\cdot\cdot}{F}}\!:\right]^-,\ \text{ionic}$$

$$2H\!\cdot\, +\, :\!\overset{\cdot\cdot}{\underset{\cdot}{S}}\!\cdot\, \rightarrow\, :\!\overset{\cdot\cdot}{\underset{\cdot\cdot}{S}}\!:H$$
$$\qquad\qquad\qquad H \quad\text{, polar covalent (frames 39–48)}$$

10. (a) Q in Group IA, Y in Group VIIA (frames 2, 8)

　　(b) ionic (frames 27–61)

　　(c) 2– (frames 12–26)

　　(d) structure 2 (It is the only one that obeys the octet rule for each atom.) (frames 17, 43–48)

11. neon (frames 18–20)

12. argon (frames 18–20)

13. linear (frame 63)

14. polar (frames 63, 67)

15. polar (frames 63, 67)

G.N. LEWIS AND CHEMISTRY

In this chapter you learned about Lewis symbols and how an understanding of these are used to make covalent bonds and molecules. This is very important concept in chemistry, especially in the field of organic chemistry. Lewis symbols helped pioneer our understanding for chemical bonding.

Gilbert Newton Lewis was born in Weymouth, Massachusetts, in 1875 and received his Ph.D. in chemistry from Harvard University. He was a prominent physical chemist who is most widely known for his work in covalent bonding.

G.N. Lewis

Lewis symbols and Lewis structures (also known as Lewis dot diagrams) all originate with G.N. Lewis's work in covalent bonding.

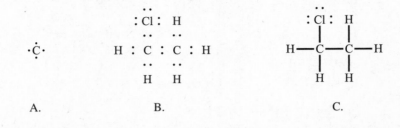

A.   B.   C.

Figure of Lewis symbol and Lewis structures: (A) Lewis symbol for carbon, (B) Lewis structure for an organic compound, C_2H_5Cl, and (C) Lewis structure for C_2H_5Cl using bars between two atoms instead of two dots.

In this figure you first see the Lewis symbol for carbon. As you learned in this chapter, the Lewis symbol for carbon has four dots around it representing its four valence electrons. It has four valence electrons and is found in Group IV of the periodic table. Figure B shows a Lewis structure (aka Lewis dot structure) for an organic compound, C_2H_5Cl (chloroethane). The Lewis structure shows the arrangement of electrons within a molecule.

Two shared electrons between two atoms represent a bond. In Figure B you also notice that chlorine has three sets of electron pairs around it. Chlorine's three sets of electron pairs are not involved in bonding. Because they are not involved in bonding, electron pairs that reside on atoms are known as lone-pair electrons or nonbonding electron pairs. Finally, Figure C shows a shorthand version of Figure B, which uses a bar in place of two shared dots between two atoms within a bond. This bar, or bond, is often drawn between two atoms instead of two dots. Both Figure B and Figure C represent the same molecule with Figure C's structure more commonly drawn especially for organic compounds as you will see in Chapter 14.

G.N. Lewis not only helped our understanding of covalent bonding but was also instrumental in the fields of thermodynamics, photochemistry, and acid–base chemistry. It is no surprise that Lewis was nominated for the Nobel Prize in Chemistry. In fact, he was nominated 41 times, but unfortunately he never won it. Nonetheless, G.N. Lewis's work remains foundational to much of our understanding in chemistry.

4 Molecular and Formula Weights

Since every molecule is made up of a definite and invariable number of component atoms, it follows that each kind of molecule has a definite and characteristic weight. You may ask, "How did anyone ever determine how many atoms of each kind are in a molecule of any pure substance?" We try to answer that question in this chapter. We may determine the molecular weight of a substance in a number of ways using instrumental techniques that are available today. Specific instrumental methods are not discussed here. Instead, we will rely upon the data that we get by analyzing substances and what you have learned so far in Chapters 1 through 3. We will depend upon the periodic table and the information it provides.

This chapter will teach you how to determine the molecular weight and percentage composition of a compound when given its molecular formula. You will also be able to reverse the process to determine the molecular formula of a compound given the percentage composition of the compound (or the weight of each element per molecule) and its molecular weight. The important thing to remember is that every molecule consists of a definite number of atoms in a fixed ratio expressed as small whole numbers.

OBJECTIVES

After completing this chapter, you will be able to

- explain the difference between a molecule of a compound and a molecule of an element;
- calculate the molecular weight of a compound when given its actual formula;
- explain the difference between an empirical formula and an actual formula;
- calculate the percentage by weight of each element in a compound when given its actual formula or empirical formula;
- determine the actual formula of a compound when given its empirical formula and molecular weight;
- calculate the actual formula of a compound when given its percentage composition and its molecular weight.

1 In Chapter 2 you learned that each atom has an average atomic weight expressed in atomic mass units (amu) and that atomic weights are included on the periodic table. You also learned that an atomic weight expressed in grams represents the weight of 6.022×10^{23} atoms.

In Chapter 3 you learned that atoms combine to form compounds of two major categories, depending upon the kind of chemical bond involved. In this chapter you will learn how to determine the weights of compounds.

Just to review a bit, the atomic weight of a Cl atom is _____ amu.

What is the weight in grams of 6.022×10^{23} atoms of Cl? _____

Answer: 35.453; 35.453

2 The chlorine gas molecule (Cl_2) is made up of two chlorine atoms. If one atom of chlorine weighs 35.45 amu (to the nearest hundredth), one molecule of Cl_2 should weigh _____. (Note: Calculate your answer to the nearest hundredth for this and further questions unless otherwise stated.)

Answer: 70.90 amu ($35.45 \times 2 = 70.90$)

MOLECULAR WEIGHT

3 Carbon monoxide (CO) is a molecule made up of one atom of carbon and one atom of oxygen. If an atom of carbon weighs 12.01 amu and an atom of oxygen weighs 16.00 amu, a molecule of CO should weigh _____ (to the nearest hundredth).

Answer: 28.01 amu ($12.01 + 16.00 = 28.01$)

4 The molecular weight of a molecule is the sum of the combined atomic weights of the atoms within the molecule. If the atomic weight of H is 1.01 amu and the atomic weight of Cl is 35.45 amu, a molecular weight of HCl is _____ amu (to the nearest hundredth).

Answer: 36.46 ($1.01 + 35.45 = 36.46$)

5 In Chapter 3 you learned that a molecule of a compound consists of two or more different types of atoms that are chemically combined by one or more chemical bonds. The molecule carbon dioxide (CO_2) is made up of one atom of carbon and two atoms of oxygen. If the atomic weight of C is 12.01 amu and the atomic weight of O is 16.00 amu, the molecular weight of CO_2 is _____ amu (to the nearest hundredth).

Answer: 44.01 (1 C + 2 O = $12.01 + 32.00 = 44.01$)

6 The term "molecule" can mean either a *molecule of an element* or a *molecule of a compound*.

A molecule of an element consists of one or more atoms of the *same* kind. Examples are H_2 (which exists as a diatomic molecule), Ar (which exists as a single argon atom), and S_8 (which exists as a molecule of eight sulfur atoms).

Since argon normally exists as a single atom, the single atom (Ar) can be called either the argon atom or the argon molecule. The atomic weight of Ar is 39.95 amu. The molecular weight of Ar is _____ amu.

Answer: 39.95 (The weight of one atom of Ar is the same as the weight of the Ar molecule because the Ar molecule is the Ar atom.)

7 A single atom molecule is an exception to the rule. Most molecules of either an element or a compound usually involve two or more atoms combined with covalent bonds.

The molecular weight of O_2 is _____ amu. The molecule O_2 is a molecule of a(n) (compound, element) _____.

Answer: 32.00 (twice the weight of a single O atom); element (since only one element, oxygen, is included in the molecule)

8 The atomic weight of S is 32.06 amu. The molecular weight of S_8 is _____ amu. S_8 is a molecule of an element because _____.

Answer: 256.48 ($8 \times 32.06 = 256.48$); only one element (sulfur) is included in the molecule

9 The following example shows a simple procedure for finding the molecular weight of sucrose sugar ($C_{12}H_{22}O_{11}$).

Element	Number of atoms	Atomic weight	Atoms × weight
C	12 C	12.01	$12 \times 12.01 = 144.12$
H	22 H	1.01	$22 \times 1.01 = 22.22$
O	11 O	16.00	$11 \times 16.00 = 176.00$

(a) The molecular weight of $C_{12}H_{22}O_{11}$ is _____ amu. (Hint: Add the results of Atoms × weight.)

(b) Sucrose ($C_{12}H_{22}O_{11}$) is a molecule of a(n) (compound, element) _____.

Answer: 342.34; compound (because more than one element is included in the molecule: carbon, oxygen, and hydrogen)

10 Determine the molecular weight of ethyl alcohol (C_2H_5OH). From now on you should use the periodic table to determine the appropriate atomic weights (to the nearest hundredth). Below are the column headings we used for our table in frame 9. Use a separate sheet of paper and make a similar table to calculate the molecular weight of C_2H_5OH.

Element	Number of atoms	Atomic weight	Atoms × weight

The molecular weight of C_2H_5OH is _____ amu.

Answer:

Element	Number of atoms	Atomic weight	Atoms × weight
C	2 C	12.01	$2 \times 12.01 = 24.02$
H	6 H	1.01	$6 \times 1.01 = 6.06$
O	1 O	16.00	$1 \times 16.00 = \underline{16.00}$
			Molecular weight is 46.08 amu

FORMULA WEIGHT

11 In Chapter 3 you learned that there were two general categories of compounds, depending upon whether the type of chemical bonding involved was covalent or ionic. Covalent bonding generally results in molecular compounds while ionic bonding results in ionic compounds. The term **molecular weight**, therefore, should properly refer only to molecular compounds and not to ionic compounds. Chemists have coined the term **formula weight** to refer to either molecular compounds or ionic compounds. The formula weight of a molecule is the same as the molecular weight. The formula weight can also refer to the weight of an ionic compound.

The weight of CO_2 (a covalent compound) is 44.01 amu. Is 44.01 amu the molecular weight, the formula weight, or either the molecular or the formula weight? _____

The weight of NaCl (an ionic compound) is 58.44 amu. Is 58.44 amu the molecular weight, the formula weight, or either the molecular or the formula weight? _____

Answer: either molecular or formula weight; formula weight

12 Determine the formula weight of copper sulfate, $CuSO_4$. Use a separate sheet of paper to set up a table similar to those in frames 9 and 10.

Answer:

Element	Number of atoms	Atomic weight	Atoms × weight
Cu	1 Cu	63.55	$1 \times 63.55 = 63.55$
S	1 S	32.06	$1 \times 32.06 = 32.06$
O	4 O	16.00	$4 \times 16.00 = \underline{64.00}$
			Formula weight is 159.61 amu

13 Answer the following questions for vitamin C ($C_6H_8O_6$).

(a) Calculate the formula weight of vitamin C. (Set up a table to calculate the amu.)

(b) Vitamin C is a covalent compound. What is its molecular weight? _____ amu

Answer:

(a)

Element	Number of atoms	Atomic weight	Atoms × weight
C	6 C	12.01	$6 \times 12.01 = 72.06$
H	8 H	1.01	$8 \times 1.01 = 8.08$
O	6 O	16.00	$6 \times 16.00 = \underline{96.00}$
			Formula weight is 176.14 amu

(b) 176.14 (Molecular weight is the same as the formula weight for a covalent compound.)

14 Carbon monoxide (CO) is a molecule with one atom of carbon and one atom of oxygen. The molecular weight of CO is 28.01 amu. The atomic weight of C is 12.01 amu. The atomic weight of O is 16.00 amu. Which atom (carbon or oxygen) accounts for more than half of the molecular weight of the CO molecule? _____

Answer: oxygen

PERCENTAGE COMPOSITION

15   To find the actual proportion by weight (percentage) of an element in a compound, the weight of the part must be divided by the weight of the whole. For example, a bolt and nut can be compared to a molecule with two atoms. If the nut weighed 100 amu and the bolt weighed 200 amu, the whole thing would weigh $100 + 200 = 300$ amu.

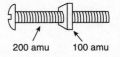

200 amu   100 amu

The nut is a part of the whole. To find the percentage by weight that the nut occupies within the whole, divide the weight of the nut by the weight of the whole and multiply by 100. Here's how this percentage is calculated.

$$(100 \text{ amu}/300 \text{ amu}) \times 100\% = 33.3\%$$

The percentage of the weight contributed by the nut is equal to 33.3% of the weight of the whole.

Assume that a bolt weighs 50 amu and a nut weighs 30 amu. What percentage by weight does the nut contribute in relation to the whole? The nut weighs _____% of the whole.

Answer:

nut = 30 amu

bolt = 50 amu

the whole = 50 + 30 = 80 amu

percentage by weight of nut = $\dfrac{\text{nut}}{\text{whole}} \times 100\% = \dfrac{30 \text{ amu}}{80 \text{ amu}} \times 100\% = 37.5\%$

16   A bolt has two nuts.

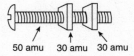

50 amu   30 amu   30 amu

Each nut weighs 30 amu and the bolt weighs 50 amu. What percentage by weight do the nuts contribute in relation to the whole? The nuts weigh _____% of the whole.

Answer:

nuts = 2 × 30 = 60 amu

bolt = 50 amu

the total whole = 60 + 50 = 110 amu

$$\text{percentage by weight of nuts} = \frac{\text{nuts}}{\text{whole}} \times 100\% = \frac{60\ \text{amu}}{110\ \text{amu}} \times 100\% = 54.5\%$$

17 The carbon monoxide molecule can be considered analogous to the bolt and nut combination. The molecule consists of one carbon atom weighing 12.01 amu (the atomic weight of carbon) and one oxygen atom weighing 16.00 amu (the atomic weight of oxygen). The whole molecule weighs 12.01 + 16.00 = 28.01 amu (formula weight). The oxygen atom weighs _____ _____% of the whole molecule.

Answer:

oxygen atom = 16.00 amu

carbon atom = 12.01 amu

whole molecule = 16.00 + 12.01 = 28.01 amu

percentage by weight of oxygen in carbon monoxide

$$= \frac{\text{oxygen atom}}{\text{formula weight}} \times 100\% = \frac{16.00\ \text{amu}}{28.01\ \text{amu}} \times 100\% = 57.1\%$$

18 To find the percentage by weight of any element within a compound, divide the total atomic weight of that element by the formula weight of the compound and multiply the result by 100%.

The molecule carbon dioxide (CO_2) has a total formula weight of 44.01 amu and consists of one carbon atom with atomic weight of 12.01 and two oxygen atoms with combined atomic weights of _____ amu.

Answer: 32.00 (16.00 × 2 = 32.00 amu)

19 Using the information in frame 18, find the percentage by weight of oxygen within the compound carbon dioxide (CO_2). _____

Answer: $\dfrac{32.00\ \text{amu}}{44.01\ \text{amu}} \times 100\% = 72.7\%$ oxygen by weight

Now calculate the percentage by weight of carbon within carbon dioxide.

_____

Answer: You could do this in two ways. Since you just found that carbon dioxide consists of 72.7% oxygen by weight, the rest must be carbon:

$$100\% - 72.7\% = 27.3\% \text{ carbon by weight}$$

Or, if you did not already know the percentage of oxygen, then:

$$\frac{12.01 \text{ amu}}{44.01 \text{ amu}} \times 100\% = 27.3\% \text{ carbon by weight}$$

 Calculate the percentage by weight of hydrogen (H) in water (H_2O). (Find the formula weight of H_2O first.)

% hydrogen = _____

Answer:

Element	Number of atoms	Atomic weight	Atoms × weight
H	2 H	1.01	$2 \times 1.01 =$ 2.02
O	1 O	16.00	$1 \times 16.00 = \underline{16.00}$
			Formula weight is 18.02 amu

To find the percentage composition by weight:

$$\text{percentage hydrogen in } H_2O = \frac{\text{weight of hydrogen}}{\text{formula weight}} \times 100\%$$

$$= \frac{2.02 \text{ amu}}{18.02 \text{ amu}} \times 100\% = 11.2\%$$

22 Determine the percentage of carbon by weight in glucose ($C_6H_{12}O_6$).

% carbon = _____

Answer:
The formula weight of glucose ($C_6H_{12}O_6$):

Element	Number of atoms	Atomic weight	Atoms × weight
C	6 C	12.01	$6 \times 12.01 = 72.06$
H	12 H	1.01	$12 \times 1.01 = 12.12$
O	6 O	16.00	$6 \times 16.00 = \underline{96.00}$
			Formula weight is 180.18 amu

$$\text{percentage carbon in glucose} = \frac{\text{weight of carbon}}{\text{formula weight of glucose}}$$

$$= \frac{72.06 \text{ amu}}{180.18 \text{ amu}} \times 100\% = 39.99\%$$

 In the preceding example (glucose, $C_6H_{12}O_6$), the percentage composition by weight of *all* elements (carbon, hydrogen, and oxygen) *totaled together* should equal _____%.

Answer: 100 (Since all the partial weights of all elements add up to the weight of the whole compound, all percentages also add up to their whole, which is 100%.)

EMPIRICAL FORMULA

 A chemist often must determine the formula of a compound when only the percentage composition by weight is known. Assume a compound is composed of only two elements, carbon and hydrogen. By analysis, the compound is found to contain 92.2% carbon by weight. What is the percentage composition by weight of hydrogen in this compound? _____

Answer: 7.8% (All partial composition percentages must add up to 100%. Since there are only two elements in the compound, their percentages must add up to 100%. If carbon accounts for 92.2%, then hydrogen must account for the other 7.8%.)

 A compound of undetermined formula contains 92.2% carbon by weight and 7.8% hydrogen by weight. This means that for every 100 amu of that compound, 92.2 amu would be contributed by carbon. How many amu would be contributed by hydrogen? _____

Answer: 7.8 amu

 Knowing that 100 amu of a compound contains 92.2 amu of carbon and 7.8 amu of hydrogen gives us the relative *weights* of the elements in a compound. In order to determine a chemical formula, we need to know the relative *number of atoms* of each element in the compound. We need to know the ratio of carbon atoms to hydrogen atoms. A 100 amu sample of a compound contains 92.2 amu of carbon and 7.8 amu of hydrogen. We know that the atomic weight of carbon is 12.01 amu and represents the weight of one carbon atom. A weight of 92.2 amu represents the weight of how many carbon atoms (to the nearest tenth)? _____

Answer: $92.2 \text{ amu} \times \dfrac{1 \text{ carbon atom}}{12.01 \text{ amu}} = 7.7 \text{ carbon atoms}$

27 A 100 amu sample of a compound contains 92.2 amu of carbon (representing 7.7 atoms of carbon) and 7.8 amu of hydrogen. How many atoms of hydrogen are contained in the sample (to the nearest tenth)? _____

Answer: $7.8 \text{ amu} \times \dfrac{1 \text{ hydrogen atom}}{1.01 \text{ amu}} = 7.7 \text{ hydrogen atoms}$

28 To arrive at a tentative formula for this unknown compound, we need to know the ratio of carbon atoms to hydrogen atoms within a single molecule. Since there are 7.7 atoms of hydrogen for 7.7 atoms of carbon, there must be how many atom(s) of hydrogen for each atom of carbon? _____

Answer: one

$$\dfrac{7.7 \text{ atoms of hydrogen}}{7.7 \text{ atoms of carbon}} = \dfrac{1 \text{ atom of hydrogen}}{1 \text{ atom of carbon}}$$

29 Since carbon and hydrogen are in a ratio of 1:1 in this unknown compound, the simplest formula is CH (or HC). The simplest possible chemical formula that represents the ratios of the atoms within an unknown molecule is called an **empirical formula**. The simple formula (CH) is called a(n) _____.

Answer: empirical formula

30 An unknown compound with an empirical formula of CH could be any of several compounds. For example, both acetylene (C_2H_2) and benzene (C_6H_6) have an empirical formula of CH. The formulas for acetylene and benzene (C_2H_2 and C_6H_6) are called **molecular formulas** because they describe the actual number of atoms contained in each molecule.

(a) A formula describing the simplest ratio between atoms in an unknown molecule is called a(n) _____.

(b) A formula describing the actual number of atoms contained in each molecule is called the _____.

Answer: (a) empirical formula; (b) molecular formula

31 An empirical formula is sometimes the same as an actual molecular formula. A formula determined by percentage weight analysis is considered to be an empirical formula since it only represents the ratio of one atom to another. By weight analysis, a compound is found to consist of carbon and oxygen in a 1:1 ratio. A formula of CO is assigned to the compound and you suspect that the compound could be carbon monoxide but have no proof. The formula CO is in this case a(n) _____ (empirical, molecular) formula.

Answer: empirical (If by simple weight analysis, the compound has a carbon and oxygen ratio of 1 : 1, then the assigned formula CO is empirical. If it is determined by additional testing that the compound is indeed carbon monoxide, then the formula CO would be a molecular formula.)

32 To determine an empirical formula if given percentage composition, follow these three steps.

1. Drop the percentage signs and replace them with amu. The result is the weight composition of a 100 amu sample.

2. Multiply each of the results of step 1 by the appropriate atoms per atomic weight conversion factor (as shown in frames 26 and 27). The result is the relative "number" of atoms in a 100 amu sample.

3. Divide each of the relative "number" of atoms obtained in step 2 by the smallest number to obtain the simplest ratio of atoms in whole numbers.

   A compound is found to be 20% Ca and 80% Br by weight. Following step 1, a 100 amu sample of this compound would be composed of _____ amu of Ca and _____ amu of Br.

Answer: 20; 80

33 Using step 2 from frame 32, determine how many atoms (rounded to the nearest tenth) of the component elements are present in the 100 amu sample.

(a) atoms of Ca = _____

(b) atoms of Br = _____

Answer:

(a) $20 \text{ amu} \times \dfrac{1 \text{ Ca atom}}{40.08 \text{ amu}} = 0.5 \text{ atom of Ca}$

(b) $80 \text{ amu} \times \dfrac{1 \text{ Br atom}}{79.90 \text{ amu}} = 1.0 \text{ atom of Br}$

34 In this 100 amu sample of an unknown compound composed of 20 amu of Ca and 80 amu of Br are 0.5 atoms of Ca and 1.0 atoms of Br. However, 0.5 atoms is not a whole number. Using step 3 from frame 32 to obtain the simplest ratio of atoms in *whole numbers*, divide both 0.5 atoms of Ca and the 1.0 atoms of Br by the smallest number, which is 0.5. (Note: If the answers obtained in step 2 are whole numbers, step 3 is unnecessary.)

$$\frac{0.5}{0.5} = \underline{\quad}\text{atom(s) of Ca}, \frac{1.0}{0.5} = \underline{\quad}\text{atom(s) of Br}$$

Answer: one; two

35 The simplest empirical formula derived from a ratio of one atom of Ca to two atoms of Br would be _____.

Answer: $CaBr_2$

36 A compound is made up of 79.9% carbon and 20.1% hydrogen by weight. What is its empirical formula? (Apply the three steps from frame 32. Round to the nearest whole number.) _____

Answer:

1. A sample of the unknown compound weighing 100 amu is made up of 79.9 amu of carbon and 20.1 amu of hydrogen (replacing "percentage" with "amu" to express weight composition).

2. The relative number of atoms in the 100 amu sample are as follows. A 79.9 amu weight of carbon contains 6.7 atoms of carbon.

$$79.9 \text{ amu} \times \frac{1 \text{ C atom}}{12.01 \text{ amu}} = 6.7 \text{ C atoms}$$

   A 20.1 amu weight of hydrogen contains 19.9 atoms of hydrogen.

$$20.1 \text{ amu} \times \frac{1 \text{ H atom}}{1.01 \text{ amu}} = 19.9 \text{ atoms}$$

3. We must divide the 6.7 atoms and the 19.9 atoms obtained in step 2 by the smallest number (6.7) to find the simplest whole number ratio of atoms.

$$\frac{6.7}{6.7} = 1 \text{ atom of carbon} \qquad \frac{19.9}{6.7} = 3 \text{ atoms of hydrogen}$$

4. The ratio of one atom of carbon to three atoms of hydrogen provides an empirical formula of CH_3 (or H_3C).

37 The same procedure can be used to determine the empirical formula of a compound with three or more elements. A sample of a compound is composed of 40% calcium (Ca), 12% carbon (C), and 48% oxygen (O) by weight. A 100 amu sample of this compound would be composed of _____ amu of Ca, _____ amu of C, and _____ amu of O.

Answer: 40; 12; 48

38 What is the empirical formula of the compound from the previous frame? (Write the formula in the order of the elements given, Ca, C, and O. Use the three-step procedure.)

Answer:

1. In a 100 amu sample, Ca = 40 amu, C = 12 amu, and O = 48 amu.
2. In that 100 amu sample there are:

$$40 \text{ amu} \times \frac{1 \text{ atom of Ca}}{48.08 \text{ amu}} = 1.0 \text{ atom of Ca}$$

$$12 \text{ amu} \times \frac{1 \text{ atom of C}}{12.01 \text{ amu}} = 1.0 \text{ atom of C}$$

$$48 \text{ amu} \times \frac{1 \text{ atom of O}}{16.00 \text{ amu}} = 3.0 \text{ atoms of O}$$

3. Dividing by the smallest number given in step 2 in order to obtain a simple whole number ratio is not necessary in this case. We already have simple whole numbers.

The empirical formula must be $CaCO_3$ (elements listed from most metallic to least metallic).

39 Determine the empirical formula of a compound made up of 28.7% K, 1.5% H, 22.8% P, and 47.0% O. Write the formula in the order of the elements given (K, H, P, O). (In step 2, calculate to the nearest thousandth.) _____

Answer:

1. In a 100 amu sample, K = 28.7 amu, H = 1.5 amu, P = 22.8 amu, and O = 47.0 amu
2. In that 100 amu sample there are:

$$28.7 \text{ amu} \times \frac{1 \text{ atom of K}}{39.10 \text{ amu}} = 0.734 \text{ atom of K}$$

$$1.5 \text{ amu} \times \frac{1 \text{ atom of H}}{1.01 \text{ amu}} = 1.485 \text{ atoms of H}$$

$$22.8 \text{ amu} \times \frac{1 \text{ atom of P}}{30.97 \text{ amu}} = 0.736 \text{ atom of P}$$

$$47.0 \text{ amu} \times \frac{1 \text{ atom of O}}{16.00 \text{ amu}} = 2.938 \text{ atoms of O}$$

3. Dividing by the smallest number given in step 2 to find the simplest whole number ratio:

$$\frac{0.734}{0.734} = 1.0 \text{ atom of K}$$

$$\frac{1.485}{0.734} = 2.0 \text{ atoms of H}$$

$$\frac{0.736}{0.734} = 1.0 \text{ atom of P}$$

$$\frac{2.938}{0.734} = 4.0 \text{ atoms of O}$$

KH_2PO_4 is the empirical formula.

40 Determine the empirical formula of a compound composed of 15.15% K, 10.45% Al, 24.80% S, and 49.60% O. Write the formula in the order of the elements given (K, Al, S, O). (In step 2, calculate to the nearest thousandth.)

_____

Answer:

1. In a 100 amu sample, K = 15.15 amu, Al = 10.45 amu, S = 24.80 amu, and O = 49.60 amu.

2. In the 100 amu sample, there are:

$$15.15 \text{ amu} \times \frac{1 \text{ atom of K}}{39.10 \text{ amu}} = 0.387 \text{ atom of K}$$

$$10.45 \text{ amu} \times \frac{1 \text{ atom of Al}}{26.98 \text{ amu}} = 0.387 \text{ atom of Al}$$

$$24.80 \text{ amu} \times \frac{1 \text{ atom of S}}{32.06 \text{ amu}} = 0.774 \text{ atom of S}$$

$$49.60 \text{ amu} \times \frac{1 \text{ atom of O}}{16.00 \text{ amu}} = 3.100 \text{ atoms of O}$$

3. Divide each of the above answers by the smallest answer to find the simplest whole number ratio.

$$\frac{0.387}{0.387} = 1.0 \text{ atom of K}$$

$$\frac{0.387}{0.387} = 1.0 \text{ atom of Al}$$

$$\frac{0.774}{0.387} = 2.0 \text{ atoms of S}$$

$$\frac{3.100}{0.387} = 8.0 \text{ atoms of O}$$

$KAlS_2O_8$ is an acceptable empirical formula.

41 To review, one atom of chlorine (Cl) weighs _____ amu and 6.022×10^{23} atoms of chlorine (Cl) weigh _____ grams.

Answer: 35.45; 35.45

42 The number 6.022×10^{23} appears often in chemical calculations. The term **mole** means 6.022×10^{23} chemical units and is much easier to write and remember than the number it represents.

One mole of atoms means 6.022×10^{23} atoms.
One mole of electrons means 6.022×10^{23} electrons.
One mole of ions means 6.022×10^{23} ions.

The current abbreviation for mole is **mol**. We will use the abbreviation in all calculations involving mole.

A gram atomic weight of chlorine (Cl) represents the weight of 1 _____ of chlorine atoms.

Answer: mole (mol) (6.022×10^{23} atoms)

43 A mole of molecules represents how many molecules? _____

Answer: 6.022×10^{23}

44 The atomic weight in grams of calcium (Ca) is 40.08 grams. A mole of Ca atoms weighs how many grams? _____

Answer: 40.08

45 The element calcium (Ca) weighs _____ grams per gram atomic weight. The element calcium (Ca) weighs _____ grams per mole of atoms.

Answer: 40.08; 40.08 (A gram atomic weight is the weight of 1 mol of atoms.)

46 The mole is a very useful quantity in chemistry and is often encountered. You will begin using the mole in the next part of this chapter.

You have already used the conversion factor of $\dfrac{1 \text{ atom}}{\text{atomic weight in amu}}$.

You will now begin to use the conversion factor of $\dfrac{1 \text{ mol of atoms}}{\text{atomic weight in grams}}$.

For example, since 1 mol of chlorine (Cl) atoms weigh 35.45 grams, the conversion factor would be $\dfrac{1 \text{ mol Cl atoms}}{35.45 \text{ grams}}$.

Instead of percentages, the composition of a compound may be given directly in weights. For example, a 5.00 gram sample of a compound (to the nearest hundredth gram) consists of *only* chlorine and phosphorus. If the 5.00 gram sample was known to contain 1.13 grams of phosphorus (P), how many grams of chlorine (Cl) would be present? _____

Answer: 3.87 (The 5.00 gram sample was of a compound with only two elements. Since one element made up 1.13 grams, the other element must account for the remainder: $5.00 - 1.13 = 3.87$ grams.)

47 To find the empirical formula of the chlorine and phosphorus compound, first multiply the weights of the sample elements by the appropriate mole per gram atomic weight conversion factor (see frame 46). The result is the number of moles of each element present in the sample.

$$1.13 \text{ grams of P} \times \frac{1 \text{ mol of P atoms}}{30.97 \text{ grams}} = 0.0365 \text{ mol of P atoms}$$

Use the same method to determine the number of moles of Cl atoms in the sample (rounded to the nearest thousandth).

$$3.87 \text{ grams of Cl} \times \frac{1 \text{ mol of Cl atoms}}{35.45 \text{ grams}} = \underline{\hspace{1cm}} \text{mol of Cl atoms}$$

Answer: 0.109

48   To determine an empirical formula for this sample, we must determine a whole number ratio of atoms. This can be done by simply dividing each mole quantity by the smallest number.

$$\frac{0.0365 \text{ mol of P atoms}}{0.0365 \text{ mol}} = 1.0 \text{ atom of P}$$

$$\frac{0.109 \text{ mol of Cl atoms}}{0.0365 \text{ mol}} = \underline{\hspace{1cm}} \text{atom(s) of Cl}$$

The resulting empirical formula is _____.

Answer: 3.0; PCl_3 or Cl_3P (One atom of P and three atoms of Cl. PCl_3 is best because the more metallic element is named first.)

49   The steps in determining an empirical formula when given a sample are similar to steps 2 and 3 encountered earlier (refer back to frame 32) when using percentage composition to determine empirical formula.

A. Determine the number of moles of atoms of each element by multiplying the sample portion of each element by its appropriate mole per gram atomic weight conversion factor.

B. Divide each answer from step A by the smallest answer to find the simplest whole number ratio of one atom to another.
    A 10-gram sample of a compound is made up of 2.73 grams of carbon and 7.27 grams of oxygen. Complete step A below. (Round to the nearest thousandth.)

$$2.73 \text{ grams of C} \times \frac{1 \text{ mol of C atoms}}{12.01 \text{ grams}} = 0.227 \text{ mol of C atoms}$$

$$7.27 \text{ grams of O} = \underline{\hspace{1cm}} \text{mol of O atoms}$$

Answer: 0.454

50 Applying step B to the same sample:

$$\frac{0.227 \text{ mol of C atoms}}{0.227 \text{ mol}} = 1 \text{ carbon (C) atom}$$

(a) Complete step B for oxygen (O) atoms. _____

(b) What is the resulting empirical formula of the sample? _____

Answer:

(a) $\dfrac{0.454 \text{ mol of O atoms}}{0.227 \text{ mol}} = 2 \text{ atoms of O}$

(b) CO_2

51 A 5.00-gram sample of a compound made up of calcium (Ca) and chlorine (Cl) is analyzed and contains 1.80 grams of Ca. The 5.00-gram sample contains how many grams of Cl? _____

Answer: 3.20 (5.00 − 1.80 = 3.20)

52 How many moles of atoms are represented by 1.80 grams of Ca and by 3.20 grams of Cl? (Calculate moles to the nearest thousandth.)

(a) moles of Ca atoms = _____

(b) moles of Cl atoms = _____

Answer:

(a) 1.80 grams Ca $\times \dfrac{1 \text{ mol of Ca}}{40.08 \text{ grams}} = 0.045 \text{ mol of Ca}$

(b) 3.20 grams Cl $\times \dfrac{1 \text{ mol of Cl}}{35.45 \text{ grams}} = 0.090 \text{ mol of Cl}$

53 Determine the simplest whole number ratio of Ca atoms to Cl atoms, and give the resulting empirical formula. _____

Answer:

$$\frac{0.045 \text{ mol of Ca atoms}}{0.045 \text{ mol}} = 1 \text{ atom of Ca}$$

$$\frac{0.090 \text{ mol of Cl atoms}}{0.045 \text{ mol}} = 2 \text{ atoms of Cl}$$

There is one atom of Ca for every two atoms of Cl, resulting in an empirical formula of $CaCl_2$.

54 A 10.00-gram sample of a compound is composed of 1.59 grams of boron (B) and 8.41 grams of fluorine (F). Determine the empirical formula of the compound. (Calculate step A to the nearest thousandth.) _____

Answer:

A. Determine the number of moles of each element within the sample.

$$1.59 \text{ grams B} \times \frac{1 \text{ mol of B}}{10.81 \text{ grams}} = 0.147 \text{ mol of B}$$

$$8.41 \text{ grams F} \times \frac{1 \text{ mol of F}}{19.00 \text{ grams}} = 0.443 \text{ mol of F}$$

B. Determine the simplest whole number ratio of B atoms to F atoms.

$$\frac{0.147 \text{ mol of B atoms}}{0.147 \text{ mol}} = 1 \text{ atom of B}$$

$$\frac{0.443 \text{ mol of F atoms}}{0.147 \text{ mol}} = 3 \text{ atoms of F}$$

The empirical formula is BF_3.

55 A 5.00-gram sample of a compound contains 3.74 grams of carbon (C) and is made up of only carbon (C) and hydrogen (H). Determine the empirical formula for this compound. (Hint: First find the weight of hydrogen in the sample. Calculate step B to the nearest thousandth.) _____

Answer: Since the sample is 5.00 grams total and carbon (C) contributes 3.74 grams, the hydrogen (H) must weigh 1.26 grams (5.00 − 3.74 = 1.26).

A. Determine the number of moles of atoms of each element within the sample.

$$3.74 \text{ grams C} \times \frac{1 \text{ mol of C}}{12.01 \text{ grams}} = 0.311 \text{ mol of C}$$

$$1.26 \text{ grams H} \times \frac{1 \text{ mol of H}}{1.01 \text{ grams}} = 1.248 \text{ mol of H}$$

B. Determine the simplest whole number ratio of C atoms to H atoms.

$$\frac{0.311 \text{ mol of C atoms}}{0.311 \text{ mol}} = 1 \text{ C atom}$$

$$\frac{1.248 \text{ mol of H atoms}}{0.311 \text{ mol}} = 4 \text{ H atoms}$$

The empirical formula is CH_4 (or H_4C).

56 An empirical formula weight can be expressed in either amu or grams. It is simply the combined atomic weights of the elements in the empirical formula. A sample of a compound has been found to have an empirical formula of CH (one carbon atom for every hydrogen atom). The empirical formula weight for CH is _____ amu.

Answer: 13.02 (C = 12.01 amu and H = 1.01 amu; thus, CH = 12.01 + 1.01 = 13.02 amu)

ACTUAL MOLECULAR FORMULA

57 The actual formula of a compound may be different from the empirical formula. For example, instead of CH, the actual formula could be C_2H_2 or C_6H_6. The empirical formula tells us only the *ratio* of one atom to another. In the case of the empirical formula CH, we know that the actual formula has one C atom for every H atom. After determining the empirical formula, a chemist must devise some other experimental means to determine the actual formula. This is often done by finding the formula weight of the actual compound. Suppose that, by experiment, it was determined that the actual formula weight of an unknown compound was double that of the empirical formula weight. The empirical formula is CH. The actual formula would be double that of CH. Which of the following formulas would you expect to be the actual formula: C_2H_2 or C_6H_6?

_____

Answer: C_2H_2

58 To determine the relationship between the empirical formula and the actual formula, divide the actual formula weight by the empirical formula weight. Suppose the actual formula weight has been found to be 78.11 amu. The empirical formula weight for CH is 13.02 amu.

(a) $\dfrac{\text{actual formula weight}}{\text{empirical formula weight}} = \dfrac{78.11 \text{ amu}}{13.02 \text{ amu}} =$ _____

(b) The actual formula is how many times larger than the empirical formula?

_____

(c) If the empirical formula is CH, the actual formula would be _____.

Answer:

(a) 6.0

(b) 6 times larger (note that the actual weight is a whole number multiple of the empirical formula)

(c) C_6H_6

59 A compound has been found to have an empirical formula of HO (one atom of hydrogen for every atom of oxygen). The actual formula weight has been found to be 34.02 amu. Determine the empirical formula weight of HO. (Remember, the empirical formula weight is the combined atomic weights of the elements in the empirical formula.)

(a) The empirical formula weight of HO is _____ amu.

(b) What is the actual formula? _____

Answer:

(a) 17.01(16.00 + 1.01 = 17.01 amu)

(b) Divide the actual formula weight by the empirical formula weight to determine the relationship between the actual formula and the empirical formula.

$$\frac{\text{actual formula weight}}{\text{empirical formula weight}} = \frac{34.02 \text{ amu}}{17.01 \text{ amu}} = 2$$

The actual formula is two times as large as the empirical formula. Since the empirical formula is HO, the actual formula is H_2O_2.

60 A compound is found to have an empirical formula of CH_3. The empirical formula weight for CH_3 is _____ amu. (Use a separate sheet of paper to set up a table of calculation for this and the following frames.)

Answer:

Element	Number of atoms	Atomic weight	Atoms × weight
C	1 C	12.01	1 × 12.01 = 12.01
H	3 H	1.01	3 × 1.01 = 3.03
			Empirical formula weight is 15.04 amu

61 The empirical formula is CH_3 and the empirical formula weight is 15.04 amu. The actual formula weight has been found to be 30.08 amu. What is the actual formula? _____

Answer: Divide the actual formula weight by the empirical formula weight:

$$\frac{\text{actual formula weight}}{\text{empirical formula weight}} = \frac{30.08}{15.04} = 2$$

The actual formula is two times as large as the empirical formula. Thus, the actual formula is C_2H_6.

62 A compound has an empirical formula of CH_2O. The actual formula weight has been found to be 180.18 amu.

(a) Determine the empirical formula weight. _____

(b) What is the actual formula? _____

Answer:

(a)

Element	Number of atoms	Atomic weight	Atoms × weight
C	1 C	12.01	1 × 12.01 = 12.01
H	2 H	1.01	2 × 1.01 = 2.02
O	1 O	16.00	1 × 16.00 = 16.00

Empirical formula weight is 30.03 amu

(b) Divide the actual formula weight by the empirical formula weight:

$$\frac{\text{actual formula weight}}{\text{empirical formula weight}} = \frac{180.18}{30.03} = 6$$

The actual formula is six times as large as the empirical formula. Thus, the actual formula is $C_6H_{12}O_6$.

63 A compound has an empirical formula of CO. The actual formula weight has been found to be 28.01 amu.

(a) The empirical formula weight of CO is _____.

(b) What is the actual formula? _____

Answer:

(a) 28.01 amu (16.00 + 12.01 = 28.01)

(b) Since the actual formula weight and the empirical formula weight are the same, the empirical formula and the actual formula must also be the same.

$$\frac{\text{actual formula weight}}{\text{empirical formula weight}} = \frac{28.01}{28.01} = 1$$

The actual formula is CO.

64 Formula weights can also be expressed in grams. A compound has an empirical formula of NH_2. The actual formula weight has been found to be 32.06 grams.

(a) Determine the empirical formula weight in grams. _____

(b) What is the actual formula? _____

Answer:

(a)

Element	Number of atoms	Atomic weight	Atoms × weight
N	1 N	14.01	$1 \times 14.01 = 14.01$
H	2 H	1.01	$2 \times 1.01 = \underline{2.02}$

Empirical formula weight is 16.03 amu

(b) Divide the actual formula weight by the empirical formula weight.

$$\frac{\text{actual formula weight}}{\text{empirical formula weight}} = \frac{32.06}{16.03} = 2$$

The actual formula is twice as large as the empirical formula. Thus, the actual formula is N_2H_4.

65 A 10.00-gram sample of a compound is made up of 3.04 grams of nitrogen (N) and the remainder is oxygen (O). The actual formula weight of this compound is 92.02 grams. Determine the actual formula of this compound. (When determining the number of moles of each element, calculate to the nearest thousandth.)

Answer:

First, calculate the amount of oxygen in the sample.

$$10.00 \text{ grams} - 3.04 \text{ grams} = 6.96 \text{ grams of oxygen}$$

Second, determine the number of moles of each element within the sample.

$$3.04 \text{ grams of N} \times \frac{1 \text{ mol of N}}{14.01 \text{ grams}} = 0.217 \text{ mol of N}$$

$$6.96 \text{ grams of O} \times \frac{1 \text{ mol of O}}{16.00 \text{ grams}} = 0.435 \text{ mol of O}$$

Third, determine the simplest whole number ratio of N atoms to O atoms to find the empirical formula.

$$\frac{0.217 \text{ mol of N atoms}}{0.217 \text{ mol}} = 1 \text{ N atom}$$

$$\frac{0.435 \text{ mol of O atoms}}{0.217 \text{ mol}} = 2 \text{ O atoms}$$

The empirical formula is NO_2.

Fourth, determine the actual formula.

Element	Number of atoms	Atomic weight	Atoms × weight
N	1 N	14.01	$1 \times 14.01 = 14.01$
O	2 O	16.00	$2 \times 16.00 = \underline{32.00}$

Empirical formula weight is 46.01 grams

Divide the actual formula weight by the empirical formula weight.

$$\frac{\text{actual formula weight}}{\text{empirical formula weight}} = \frac{92.02 \text{ grams}}{46.01 \text{ grams}} = 2$$

The actual formula is twice as large as the empirical formula. Thus, the actual formula is N_2O_4.

What you have just learned is very important to a research chemist who is trying to determine the formula of a compound that has just been produced. The chemist can utilize a variety of instrumental and chemical processes to determine the percentage composition of the compound. In later chapters, you will learn that a chemist can also utilize several different ways to experimentally determine the formula weight of a compound. The combination of these methods permits the researcher to calculate the actual formula of the compound.

You will need to be able to determine formula weights in several of the remaining chapters of this book and in any chemistry course you take. It is important that you have a good understanding of Chapters 1 through 4 before continuing your study.

In Chapter 5 you will learn how chemists name the compounds they produce based upon an internationally accepted scheme.

SELF-TEST

This self-test is designed to show how well you have mastered this chapter's objectives. Correct answers and review instructions follow the test. Round answers to the nearest hundredth.

1. What is the formula weight of dihydrogen sulfide, H_2S, to the nearest hundredth? _____

2. What is the percentage by weight of sulfur in dihydrogen sulfide? _____

3. What is the formula weight of aspirin (acetylsalicylic acid), $C_9H_8O_4$?
   _____

4. What is the percentage by weight of oxygen in aspirin? _____

5. What is the formula weight of acetic acid, $C_2H_4O_2$? _____

6. What is the percentage by weight of carbon in acetic acid? _____

7. What is the actual formula of a compound that is 39.99% carbon, 6.67% hydrogen, and 53.34% oxygen if its actual formula weight is 150.15 amu? (Use a separate sheet of paper for your calculations.) _____

8. How many moles of atoms are represented by 2.50 grams of Na and 4.44 grams of Br? (Calculate moles to the nearest hundredth.)

   (a) moles of Na atoms = _____

   (b) moles of Br atoms = _____

9. How many moles of atoms are represented by 0.35 grams of Mg and 0.98 grams of O? (Calculate moles to the nearest hundredth.)

   (a) moles of Mg atoms = _____

   (b) moles of O atoms = _____

10. What is the actual formula of a compound that is 92.26% carbon and 7.74% hydrogen, if its actual formula weight is 78.11 amu? (Use a separate sheet of paper for your calculations.)_____

11. What is the actual formula of a compound that is 85.63% carbon and 14.37% hydrogen, if its actual formula weight is 56.11 amu? (Use a separate sheet of paper for your calculations.)

12. Lisinopril ($C_{21}H_{31}N_3O_5$) is a medication that is often prescribed to treat high blood pressure. The formula weight for lisinopril is 405.495 amu. Calculate the % carbon and % hydrogen for this compound.

13. Cadaverine ($C_5H_{14}N_2$) is a noxious compound produced from the putrefaction of animal tissue. The formula weight for cadaverine is 102.178 amu. Calculate the % N and % H of this compound.

14. Explain the difference between an empirical formula and the actual formula of a compound._____

15. How is a molecule of a compound different from a molecule of an element?_____

ANSWERS

Compare your answers to the self-test with those given below. If you answer all questions correctly, you are ready to proceed to the next chapter. If you miss any, review the frames indicated in parentheses following the answers. If you miss several questions, you should probably reread the chapter carefully.

1. 34.08 amu (frames 4, 9, 13)

2. %sulfur = 32.06 amu of sulfur/34.08 amu of dihydrogen sulfide × 100 % = 94.07 % S (frames 18–22)

3. 180.17 amu (frames 4, 9, 13)

4. %oxygen = $\dfrac{64.00 \text{ amu of oxygen}}{180.17 \text{ amu of aspirin}} \times 100\% = 35.52\%$ (frames 18–22)

5. 60.06 amu (frames 4, 9, 13)

6. %carbon = [2 × (1 atom of carbon × 12.01 amu of carbon)/60.06 amu acetic acid] × 100 % = 24.02/60.06 × 100% = 40%

7. Step 1:

$$\text{atoms of C} = \frac{39.99 \text{ amu of C}}{12.01 \text{ amu per atom of C}} = 3.33 \text{ atoms of C}$$

$$\text{atoms of H} = \frac{6.67 \text{ amu of H}}{1.01 \text{ amu per atom of H}} = 6.60 \text{ atoms of H}$$

$$\text{atoms of O} = \frac{53.34 \text{ amu of O}}{16.00 \text{ amu per atom of O}} = 3.33 \text{ atoms of O}$$

Step 2: $\dfrac{3.33 \text{ atoms of C}}{3.33} = 1.00 \text{ atom of C}$

$\dfrac{6.60 \text{ atoms of H}}{3.33} = 2.00 \text{ atom of H}$

$\dfrac{3.33 \text{ atoms of O}}{3.33} = 1.00 \text{ atom of O}$

Step 3: The empirical formula is CH_2O

Step 4: $\dfrac{\text{actual formula weight}}{\text{empirical formula weight}} = \dfrac{150.15}{30.03} = 5$

Therefore, the actual formula is $C_5H_{10}O_5$. (frames 24–39, 56–59)

8. (a) 2.5 grams Na $\times \dfrac{1 \text{ mol of Na}}{22.99 \text{ grams Na}} = 0.11 \text{ mol of Na}$ (frame 52)

    (b) 4.4 grams Br $\times \dfrac{1 \text{ mol of Br}}{79.904 \text{ grams Br}} = 0.055 \text{ mol of Br}$ (frame 52)

9. (a) 0.35 grams Mg $\times \dfrac{1 \text{ mol of Mg}}{24.305 \text{ grams Mg}} = 0.014 \text{ mol of Mg}$ (frame 52)

    (b) 0.98 grams O $\times \dfrac{1 \text{ mol of O}}{16.00 \text{ grams O}} = 0.061 \text{ mol of O}$ (frame 52)

10. Step 1: atoms of C = $\dfrac{92.26 \text{ amu C}}{12.011 \text{ amu C}} = 7.68 \text{ atoms C}$

    atoms of H = $\dfrac{7.74 \text{ amu H}}{1.008 \text{ amu H}} = 7.68 \text{ atoms H}$

Step 2: $\dfrac{7.68 \text{ atoms C}}{7.68} = 1$ atom of C

$\dfrac{7.68 \text{ atoms H}}{7.68} = 1$ atom of H

Step 3: The empirical formula is CH.

Step 4: $\dfrac{\text{actual formula weight}}{\text{empirical formula weight}} = \dfrac{78.11}{13.02} = 6$

Therefore, the actual formula is C_6H_6. (frames 24–39, 56–59)

11. Step 1: atoms of $C = \dfrac{85.63 \text{ amu C}}{12.011 \text{ amu C}} = 7.13$ atoms C

atoms of $H = \dfrac{14.37 \text{ amu H}}{1.008 \text{ amu H}} = 14.26$ atoms H

Step 2: $\dfrac{7.13 \text{ atoms C}}{7.13} = 1$ atom of C

$\dfrac{14.26 \text{ atoms H}}{7.13} = 2$ atoms of H

Step 3: The empirical formula is CH_2.

Step 4: $\dfrac{\text{actual formula weight}}{\text{empirical formula weight}} = \dfrac{56.11}{14.03} = 4$

Therefore, the actual formula is C_4H_8. (frames 24–39, 56–59)

12.

$\%\text{carbon} = \dfrac{(21 \text{ atoms C} \times 12.011 \text{ amu C})}{405.495 \text{ amu lisinopril}} = 62.203\%\text{C}$

$\%\text{hydrogen} = \dfrac{(31 \text{ atoms H} \times 1.008 \text{ amu H})}{405.495 \text{ amu lisinopril}} = 7.706\%\text{H}$

(frames 15–23)

13.

$\%\text{nitrogen} = \dfrac{(2 \text{ atoms N} \times 14.007 \text{ amu N})}{102.178 \text{ amu cadaverine}} = 27.416\%\text{C}$

$\%\text{hydrogen} = \dfrac{(14 \text{ atoms H} \times 1.008 \text{ amu H})}{102.178 \text{ amu cadaverine}} = 13.811\%\text{H}$

(frames 15–23)

14. An empirical formula merely indicates the combining *ratio* of the atoms. The actual formula has the same atomic ratio as the empirical formula but includes the *actual number* of each atom present according to the measured formula weight. (frames 29–32)

15. A molecule of a compound consists of atoms of *different* elements, whereas a molecule of an element consists of atoms of the *same* element. (frame 6)

5 Nomenclature

Just as you have a given name by which you are recognized, every chemical compound has a name that makes it readily identifiable to every chemist.

In Chapter 4 you were introduced to molecular formulas that indicate the ratio of the combining elements. Our task in this chapter is to teach you how to name a compound when given its molecular formula, and vice versa. To be proficient in naming compounds and writing formulas (and later in writing and balancing chemical equations, Chapter 6), you should begin to memorize the oxidation numbers (valence charge) of the more common elements and groups of atoms that form what we call **polyatomic ions**. A table of common oxidation numbers is found on page 107.

Just as we are given nicknames by friends and become known by names that are not really our own names, the designation of chemical compounds is sometimes complicated by the use of common names that have been carried through the years. The establishment of an international nomenclature scheme, which we will study in this chapter, is an attempt to clear up this difficulty.

OBJECTIVE

After completing this chapter, you will be able to correctly name binary and ternary compounds, including binary and oxyacids and their salts, when given the correct chemical formula, and vice versa.

Chemical formulas are the chemist's international shorthand for describing a substance being studied. A chemical formula specifies what elements and how many atoms of each element are present in any substance.

Before we start naming compounds and writing formulas, we must cover a few ground rules. We present them here collectively and then separately as we work through the chapter. The scheme that we present is used for **inorganic compounds** composed of metals and nonmetals, as opposed to **organic compounds** that consist mainly of C, H, and O and a few other elements. The nomenclature of organic compounds is entirely different. You will be introduced to organic compounds in Chapter 14.

The first compounds we will discuss are **binary compounds** that contain only two elements. One element is usually a metal and the second element is a nonmetal. Remember that metals are found in Groups IA, IIA, IIIA, all the

B groups, and toward the bottom of Groups IVA, VA, and VIA in the periodic table. Metals form positive charges by giving up electrons. The number of metals is much larger than the number of nonmetals. Nonmetals are located in the upper right-hand portion of the periodic table and form negative charges by gaining electrons.

The rules for naming binary compounds are:

- The symbol of the metallic or positively charged element is written first in the formula and named first.

- The symbol of the nonmetallic or negatively charged element is written and named second.

- The name of the metal remains unchanged, but the name of the nonmetal carries the suffix *ide*.

$$CaO \text{ — calcium oxide}$$

$$LiF \text{ — lithium fluoride}$$

- Common groups of elements (polyatomic ions) are treated as a single element.

$$NH_4^+ \text{ — ammonium ion}$$

$$OH^- \text{ — hydroxide ion}$$

$$CN^- \text{ — cyanide ion}$$

These are specified in this way.

$$NH_4Br \text{ – ammonium bromide}$$

$$KOH \text{ – potassium hydroxide}$$

$$NH_4CN \text{ – ammonium cyanide}$$

- If you check the table of charges on page 107, you will note that each formula we have written above is electrically neutral. That is, the sum of all the positive and negative charges is zero. We chose elements and polyatomic ions that carried numerically the same charge. What if we had chosen Li^+ and O^{2-} to form a compound? The formula would be Li_2O because it must be electrically neutral ($2Li^+ = 2+$, $1O^{2-} = 2-$, $2+$ and $2-$ add up to zero). Li_2O means two atoms of Li combined with one atom of O. We use small whole numbers as subscripts to make the formula electrically neutral.

$$Ba^{2+} \text{ and } I^- \text{ form } BaI_2, \text{ barium iodide}$$

$$Al^{3+} \text{ and } OH^- \text{ form } Al(OH)_3, \text{ aluminum hydroxide}$$

$$Rb \text{ and } S^{2-} \text{ form } Rb_2S, \text{ rubidium sulfide}$$

In the next section we will learn to name and write the formulas for some binary compounds.

TABLE OF COMMON OXIDATION NUMBERS

Names and charges of some positive ions:

Name	Formula and charge	Name	Formula and charge
aluminum	Al^{3+}	lead(II), plumbous	Pb^{2+}
ammonium	NH_4^+	lead(IV), plumbic	Pb^{4+}
barium	Ba^{2+}	lithium	Li^+
calcium	Ca^{2+}	magnesium	Mg^{2+}
chromium(II), chromous	Cr^{2+}	manganese(II), manganous	Mn^{2+}
chromium(III), chromic	Cr^{3+}	manganese(III), manganic	Mn^{3+}
chromium(VI), perchromic	Cr^{6+}	mercury(I), mercurous	Hg_2^{2+}
copper(I), cuprous	Cu^+	mercury(II), mercuric	Hg^{2+}
copper(II), cupric	Cu^{2+}	potassium	K^+
gold(I), aurous	Au^+	silver	Ag^+
gold(III), auric	Au^{3+}	sodium	Na^+
hydrogen	H^+	tin(II), stannous	Sn^{2+}
iron(II), ferrous	Fe^{2+}	tin(IV), stannic	Sn^{4+}
iron(III), ferric	Fe^{3+}	zinc	Zn^{2+}

Names and charges of some negative ions:

Name	Formula and charge	Name	Formula and charge
bromide	Br^-	nitride	N^{3-}
chloride	Cl^-	oxide	O^{2-}
fluoride	F^-	sulfide	S^{2-}
iodide	I^-		

Names and charges of some polyatomic ions (all are negative except ammonium):

Name	Formula and charge	Name	Formula and charge
acetate	$C_2H_3O_2^-$	hypoiodite	IO^-
ammonium	NH_4^+	iodate	IO_3^-
bromate	BrO_3^-	nitrate	NO_3^-
bromite	BrO_2^-	nitrite	NO_2^-
carbonate	CO_3^{2-}	perchlorate	ClO_4^-
chlorate	ClO_3^-	periodate	IO_4^-
chlorite	ClO_2^-	phosphate	PO_4^{3-}
cyanide	CN^-	selenate	SeO_4^{2-}
hydroxide	OH^-	selenide	Se^{2-}
hypobromite	BrO^-	sulfate	SO_4^{2-}
hypochlorite	ClO^-	sulfite	SO_3^{2-}

BINARY COMPOUNDS

1 Common table salt is called sodium chloride by chemists. Knowing that NaCl is sodium chloride will help you to remember the names of a whole class of chemical compounds. Two-element compounds are classified as binary compounds. Because NaCl contains two elements, it is classified as a _____ compound.

Answer: binary

2 The names of all binary compounds end in the letters *ide*. Note that the written name of NaCl is sodium chloride. The compound KCl is called potassium _____.

Answer: chloride

3 The compound CaS is called _____.

Answer: calcium sulfide

4 Na_2O is called _____.

Answer: sodium oxide

5 Several common ions, called polyatomic ions, are treated as single elements. The ammonium ion (NH_4^+) is treated as a single element and is named *first* in a compound. The hydroxide ion (OH^-) and the cyanide ion (CN^-) are also treated as single elements in naming. These two are named *last*.

   Give the name for each of the following.

   (a) NH_4Cl _____

   (b) KCN _____

   (c) NH_4I _____

   (d) HCl _____

Answer: (a) ammonium chloride; (b) potassium cyanide; (c) ammonium iodide; (d) hydrogen chloride

6 Give the formula for each of the following.

(a) lithium hydroxide _____

(b) ammonium hydroxide _____

(c) barium hydroxide _____

(d) hydrogen bromide _____

Answer: (a) LiOH; (b) NH_4OH; (c) $Ba(OH)_2$; (d) HBr

7 Now give the formula for each of the following.

(a) potassium sulfide _____

(b) magnesium sulfide _____

(c) aluminum sulfide _____

Answer: (a) K_2S; (b) MgS; (c) Al_2S_3 ($2Al^{3+} = 6+$, $3S^{2-} = 6-$,   6+ and 6− = 0)

You may have noticed that some common metallic elements, such as iron, copper, gold, chromium, mercury, tin, and lead, can have more than one oxidation number (other than zero). This means that iron can combine with chlorine to form two different compounds: $FeCl_2$ and $FeCl_3$. We can't call them both iron chloride so we must do something else.

There are two accepted methods of naming compounds containing these metals. One method is to write the oxidation number of the metal in Roman numerals to the right of the metal name followed by the name of the nonmetal with its suffix *ide*. For example, $FeCl_2$ is named "iron(II) chloride" because iron in this compound has an oxidation number of 2+ and Cl is 1−. $FeCl_3$ is named "iron(III) chloride" because iron in this compound has an oxidation number of 3+.

8 In the compound FeI_2, iron has an oxidation number of 2+. FeI_2 is named iron(_____) iodide.

Answer: II

9 In the compound $Fe(OH)_3$, iron has an oxidation number of 3+. $Fe(OH)_3$ is named _____

Answer: iron(III) hydroxide

10   Iron, copper, gold, chromium, tin, lead, and mercury can form ions with two oxidation numbers.

(a) In the compound CuO, the oxidation number of O is 2−. Thus, the oxidation number of copper is _____.

(b) The name of CuO is _____.

Answer: (a) 2+ (charges must add up to zero); (b) copper(II) oxide

11   Given the compound Cu_2O:

(a) The oxidation number of copper in the compound is _____.

(b) The name of the compound is _____.

Answer: (a) 1+; (b) copper(I) oxide

12   As we mentioned, iron, copper, gold, chromium, mercury, lead, and tin are metals that can form ions with two (or more) oxidation numbers.

Metal	Possible oxidation numbers
iron	2+ and 3+
copper	1+ and 2+
gold	1+ and 3+
tin	2+ and 4+
chromium	2+, 3+, and 6+
mercury	1+ and 2+
lead	2+ and 4+

What is the oxidation number of the metal and the name of the compound for the following?

(a) AuF_3 if F = 1− _____, _____

(b) $SnCl_4$ if Cl = 1− _____, _____

(c) PbO_2 if O = 2− _____, _____

(d) Cr_2O_3 if O = 2− _____, _____

Answer: (a) 3+, gold(III) fluoride; (b) 4+, tin(IV) chloride; (c) 4+, lead(IV) oxide; (d) 3+, chromium(III) oxide

13 What are the formulas for the following compounds?

(a) chromium(II) bromide_____

(b) lead(II) hydroxide_____

(c) gold(I) cyanide_____

(d) mercury(II) oxide_____

Answer: (a) $CrBr_2$; (b) $Pb(OH)_2$; (c) AuCN; (d) HgO

14 An older method of naming compounds made from these metals (iron, copper, gold, chromium, tin, mercury, and lead) is to add *ic* or *ous* to their Latin names of these metals.

The suffix *ic* is added to the name of the element when the *higher* oxidation state is used. The suffix *ous* is added to the element name when the *lower* oxidation state is displayed in a compound. For example:

Gold(I) is the same as aurous. Gold(III) is the same as auric. Therefore, AuCl could be named gold(I) chloride or aurous chloride.

Here are more applications of the naming method to our group of metals.

Lead(II) is the same as plumbous. Lead(IV) is the same as plumbic.
Mercury(I) is the same as mercurous. Mercury(II) is the same as mercuric.
Tin(II) is the same as stannous. Tin(IV) is the same as stannic.
Copper(I) is the same as cuprous. Copper(II) is the same as cupric.
Iron(II) is the same as ferrous. Iron(III) is the same as ferric.
Chromium(II) is the same as chromous. Chromium(III) is the same as chromic.
Chromium(VI) is an exception, since chromium can have three different oxidation states.
Chromium(VI) is also called perchromic, a name you will learn more about later.

(a) Two names for Fe_2O_3 are _____ and
_____.

(b) Two names for CuOH are _____ and
_____.

(c) Two names for Cr_2S_3 are _____ and
_____.

(d) Two names for $PbCl_2$ are _____ and
_____.

Answer: (a) iron(III) oxide, ferric oxide; (b) copper(I) hydroxide, cuprous hydroxide; (c) chromium(III) sulfide, chromic sulfide; (d) lead(II) chloride, plumbous chloride

15 Write the formulas for these compounds.

(a) stannous fluoride _____

(b) mercuric iodide _____

(c) plumbous oxide _____

(d) stannic sulfide _____

Answer: (a) SnF_2; (b) HgI_2; (c) PbO; (d) SnS_2

The most common metals capable of two (or more) oxidation states have been listed in frame 12. In binary compounds, the nonmetals have only one oxidation number. The halogens are all 1−; oxygen and sulfur are each 2−. The Group IA metals are all 1+, and the Group IIA metals are 2+.

Before you continue you should memorize the oxidation states of the metals, non-metals, and polyatomic ions as listed on the Table of Common Oxidation Numbers on page 107.

16 Certain pairs of nonmetallic elements form more than one binary compound. Good examples are carbon and oxygen. They combine to form CO and CO_2. Because carbon is more metallic than oxygen, it is written first. Based upon what you have just learned about naming compounds, their names using Roman numerals would be, respectively, _____ and _____.

Answer: carbon(II) oxide; carbon(IV) oxide

Through previous experience you probably know CO is also commonly called carbon monoxide and CO_2 is carbon dioxide.

The prefixes *mono* (1), *di* (2), *tri* (3), *tetra* (4), *penta* (5), *hexa* (6), *hepta* (7), and *octa* (8) indicate the number of atoms of each element present in the compound. Many binary compounds of carbon, nitrogen, sulfur, and oxygen are named using prefixes instead of the Roman numeral scheme. The next few frames will illustrate how prefixes are used.

17 SO_2 is called sulfur dioxide. You might have called it monosulfur dioxide, but when naming binary compounds that have only one positive atom in the compound, the *mono* prefix is dropped. For example, CO_2 is carbon dioxide not monocarbon dioxide. CO is carbon monoxide, not monocarbon monoxide.

What is the name for SO_3? _____

Answer: sulfur trioxide (one atom of sulfur, three atoms of oxygen)

18 Several binary compounds of oxygen and nitrogen are N_2O, NO, NO_2, N_2O_3, and N_2O_5. N_2O_3 is named *di* nitrogen *trioxide*. NO_2 is named nitrogen *di*oxide.

(a) NO is named _____

(b) N_2O is named _____

(c) N_2O_5 is named _____

Answer: (a) nitrogen monoxide; (b) dinitrogen monoxide;(c) dinitrogen pentoxide

19 Use the same system to name the following compounds.

(a) CS_2 _____

(b) BF_3 _____

(c) CCl_4 _____

Answer: (a) carbon disulfide; (b) boron trifluoride; (c) carbon tetrachloride

20 Write the formula for the following compounds.

(a) nitrogen triiodide _____

(b) nitrogen dioxide _____

(c) diphosphorus trioxide _____

(d) diphosphorus pentoxide _____

(e) dinitrogen tetroxide _____

Answer: (a) NI_3; (b) NO_2; (c) P_2O_3; (d) P_2O_5; (e) N_2O_4

21 Previously we have seen prefixes in parentheses used instead of oxidation numbers for binary compounds of metals when the metal has more than one possible oxidation number. For example, MnO_2 may be called manganese(IV) oxide or manganese dioxide. SeO_3 may be called selenium(VI) oxide or selenium trioxide. Likewise, N_2O_3 may be called dinitrogen trioxide or nitrogen(III) oxide.

What other name could be given for the following?

(a) NO_2: nitrogen dioxide or _____

(b) CCl_4: carbon tetrachloride or _____

(c) P_2O_3: diphosphorus trioxide or _____

(d) PbO_2: plumbic oxide, lead(IV) oxide, or _____

Answer: (a) nitrogen(IV) oxide; (b) carbon(IV) chloride; (c) phosphorus(III) oxide; (d) lead dioxide

22 What would be the formula for the following?

(a) lead monoxide _____

(b) carbon(IV) sulfide _____

Answer: (a) PbO; (b) CS_2

23 Placing the Roman numeral of the oxidation number of the first named element in the compound name, such as carbon(IV) oxide, is the method recommended by the International Union of Pure and Applied Chemistry (IUPAC) and is less confusing than other methods.

However, as long as other methods of naming exist, you should be able to translate back and forth between methods.

Remember that placing the Roman numeral of the oxidation number of the first element in a compound is valid only if the first–named element can have *more than one* oxidation number. If the first element can have only one oxidation number, we do not use the Roman numeral or a prefix on either name.

Name these compounds.

(a) CaO _____

(b) CO _____

(c) K_3N _____

(d) $MnCl_3$ _____

Answer: (a) calcium oxide (calcium has only one oxidation number); (b) carbon(II) oxide or carbon monoxide; (c) potassium nitride; (d) manganese(III) chloride or manganese trichloride

24 Write the formulas for these compounds.

(a) manganese difluoride _____

(b) chromium(VI) oxide _____

(c) silicon dioxide _____

Answer: (a) MnF_2; (b) CrO_2; (c) SiO_2

BINARY ACIDS

25 One special class of compounds has the H^+ ion serving in place of a metal ion in the compound. These compounds have special properties, which we will identify as acidic properties in Chapter 13. For now we will simply call them acids. The first names of these acids are combinations of prefixes and suffixes attached to the name of another nonmetal, and the second name is "acid."

In these frames we will look at the scheme for naming **binary acids**. Remember, "binary" means only *two* elements are present. The compound HF can be named as a binary compound with the name _____.

Answer: hydrogen fluoride

26 The compound HF is an example of a binary acid. It can be named as an ordinary binary compound, or it can be named hydrofluoric acid. This method of naming substitutes *hydro* for hydrogen and adds the suffix *ic* to the stem or full name of the second element.

Name the binary acid HCl in two ways: _____ or _____.

Answer: hydrogen chloride; hydrochloric acid

27 You may already be familiar with HCl as hydrochloric acid. If you remember this name for HCl, you can remember the method for naming all binary acids.

HBr can be named _____ or _____.

Answer: hydrogen bromide; hydrobromic acid

28 Sulfur (S) and selenium (Se) both have only one negative oxidation number, 2−.

(a) What would be the formula of the binary compounds formed between hydrogen and sulfur? _____ Between hydrogen and selenium? _____

(b) What are the names of those compounds? _____

Answer: (a) H_2S; H_2Se; (b) hydrogen sulfide or hydrosulfuric acid; hydrogen selenide or hydroselenic acid

29 Name H_2Te as an acid. _____

Answer: hydrotelluric acid

OXYACIDS

30 Some very common useful acids contain hydrogen, a nonmetal, and oxygen. Acids that contain these three elements are called **oxyacids**. The *oxy* in oxygen may help you to remember the term "oxyacid."

The most common oxyacid is H_2SO_4, sulfuric acid. Another oxyacid is HNO_3, nitric acid. You may already be familiar with these acids.

Which of the following compounds are oxyacids: HCl, H_2SO_3, HI, H_2O, H_3PO_4? _____

Answer: H_2SO_3, H_3PO_4

Oxyacids are probably the most difficult compounds to name because the same three elements may combine in different ratios. We know, for example, that H, S, and O combine to form H_2SO_4 and H_2SO_3; H, N, and O form HNO_3 and HNO_2. Phosphorus and the halogens form several oxyacids. The next few frames will present a scheme by which oxyacids may be named.

31 The most common forms of oxyacids are named by adding *ic* to the stem or name of the nonmetal in each acid. The *ic* acids *usually* contain either three or four oxygen atoms in their formulas, but there is no systematic way to know whether an *ic* acid contains three or four oxygen atoms. A handy way to remember whether an *ic* acid contains three or four oxygen atoms is to list several *ic* oxyacids alphabetically according to the first letter of the nonmetals symbol, as shown below. Note that as the alphabet reaches the letter O, the number of oxygen atoms in the *ic* acid increases from three to four.

Acid	Name
$HBrO_3$	bromic acid
H_2CO_3	carbonic acid
$HClO_3$	chloric acid
HNO_3	nitric acid
H_3PO_4	phosphoric acid
H_2SO_4	sulfuric acid

HIO_3 follows the rule of thumb. What is the name of HIO_3? _____

Answer: iodic acid

32 H_2SeO_4 follows the rule of thumb. What is the name of H_2SeO_4? _____

Answer: selenic acid

33 Note that we have not included any reference to hydrogen in the names of oxyacids as we did with binary acids. Nor have we included prefixes (*mono, di,* or others). There is no need to do so since the name "acid" tells us there is hydrogen present, and the number of hydrogens present is a function of the oxidation number (charge) of the polyatomic ion with which it is combined.

Look at the table of common polyatomic ions on page 107 (if you need to) and see if you can find the polyatomic ion and its charge for the following *ic* acids.

(a) H_3PO_4 _____

(b) H_2SO_4 _____

(c) HNO_3 _____

Answer: (a) PO_4^{3-}; (b) SO_4^{2-}; (c) NO_3^-

34 Bromic acid follows the rule of thumb. What is the formula of bromic acid? _____ What is the polyatomic ion and its charge? _____

Answer: $HBrO_3$; BrO_3^- (the charge is 1–)

35 Chloric acid follows the rule of thumb. What is the polyatomic ion and its charge for chloric acid?_____

Answer: ClO_3^- (the charge is 1–)

36 Once you have learned the name and formula of the *ic* acid for any nonmetal, all other oxyacids with the same nonmetal are systematically named (e.g., H_2SO_4, H_2SO_3). If an oxyacid contains *one less* oxygen than the *ic* acid, it is named by using the same stem but changing the *ic* to *ous*. Since H_2SO_4 is sulfuric acid, H_2SO_3 is *sulfurous* acid.

HClO$_3$ is chloric acid. What is the formula for chlorous acid? _____

Answer: $HClO_2$ (one less oxygen than the "ic" acid)

37 Write the formulas for the following acids.

(a) Nitrous acid _____

(b) Phosphorous acid _____

Answer: (a) HNO_2; (b) H_3PO_3

38 Name the following compounds.

(a) $HBrO_2$ _____

(b) HIO_2 _____

(c) H_2SeO_3 _____

Answer: (a) bromous acid; (b) iodous acid; (c) selenous acid

39 An acid with *two less* oxygens than the *ic* acid is named by dropping the *ic* and substituting *ous,* and placing the prefix *hypo* before the nonmetal. "Hypo" means under or beneath (such as in hypodermic, meaning under the skin). A "hypo-ous" acid has one less oxygen than an "*ous*" acid.

(a) $HClO_3$ is _____ acid.

(b) $HClO_2$ is _____ acid.

(c) $HClO$ is _____ acid.

Answer: (a) chloric; (b) chlorous; (c) hypochlorous

40 Write the formulas for these acids.

(a) hypobromous acid _____

(b) hypoiodous acid _____

(c) hypophosphorous acid _____

Answer: (a) HBrO; (b) HIO; (c) H_3PO_2

41 In some cases it may be possible for an acid to have *one more* oxygen atom than the corresponding *ic* acid. Such an acid is named by placing the prefix *per* before the nonmetal and keeping the *ic* in the name.

(a) $HClO_3$ is _____ acid.

(b) $HClO_4$ is _____ acid.

Answer: (a) chloric; (b) perchloric

42   Use the naming rule for the following acids.

(a) $HBrO_3$ is named _____.

(b) $HBrO_2$ is named _____.

(c) $HBrO_4$ is named _____.

Answer: (a) bromic acid; (c) bromous acid; (c) perbromic acid

43   Oxyacids really behave as though they have only two parts: the hydrogen ion(s) (H+) and the polyatomic ion. The polyatomic ion that we have discussed is called the **acid polyatomic ion**. For example, the acid polyatomic ion of HNO_3 is $NO_3{}^-$, in H_2SO_4 it is $SO_4{}^{2-}$, and in H_3PO_4 it is $PO_4{}^{3-}$. The acid polyatomic ion carries its own name, as you may have noticed in the table on page 107. The names of acid polyatomic ions for *ic* acids are derived by dropping the *ic* ending and replacing it with *ate*. For example, the $NO_3{}^-$ polyatomic ion is called the *nitrate* polyatomic ion. The $SO_4{}^{2-}$ polyatomic ion is called the *sulfate* polyatomic ion. The $PO_4{}^{3-}$ polyatomic ion is called the _____ polyatomic ion.

Answer: phosphate

44   What are the polyatomic ion names for the following acids?

(a) $HClO_3$ _____

(b) $HBrO_3$ _____

Answer: (a) chlorate; (b) bromate

45   Similarly, *ous* acids have *ite* acid polyatomic ion names. For example, in HNO_2, nitrous acid, the $NO_2{}^-$ is called the *nitrite* polyatomic ion. What is the acid polyatomic ion name for each of the following?

(a) H_2SO_3 _____

(b) $HClO_2$ _____

(c) H_3PO_3 _____

Answer: (a) sulfite; (b) chlorite; (c) phosphite

46 "Hypo-ous" acids have "hypo-ite" polyatomic ions.

(a) ClO^- is the _____ polyatomic ion.

(b) IO^- is the _____ polyatomic ion.

(c) PO_2^{3-} is the _____ polyatomic ion.

Answer: (a) hypochlorite; (b) hypoiodite; (c) hypophosphite

47 "Per-ic" acids have "per-ate" polyatomic ions.

(a) ClO_4^- is the _____ polyatomic ion.

(b) BrO_4^- is the _____ polyatomic ion.

Answer: (a) perchlorate; (b) perbromate

SALTS

48 A very large number of compounds are formed by combining acid polyatomic ions and metal ions. For example, Na^+ and SO_4^{2-} form Na_2SO_4. Remember, the metal or most positive element is written first, the negative part is written second, and subscripts are used to make the positive charges and negative charges equal zero. These compounds formed from metal ions and acid polyatomic ions are called **salts**. In Chapter 13 you will see one way in which salts may be prepared in the laboratory. Salts are named by using the name of the metal followed by the name of the acid polyatomic ion. Thus, Na_2SO_4 is called sodium sulfate. It is derived from Na^+ and SO_4^{2-}.

Salts derived from *ic* acids are called *ate* salts. The acid $HClO_3$ is named chloric acid. The salt $KClO_3$ is named _____.

Answer: potassium chlorate

49 H_2SO_4 is _____ acid. $CaSO_4$ is _____.

Answer: sulfuric; calcium sulfate

50 Try your hand with the following.

(a) HNO_3 is _____ acid.

(b) The formula of the compound formed between Ba^{2+} and NO_3^- is _____.

(c) The compound of Ba^{2+} and NO_3^- is named _____. (Remember, the NO_3^- acts as a single unit. It takes two polyatomic ions to balance Ba^{2+}.)

Answer: (a) nitric; (b) $Ba(NO_3)_2$; (c) barium nitrate

51 HNO_3 is nitric acid. $Al(NO_3)_3$ is _____.

Answer: aluminum nitrate

52 $HBrO_3$ is _____ acid. $Ca(BrO_3)_2$ is _____.

Answer: bromic; calcium bromate

53 What are the formulas of the following compounds?

(a) sodium iodate _____

(b) magnesium sulfate _____

(c) lithium phosphate _____

(d) barium phosphate _____

Answer: (a) $NaIO_3$; (b) $MgSO_4$; (c) Li_3PO_4; (d) $Ba_3(PO_4)_2$

54 Similarly, *ous* acids form *ite* salts, "hypo-ous" acids form "hypo-ite" salts, and "per-ic" acids form "per-ate" salts. HIO is _____ acid. LiIO is _____.

Answer: hypoiodous; lithium hypoiodite

55 HIO_4 is _____ acid. KIO_4 is _____.

Answer: periodic; potassium periodate

56 $HBrO_2$ is _____ acid. $Mg(BrO_2)_2$ is _____.

Answer: bromous; magnesium bromite

SUMMARY

You should now be able to write formulas and name compounds with a reasonable degree of proficiency. As with most rules there are exceptions, but you will not encounter them in this book. The decision tree on page 123 is a useful summary for naming compounds.

I. Binary compounds

  (a) Contain only two different elements.

  (b) Are named by:

   1. Writing the name of the metal followed by the name of the nonmetal with an *ide* ending, or

   2. Writing the name of the metal followed by the charge of the metal in Roman numerals in parentheses, then the nonmetal name with the *ide* ending, or

   3. Using prefixes before the names of the elements with the *ide* ending on the name of the second element.

  (c) Formulas are written by using the symbol of the more metallic element followed by the symbol of the more nonmetallic element, with the necessary subscripts to make the charges equal zero.

II. Binary acids

  (a) Contain hydrogen and another nonmetal.

  (b) Are named by using a *hydro* prefix and an *ic* suffix with the name of the nonmetal.

  (c) Formulas are written with the hydrogen first and the other nonmetal second, with the appropriate subscripts to make the charges equal zero.

III. Oxyacids

  (a) Contain hydrogen, a nonmetal, and oxygen.

  (b) Are named according to the most commonly occurring acid, the *ic* acid. The suffix *ic* is added to the name of the nonmetal.

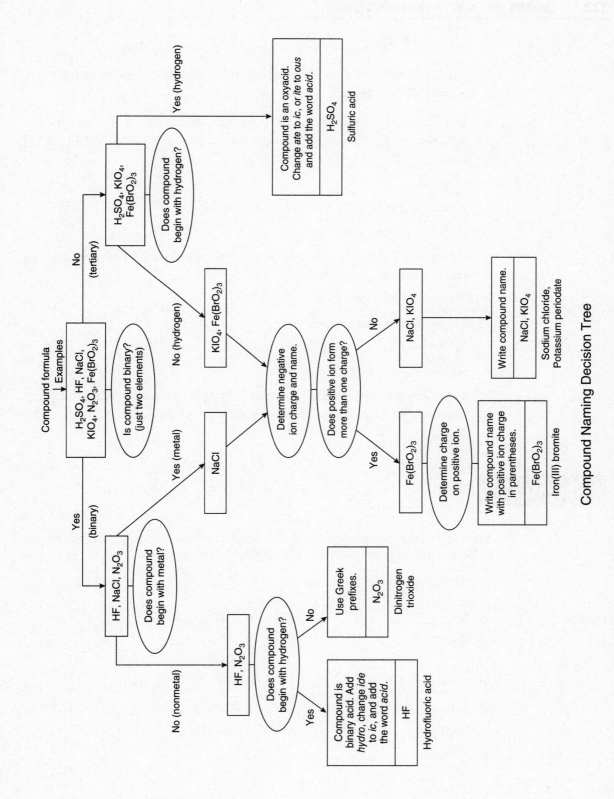

Compound Naming Decision Tree

1. Acids with *one less* oxygen than the *ic* acid are *ous* acids.

2. Acids with *two less* oxygens than the *ic* acid are "hypo-ous" acids.

3. Acids with *one more* oxygen than the *ic* acid are "per-ic" acids.

(c) Formulas are written with the hydrogen followed by the polyatomic acid ion, with necessary subscripts included to make the charges equal zero.

IV. Oxyacid salts

(a) Contain a metal, a nonmetal, and oxygen.

(b) Are derived from oxyacids.

(c) Are named by writing the name of the metal followed by the name of the polyatomic acid ion.

1. *ic* acids form *ate* salts.

2. *ous* acids form *ite* salts.

3. *Hypo-ous* acids form *hypo-ite* salts.

4. *Per-ic* acids form *per-ate* salts.

(d) Formulas are written with the metal symbol first followed by the polyatomic acid ion, with necessary subscripts to make the charges equal zero.

SELF-TEST

This self-test is designed to show how well you have mastered this chapter's objectives. Correct answers and review instructions follow the test.

1. Name the following compounds.

(a) KI _____

(b) NaCN _____

(c) $(NH_4)_2SO_4$ _____

(d) CCl_4 _____

(e) SeO_3 _____

(f) Fe_2O_3 _____

(g) N_2O_5 _____

(h) HCl _____

(i) HNO_2 _____

(j) KClO _____

(k) HIO_4 _____

(l) Na_2SO_3 _____

(m) $Ba(OH)_2$ _____

2. Write the formulas for the following compounds.

(a) lithium bromide _____

(b) carbon monoxide _____

(c) copper(I) oxide _____

(d) carbon(IV) disulfide _____

(e) cupric hydroxide _____

(f) ammonium carbonate _____

(g) hypobromous acid _____

(h) hydrosulfuric acid _____

(i) sodium perchlorate _____

(j) ammonium phosphate_____

(k) ammonium nitrite _____

(l) iron(II) iodide or ferrous iodide _____

(m) dinitrogen trioxide _____

3. What are the oxidation numbers for Cr, Sn, and Fe in the compounds $CrCl_3$, SnF_4, and Fe_2O_3?

4. What are the oxidation numbers for Cu, Au, and Pb in the compounds $CuCl_2$, $AuCl$, and PbO_2?

5. The two names for CuF are _____ and _____.

6. The two names for PbO_2 are _____ and _____.

7. The two names for HI are _____ and _____.

8. The two names for HBr are _____ and _____.

9. Which of the following compounds are binary acids: HF, H_2SO_4, HBr, H_2O, HNO_3?

10. Which of the following compounds are oxyacids: HBr, H_2SO_4, HCl, H_2O, H_3PO_4?

11. What are the names for the following oxyacids: $HClO_3$, $HBrO_3$, HIO_3?

12. What are the names for the following oxyacids: $HClO_4$, $HBrO_4$, HIO_4?

13. Identify the compound that is chlorous acid: $HClO_3$, $HClO_2$, $HClO$.

14. Identify the compound that is hypobromous acid: $HBrO_3$, $HBrO_2$, $HBrO$.

15. The salt $Hg_2(NO_3)_2$ which has a formula of $HgNO_3$ but usually exists as the dihydrate and is thus written as $Hg_2(NO_3)_2$. The dihydrate formula does not change its name. Which of the following are correct names?: mercury(II) nitrate, mercury nitrate, mercurous nitrate, mercury(I) nitrate, mercuric nitrate.

ANSWERS

Compare your answers to the self-test with those given below. If you answer all questions correctly, you are ready to proceed to the next chapter. If you miss any, review the frames indicated in parentheses following the answers. If you miss several questions, you should probably reread the chapter carefully.

1. (a) potassium iodide (frames 1–24)

   (b) sodium cyanide (frames 1–24)

   (c) ammonium sulfate (frames 48–56)

   (d) carbon tetrachloride (frames 1–24)

   (e) selenium trioxide or selenium(VI) oxide (frames 1–24)

   (f) iron(III) oxide or ferric oxide (frames 1–24)

   (g) dinitrogen pentoxide or nitrogen(V) oxide (frames 1–24)

   (h) hydrogen chloride or hydrochloric acid (frames 25–29)

   (i) nitrous acid (frames 30–47)

   (j) potassium hypochlorite (frames 48–56)

   (k) periodic acid (frames 30–47)

   (l) sodium sulfite (frames 48–56)

   (m) barium hydroxide (frames 1–24)

2. (a) LiBr (frames 1–24)

   (b) CO (frames 1–24)

   (c) Cu_2O (frames 1–24)

   (d) CS_2 (frames 1–24)

   (e) $Cu(OH)_2$ (frames 1–24)

   (f) $(NH_4)_2CO_3$ (frames 48–56)

   (g) HBrO (frames 30–47)

   (h) H_2S (frames 25–29)

   (i) $NaClO_4$ (frames 43–56)

   (j) $(NH_4)_3PO_3$ (frames 48–56)

   (k) NH_4NO_2 (frames 48–56)

   (l) FeI_2 (frames 1–24)

   (m) N_2O_3 (frames 1–24)

3. $Cr = +3$ since each $Cl = -1$; $Sn = +4$ since each $F = -1$; $Fe = +3$ since each $O = -2$ (frame 12)

4. $Cu = +2$ since each $Cl = -1$; $Au = +1$ since each $Cl = -1$; $Pb = +4$ since each $O = -2$ (frame 12)

5. Copper (I) fluoride and cuprous fluoride (frame 14)

6. Lead (IV) oxide and plumbic oxide (frame 14)

7. Hydrogen iodide and hydroiodic acid (frames 25–27)

8. Hydrogen bromide and hydrobromic acid (fames 25–27)

9. HF and HBr (frames 25–29)

10. H_2SO_4 and H_3PO_4 (frames 30–47)

11. $HClO_3$ = chloric acid; $HBrO_3$ = bromic acid; HIO_3 = iodic acid (frames 30–47)

12. $HClO_4$ = perchloric acid; $HBrO_4$ = perbromic acid; HIO_4 = periodic acid (frames 30–47)

13. $HClO_2$ (frames 30–47)

14. HBrO (frames 30–47)

15. Mercurous nitrate and mercury (I) nitrate (frames 12 and 14)

6 Chemical Equations

Now that you are familiar with atoms, symbols, molecules, formulas, and nomenclature, let's look at what happens when we mix substances together. The most important result of your efforts with this chapter will be your ability to write a balanced chemical equation that represents the reaction between two or more different substances that produces at least one new substance.

Chemical equations are the chemist's shorthand. They show at a glance what substances have been mixed together and what new substance(s) have been produced. Chemists are able to predict the products of a mixture of substances even though they may never have actually mixed the substances in the laboratory. This is very important to research chemists trying to prepare new products that are useful and beneficial to mankind.

You will learn how to complete and balance several kinds of chemical equations and how chemists recognize whether or not a reaction does indeed occur when substances are mixed.

You will discover that some things remain unchanged during a chemical reaction. Chemists are more concerned with the things that change, so you will learn a scheme that shows what does change during certain chemical reactions.

OBJECTIVES

After completing this chapter, you will be able to

- recognize and apply or illustrate: reactant, product, electrolyte (weak or strong), precipitate, soluble, insoluble, solid, liquid, gas, and aqueous;

- balance a molecular equation when given the formulas of the reactants and products;

- convert word equations to balanced molecular equations;

- convert balanced molecular equations to complete ionic equations and then to net ionic equations;

- identify a chemical equation as an example of a combination, decomposition, single displacement, or double displacement type of reaction;

- explain why or how we recognize whether or not a chemical reaction occurs;

- use a solubility table to determine if a compound is soluble or insoluble.

1   Chemical equations are useful to the chemist because they represent chemical reactions. You have already learned that certain chemical symbols represent molecules, elements, ions, or compounds. These symbols are also used in chemical equations. The substance or substances that are the starting materials in a chemical reaction are called **reactants** and are located on the left side of a chemical equation. The substance or substances produced as the result of a reaction are called **products** and are located on the right side of a chemical equation.

Hydrogen gas and oxygen gas react to produce water. The chemical equation for this reaction is $2H_2 + O_2 \rightarrow 2H_2O$.

(a) Identify the symbol(s) representing the reactant(s).

(b) Identify the symbol(s) representing the product(s).

Answer: (a) H_2 and O_2; (b) H_2O

2   Some reactions require continuing energy in order to occur. Some reactions occur spontaneously when the reactants are mixed together. Other reactions just need energy to get started and will then continue until one or more reactant is used up.

Later in this chapter we will deal with predicting whether or not chemical reactions will actually occur as written.

The following are examples of two types of chemical reactions. The equation $2H_2 + O_2 \rightarrow 2H_2O$ represents a **combination reaction** because two or more different reactants combine to form a single product. The equation $2H_3PO_4 \rightarrow H_4P_2O_7 + H_2O$ represents a **decomposition reaction** because a single reactant is decomposed to form two or more different products.

(a) The equation $2H_2O_2 \rightarrow 2H_2O + O_2$ represents what type of reaction? _____

(b) Identify the reactant(s) and product(s) in the equation. _____

Answer: (a) decomposition reaction; (b) H_2O_2 is the reactant. It is on the left side of the equation. H_2O and O_2 are the products, on the right side of the equation.

3   Identify each of the following equations as either decomposition or combination.

(a) $2C + O_2 \rightarrow 2CO$

(b) $2HO \rightarrow 2Hg + O_2$

(c) $H_2 + Cl_2 \rightarrow 2HCl$

Answer: (a) combination; (b) decomposition; (c) combination

4 The following equation represents a **single displacement reaction**.

$$Mg + 2HCl \rightarrow MgCl_2 + H_2$$

In the reaction, the metal magnesium (Mg) has displaced the hydrogen atom (H) in HCl and formed magnesium chloride and hydrogen gas (H_2) as products. In a single displacement reaction, an element and a compound react to form another element and compound.

In contrast with the single displacement reaction, the following equation is called a **double displacement reaction**; two compounds react to form two different compounds.

$$BaCl_2 + Na_2SO_4 \rightarrow BaSO_4 + 2NaCl$$

In this reaction, the Ba displaces the Na in Na_2SO_4 to form $BaSO_4$. The Na displaces Ba in $BaCl_2$ to form NaCl. In the following examples, two substances react to form two different substances.

(1) $2KCl + 2HgNO_3 \rightarrow 2KNO_3 + Hg_2Cl_2$

(2) $2Na + H_2SO_4 \rightarrow Na_2SO_4 + H_2$

(a) What type of reaction does equation 1 represent?

(b) What type of reaction does equation 2 represent?

Answer: (a) double displacement (involves two compounds reacting to form two different compounds); (b) single displacement (involves a compound and an element reacting to form a different compound and an element)

5 Identify the following reaction equations as combination, decomposition, single displacement, or double displacement.

(a) $2KClO_3 \rightarrow 2KCl + 3O_2$_____

(b) $H_2 + Cl_2 \rightarrow 2HCl$_____

(c) $Zn + Cu(NO_3)_2 \rightarrow Cu + Zn(NO_3)_2$_____

(d) $AgNO_3 + KCl \rightarrow AgCl + KNO_3$_____

(e) $FeCl_3 + 3KOH \rightarrow Fe(OH)_3 + 3KCl$_____

Answer: (a) decomposition; (b) combination; (c) single displacement; (d) double displacement; (e) double displacement

6 Double displacement reactions are especially important because a large number of reactions are of this type. Think of double displacement reactions as compounds exchanging partners. Later in this chapter you will need to predict the products of double displacement reactions when given the reactants. In the following generalized reaction equation, A and B are positive ions (e.g., metal, NH_4^+, or H^+) and X and Y are negative ions (one or more nonmetal atoms of an ionic compound that remain after the positive ion is removed).

$$AX + BY \rightarrow BX + AY$$

In this equation, A has displaced B and B has displaced A.

In a double displacement reaction, the metal, or NH_4^+, or H^+ ion of one reactant replaces the metal, or NH_4^+, or H^+ ion in the other reactant, and vice versa.

In the following double displacement reaction, the products can be predicted.

$$AgNO_3 + NaCl \rightarrow ?$$

The positive metal ions are Ag^+ and Na^+. The negative ions are NO_3^- and Cl^-. Predict the products of this reaction. _____

Answer: AgCl and $NaNO_3$

7 Predict the products of the following double displacement reactions.

(a) $CuSO_4 + Ca(OH)_2 \rightarrow$ _____

(b) $NaC_2H_3O_2 + HCl \rightarrow$ _____

Answer: (a) $CaSO_4$ and $Cu(OH)_2$; (b) NaCl and $HC_2H_3O_2$

BALANCING CHEMICAL EQUATIONS

8 In an actual reaction, all atoms must be accounted for. Atoms are neither gained nor lost. All of the equations thus far have been **balanced**: that is, the number of each type of atom on the reactant side of the equation is equal to the number of each type of atom on the product side. The balancing of chemical equations is very important to chemists, since only a balanced equation can adequately describe the ratios of reactants and products in a reaction. An equation that is not balanced does not truly represent a reaction. Look at the following chemical equation:

$$H_2 + Cl_2 \rightarrow HCl$$

(a) How many atoms are on the reactant side? H _____, Cl _____

(b) How many atoms are on the product side? H _____, Cl _____

(b) Is the equation balanced? _____

Answer: (a) two, two; (b) one, one; (c) no (The atoms on the reactant side do not equal those on the product side.)

9 Balancing equations is often a matter of trial and error. Coefficients (usually of small whole numbers) can be placed in front of reactants or products, when necessary, to obtain equal numbers of atoms on both sides of the equation. *Do not change the subscripts* within the formulas of the reactants or products.

The molecular formula for water is H_2O. To double the formula for water, which of the following is the correct expression: H_4O_4, H_4O_2, $2H_2O$?

_____

Answer: $2H_2O$ (The other two expressions have changed the actual molecular formula.)

10 The expression $2H_2O$ indicates two molecules of H_2O. This results in twice as many hydrogen atoms (and doubles the oxygen atoms also). H_2SO_4 is the formula for sulfuric acid and contains two atoms of H, one atom of S, and four atoms of O. The expression $3H_2SO_4$ indicates three molecules of H_2SO_4 containing _____ atoms of H, _____ atoms of S, and _____ atoms of O.

Answer: six; three; 12

11 Let's consider what is meant by the word "equation." The following mathematical expression is an equation because both sides are equal. It is "balanced" because each side is equal to 12.

$$3 \times 4 = 6 \times 2$$

In the purest sense of the word, a chemical "equation" that is not balanced is not a true equation because the two sides are not equal. Only a balanced equation is a true mathematical equation or true chemical equation. To balance a chemical equation, you may place coefficients (smallest possible whole numbers) in front of any substance in the equation on a trial and error basis until the equation is balanced with equal numbers of atoms on each side.

Balance the following equation by placing coefficients in front of appropriate substances. The coefficients should be the *smallest possible* whole numbers.

$$H_2 + Cl_2 \rightarrow HCl$$

Answer: $H_2 + Cl_2 \rightarrow 2HCl$

Since the reactant side has two H atoms and two Cl atoms and the product side has only one H atom and one Cl atom, placing a coefficient of 2 before HCl makes the atoms equal on both sides. It is now a true equation. The following answers are also true equations, but the coefficients are not the *smallest possible* whole numbers. Therefore, they are not correct answers.

$$2H_2 + 2Cl_2 \rightarrow 4HCl, \text{ or}$$

$$3H_2 + 3Cl_2 \rightarrow 6HCl, \text{ or}$$

$$5H_2 + 5Cl_2 \rightarrow 10HCl$$

12 The following unbalanced equation represents a decomposition reaction sometimes used in the laboratory to obtain oxygen gas.

$$HgO \rightarrow Hg + O_2$$

(a) The reactant side has _____ atom(s) of Hg and _____ atom(s) of O.

(b) The product side has _____ atom(s) of Hg and _____ atom(s) of O.

Answer: (a) one, one; (b) one, two

13 The first step in balancing the equation in frame 12 is to make the oxygen atoms on the reactant side equal to the oxygen atoms on the product side. This can be accomplished by placing a coefficient of 2 in front of HgO.

$$2HgO \rightarrow Hg + O_2$$

The resulting numbers of atoms are:

Reactants	Products
2 atoms of Hg	1 atom of Hg
2 atoms of O	2 atoms of O

Place a coefficient in front of another substance in order to complete the balancing of this equation. _____

Answer: $2HgO \rightarrow 2Hg + O_2$ (The equation now indicates two atoms of Hg on the product side as well as on the reactant side.)

 14 The chemical equation $2HgO \rightarrow 2Hg + O_2$ is a true equation and is balanced because _____.

Answer: the number of each kind of atom is equal on both reactant and product sides of the equation.

15 The following unbalanced equation represents a single displacement reaction. Add up the atoms on each side on the equation.

$$Mg + HCl \rightarrow MgCl_2 + H_2$$

Reactants	Products
_____ atom(s) of Mg	_____ atom(s) of Mg
_____ atom(s) of H	_____ atom(s) of H
_____ atom(s) of Cl	_____ atom(s) of Cl

Answer:

Reactants	Products
1 atom of Mg	1 atom of Mg
1 atom of H	2 atoms of H
1 atom of Cl	2 atoms of Cl

16 Use a coefficient to make the chlorine atoms on the reactant side equal to the chlorine atoms on the product side. _____

Answer: $Mg + 2HCl \rightarrow MgCl_2 + H_2$ (Place the coefficient of 2 before the HCl.)

17 Is the equation in frame 16 balanced? (Add the atoms on each side to determine if it is balanced.) _____

Answer: yes (Doubling the HCl resulted in doubling both H and Cl atoms on the reactant side; therefore, the equation is balanced.)

 18 The following unbalanced equation represents a combination reaction. Add up the atoms on each side of the equation.

$$Al + O_2 \rightarrow Al_2O_3$$

Reactants	Products
_____atom(s) of Al	_____atom(s) of Al
_____atom(s) of O	_____atom(s) of O

Answer:

Reactants	Products
1 atom of Al	2 atoms of Al
2 atoms of O	3 atoms of O

19 At this point, we can use a coefficient to equalize either the Al atoms or the O atoms. A good rule of thumb to follow is to leave any oxygen or hydrogen atoms until last, placing coefficients elsewhere first. Use a coefficient to equalize the number of atoms of Al on both the reactant and the product side.

$$Al + O_2 \rightarrow Al_2O_3 \underline{\hspace{4cm}}$$

Answer: $2Al + O_2 \rightarrow Al_2O_3$ (The numbers of Al atoms are now equal on both sides of the equation, but the equation is not yet balanced.)

20 Our equation from frame 19 now represents the following number of atoms.

$$2Al + O_2 \rightarrow Al_2O_3$$

Reactants	Products
2 atoms of Al	2 atoms of Al
2 atoms of O	3 atoms of O

(a) By what number must the oxygen atoms on the reactant side of the equation be multiplied to equal the oxygen atoms on the product side? _____

(b) Is that number a small whole number? _____

Answer: (a) 1 ½ (which would result in an equation of $2Al + 1\frac{1}{2}O_2 \rightarrow Al_2O_3$); (b) no (1 ½ is not a small whole number.)

21 The equation in the answer to frame 20 is now balanced. However, the coefficients in chemical equations should normally be *whole* numbers. We should multiply by the smallest possible whole number to eliminate the fraction. Here, we can multiply every item in the equation by what single whole number? _____

Answer: 2 (Multiplying every item by a factor of 2 gives a balanced equation with no fraction.)

22 Rewrite the equation, multiplying every item by 2. _____

Answer: $4Al + 3O_2 \rightarrow 2Al_2O_3$

23 The following unbalanced equation represents a single displacement reaction. Determine the number of each kind of atom in the unbalanced equation.

$$Al + HCl \rightarrow AlCl_3 + H_2$$

Reactants	Products
_____atom(s) of Al	_____atom(s) of Al
_____atom(s) of H	_____atom(s) of H
_____atom(s) of Cl	_____atom(s) of Cl

Answer:

Reactants	Products
1 atom of Al	1 atoms of Al
1 atom of H	2 atoms of H
1 atom of Cl	3 atoms of Cl

24 Remember, in balancing equations, it often is useful to leave the H atoms and the O atoms (if any) until last. Rewrite the equation below to equalize the Cl atoms on both sides. _____

$$Al + HCl \rightarrow AlCl_3 + H_2$$

Answer: $Al + 3HCl \rightarrow AlCl_3 + H_2$

25 Now that the Cl atoms are equal, is the above equation balanced? _____

Answer: no (The reactant side has three atoms of H and the product has two atoms of H.)

26 Look at the following equation.

$$Al + 3HCl \rightarrow AlCl_3 + H_2$$

What compound or element should be multiplied to make equal numbers of H atoms on both sides of the equation? _____ By what number? _____

Answer: H_2; $1\frac{1}{2}$ or $\frac{3}{2}$

27 It is usually not acceptable to write a fraction in a chemical equation.

$$Al + 3HCl \rightarrow AlCl_3 + \frac{3}{2}H_2$$

The fraction can be eliminated with simple math. Multiply every item in the equation by a factor of 2.

Answer: $2Al + 6HCl \rightarrow 2AlCl_3 + 3H_2$

28 The above equation is balanced because _____.

Answer: it has equal numbers of Al, H, and Cl atoms on each side

29 The following unbalanced equation represents a combination reaction.

$$P_4O_6 + H_2O \rightarrow H_3PO_3$$

(a) Determine the number of atoms of each element for each side

Reactants	Products
____ atom(s) of P	____ atom(s) of P
____ atom(s) of O	____ atom(s) of O
____ atom(s) of H	____ atom(s) of H

(b) Use a coefficient to make the number of P atoms equal on both sides.

Answer:
(a)

Reactants	Products
4 atoms of P	1 atom of P
7 atoms of O	3 atoms of O
2 atoms of H	3 atoms of H

(b) $P_4O_6 + H_2O \rightarrow 4H_3PO_3$ (representing four molecules of H_3PO_3)

Now that the P atoms are equal, count the number of atoms on each side of the equation.

$$P_4O_6 + H_2O \rightarrow 4H_3PO_3$$

Reactants	Products
____ atom(s) of P	____ atom(s) of P
____ atom(s) of O	____ atom(s) of O
____ atom(s) of H	____ atom(s) of H

Answer:

Reactants	Products
4 atoms of P	4 atoms of P
7 atoms of O	12 atoms of O
2 atoms of H	12 atoms of H

 Completely balance the equation. (Hint: Equalize the H atoms or the O atoms next.)

$$P_4O_6 + H_2O \rightarrow 4H_3PO_3 \underline{\hspace{6cm}}$$

Answer: $P_4O_6 + 6H_2O \rightarrow 4H_3PO_3$

32 Balance the following equation representing a combination reaction.

$$Al + Cl_2 \rightarrow AlCl_3$$

Answer:
The first step is to equalize the Cl atoms on both sides.

$$Al + \frac{3}{2}Cl_2 \rightarrow AlCl_3$$

Then eliminate the fraction.

$$2Al + 3Cl_2 \rightarrow 2AlCl_3$$

33 Balance the following equation representing a combination reaction. Leave the H and the O atoms until last.

$$P_4O_{10} + H_2O \rightarrow H_3PO_4 \underline{\hspace{5cm}}$$

Answer: $P_4O_{10} + 6H_2O \rightarrow 4H_3PO_4$

The following unbalanced equation represents a double displacement reaction.

$$BaCl_2 + Na_2SO_4 \rightarrow BaSO_4 + NaCl$$

Balance this equation. _____

Answer: $BaCl_2 + Na_2SO_4 \rightarrow BaSO_4 + 2NaCl$ (The only change necessary to balance the equation is placing a coefficient of 2 before NaCl.)

 The unbalanced equation below also represents a double displacement reaction.

$$FeCl_3 + KOH \rightarrow Fe(OH)_3 + KCl$$

Balance the equation. _____

Answer: $FeCl_3 + 3KOH \rightarrow Fe(OH)_3 + 3KCl$

WORD EQUATIONS

36 Each chemical equation represents a chemical reaction. You have already learned how to name chemical compounds in Chapter 5. Let's apply what you have already learned about naming compounds to reading chemical equations.

The reaction equation below can read as "sodium hydroxide plus hydrogen chloride (or hydrochloric acid) yields sodium chloride and hydrogen oxide (or water)."

$$NaOH + HCl \rightarrow NaCl + H_2O$$

The reaction equation below reads as: _____ plus _____ yields _____ plus _____.

$$AgNO_3 + KCl \rightarrow AgCl + KNO_3$$

Answer: silver nitrate; potassium chloride; silver chloride; potassium nitrate

37 Here are several equations of various types.

(1) $2P + 5O_2 \rightarrow 2P_2O_5$

(2) $C + O_2 \rightarrow CO_2$

(3) $FeCl_3 + 3KOH \rightarrow Fe(OH)_3 + 3KCl$

(a) Equation 1 reads as _____ plus _____ yields _____.

(b) Equation 2 reads as _____ plus _____ yields _____.

(c) Equation 3 reads as _____ plus _____ yields _____.

Answer:

(a) phosphorus; oxygen; phosphorus(V) oxide (or diphosphorus pentoxide)

(b) carbon; oxygen; carbon dioxide (or carbon(IV) oxide)

(c) iron(III) chloride (or ferric chloride); potassium hydroxide; iron(III) hydroxide (or ferric hydroxide); potassium chloride

38 Write the balanced equation for sodium plus water yields sodium hydroxide plus hydrogen gas. (Hydrogen gas is H_2.) _____

Answer:
The unbalanced equation is $Na + H_2O \rightarrow NaOH + H_2$
The balanced equation is $2Na + 2H_2O \rightarrow 2NaOH + H_2$

39 Write the balanced equation for potassium hydroxide plus hydrochloric acid yields potassium chloride plus water. _____

Answer: $KOH + HCl \rightarrow KCl + H_2O$

You have had some practice in balancing equations that represent several types of reactions. In the next section you will learn how to predict whether or not a reaction will actually occur as written in the equation.

REACTIONS: GO OR NO GO

40 Many of the reaction equations you have balanced thus far involve ionic compounds and take place in water (aqueous) solutions. The ionic compounds in aqueous solution actually **dissociate**, meaning that the negative and positive ions separate and move about freely in the solution.

The reaction equation from frame 39 can also be written in complete ionic form.

$$K^+ + OH^- + H^+ + Cl^- \rightarrow K^+ + Cl^- + H_2O$$

Which item in the complete ionic equation is obviously *not* an ion but is a covalently bonded molecule? _____

Answer: H_2O

41   In the ionic equation in frame 40, the whole reaction takes place in water (aqueous) solution. All of the ions are completely dissolved in water. We can show that the reaction takes place in water by writing (aq), which means "aqueous," behind every ion in the equation.

$$K^+(aq) + OH^-(aq) + H^+(aq) + Cl^-(aq) \rightarrow K^+(aq) + Cl^-(aq) + H_2O(l)$$

Not only did the reaction take place in water (aqueous) solution, but water is also a product. Note that (*l*) is placed after H_2O to indicate that it is a liquid. The formation of water as a product is one way to know whether a reaction actually occurs.

A second way to know that a reaction has occurred is if a solid precipitate is made. A **precipitate** is a compound that does not dissolve in water. If a precipitate is formed as a product in a reaction that is taking place in water solution, the precipitate will come out of the solution and will normally settle to the bottom of the solution container. In an experimental situation, it is easy to see if a precipitate is formed. We can also use a solubility table, such as the table (see page 143) to predict whether or not a precipitate will form.

What are two ways to tell whether or not a reaction will take place?

Answer: A reaction will take place if either of the following is true: (1) if water (H_2O) is a product or (2) if a precipitate is a product.

42   If H_2O or a precipitate is not formed as a product, or if other products that will be covered later are not formed, we may just simply have an aqueous solution of the various ions with no reaction taking place at all. It is, therefore, important to determine if a reaction is likely to produce a precipitate. The table above is very useful for determining whether or not a compound formed from a pair of ions is soluble or insoluble in water. Insoluble compounds are precipitates. To use the table, first determine which ions make up the compound. As you learned in Chapter 3, an ionic compound is made up of a positive ion, which is usually a metal or H^+ or NH_4^+, and a negative ion, which is usually the remainder of the compound and is made up of one or more nonmetal atoms.

After determining the ion pair that makes up the compound, locate the ions on the table and determine if the ion pair will form a precipitate or if the compound is soluble. For example, the compound $CaCl_2$ is made up of Ca^{2+} and Cl^- ions. According to the fourth statement on the table, all compounds made up of Cl^- ions are soluble except when combined with Ag^+, Hg^{2+}, or Pb^{2+}. Since Cl^- is combined with Ca^{2+}, the compound is soluble.

Using the table, underline those compounds that are insoluble and would therefore precipitate in aqueous solution.

$$AgNO_3, AgCl, NaCl, CaCO_3, Na_2CO_3, Mg(OH)_2$$

Answer: AgCl, CaCO₃, and Mg(OH)₂

TABLE OF SOLUBILITY OF SOME COMMON COMPOUNDS

- All common compounds made up of alkali metal ions (Li^+, Na^+, K^+, Rb^+, Cs^+, or Fr^+) or NH_4^+ and negative ions are *soluble*.

- All compounds containing $C_2H_3O_2^-$ or NO_3^- and a positive ion are *soluble*.

- All compounds containing SO_4^{2-} are *soluble except* when combined with ions of Ba^{2+}, Sr^{2+}, or Pb^{2+}.

- All compounds containing Br^-, Cl^-, or I^- are *soluble except* when combined with ions of Ag^+, Hg_2^{2+}, or Pb^{2+}.

- Compounds of HCO_3^- with alkaline earth metal ions (Be^{2+}, Mg^{2+}, Ca^{2+}, Sr^{2+}, Ba^{2+}, or Ra^{2+}) and alkali metal ions (Li^+, Na^+, K^+, Rb^+, Cs^+, or Fr^+) or NH_4^+ are *soluble*. Compounds of HCO_3^- are *insoluble* with all other positive ions.

- Compounds with OH^- are *soluble only* with Ba^{2+}, Li^+, Na^+, K^+, Rb^+, Cs^+, Fr^+, or NH_4^+. Compounds with OH^- are *insoluble* with all other positive ions.

- All compounds with CO_3^{2-}, PO_4^{3-}, or S^{2-} are *insoluble except* with alkali metal ions (Li^+, Na^+, K^+, Rb^+, Cs^+, or Fr^+) or NH_4^+.

 Underline those compounds listed below that are insoluble and will form precipitates in aqueous solution.

$$Na_2S, Mg(HCO_3)_2, PbSO_4, Hg_2I_2, MgCO_3$$

Answer: PbSO₄, Hg₂I₂, and MgCO₃

44 Does the following reaction actually occur? Why or why not? (Hint: Use the table to check to see if one of the products is a precipitate.)

$$AgNO_3 + KI \rightarrow AgI + KNO_3$$

Answer: Yes, the reaction occurs because AgI will precipitate.

45 Does the following reaction occur? Why or why not? _____

_____

$$Ba(OH)_2 + MgSO_4 \rightarrow BaSO_4 + Mg(OH)_2$$

Answer: Yes, the reaction occurs because both products are precipitates.

46 You have already learned two means for determining whether or not a reaction occurs. If either water or a precipitate is a product, a reaction occurs. Other means for determining if a reaction actually occurs are the formation of a gas, a weak electrolyte, or a covalent compound. In an experiment, it is easy to detect a gas as a product since it usually bubbles up from an aqueous solution. Typical gases that you have encountered several times in this and previous chapters are H_2 (hydrogen gas), O_2 (oxygen gas), and CO_2 (carbon dioxide gas). We will specify if gases other than these are products.

Weak electrolytes are ionic compounds that dissociate only partially and, therefore, only conduct a weak electric current in aqueous solutions. We will specify if any products are weak electrolytes. (Later chapters will deal more specifically with weak electrolytes.)

Covalent compounds were described in Chapter 3 as compounds that are not ionic and typically are composed of nonmetal atoms. Typical covalent compounds include both H_2O and the gases listed above. We will specify if covalent compounds other than these are reaction products.

If none of these products is formed, assume that a reaction does *not* occur at all. Does the following reaction occur? Why or why not?

_____

$$Mg + 2HCl \rightarrow MgCl_2 + H_2$$

Answer: Yes, the reaction occurs because a gas (H_2) is formed.

47 In the following equation, the products listed are both strong electrolytes (therefore not weak electrolytes). There are no gases or covalent compounds produced. Does the reaction occur? Why or why not?

_____

$$2NaNO_3 + MgCl_2 \rightarrow Mg(NO_3)_2 + 2NaCl$$

Answer: No, the reaction does not occur. The question has already eliminated gases and covalent compounds. Since the products are strong electrolytes, they cannot be weak electrolytes. Water is not a product. Neither of the possible products is a precipitate (according to the solubility table). Therefore, the reaction does not occur.

Here is another double displacement chemical equation. No weak electrolytes, gases, or covalent compounds are produced. Does the reaction occur? Why or why not?

$$2NH_4Br + Pb(NO_3)_2 \rightarrow 2NH_4NO_3 + PbBr_2$$

Answer: Yes, the reaction occurs. Although the products are not weak electrolytes, gases, water, or other covalent compounds, one of the products ($PbBr_2$) is a precipitate.

IONIC EQUATIONS

 Although most of the previous equations involving ions in aqueous solution have been written in *molecular* form (showing complete chemical formulas), a more correct method would be to write an equation showing the ions as *dissociated*. The following equation is in molecular form.

$$AgNO_3 + NaCl \rightarrow AgCl + NaNO_3$$

A more correct method of writing the same equation would show the reactants as ions. In the following equation (reactant side only), note that water is neither a reactant nor a product. Although the reaction takes place in water solution, it does not participate as either a reactant or a product; therefore it can either be left out of the equation or the abbreviation (aq) can be placed beside each ion. Complete the reactant side of this equation in ionic form.

$$Ag^+(aq) + NO_3^-(aq) + \underline{\hspace{1cm}}(aq) + \underline{\hspace{1cm}}(aq) \rightarrow$$

Answer: $Ag^+(aq) + NO_3^-(aq) + Na^+(aq) + Cl^-(aq) \rightarrow$

 The complete molecular equation is $AgNO_3 + NaCl \rightarrow AgCl + NaNO_3$. Since the AgCl is a precipitate and remains an undissolved solid, we place (s) beside its formula (which stands for solid). Write the complete ionic equation representing the precipitate and all ions in aqueous solution.

$$\underline{\hspace{0.7cm}}(aq) + \underline{\hspace{0.7cm}}(aq) + \underline{\hspace{0.7cm}}(aq) + \underline{\hspace{0.7cm}}(aq) \rightarrow AgCl(s) + \underline{\hspace{0.7cm}}(aq) + \underline{\hspace{0.7cm}}(aq)$$

Answer: $Ag^+(aq) + NO_3^-(aq) + Na^+(aq) + Cl^-(aq) \rightarrow AgCl(s) + Na^+(aq) + NO_3^-(aq)$

51 The equation in frame 50 is called a **complete ionic equation**. A complete ionic equation usually includes some ions that did not take part in the reaction and can be found on both sides of the equation. Those ions that do not take part in the reaction are called **spectator ions**.

Look at the complete ionic equation again.

$$Ag^+(aq) + NO_3^-(aq) + Na^+(aq) + Cl^-(aq) \rightarrow AgCl(s) + Na^+(aq) + NO_3^-(aq)$$

The spectator ions are ____ and ____.

Answer: $Na^+(aq), NO_3^-(aq)$

52 A complete ionic equation contains all ions including spectator ions. If the spectator ions are eliminated from both sides of a complete ionic equation, the result is called a **net ionic equation**.

For the same complete ionic equation, write the net ionic equation.

$$Ag^+(aq) + NO_3^-(aq) + Na^+(aq) + Cl^-(aq) \rightarrow AgCl(s) + Na^+(aq) + NO_3^-(aq)$$

Answer: $Ag^+(aq) + Cl^-(aq) \rightarrow AgCl(s)$

53 A net ionic equation contains no spectator ions. A complete ionic equation includes spectator ions. When NaOH is added to $FeCl_3$ in aqueous solution, a precipitate $Fe(OH)_3$ forms. What ions would be included in the complete ionic equation for this reaction? (Forget about balancing the equation for the moment.)

$$\_\_(aq) + \_\_(aq) + \_\_(aq) + \_\_(aq) \rightarrow Fe(OH)_3(s) + \_\_(aq) + \_\_(aq)$$

Answer: $Na^+(aq) + OH^-(aq) + Fe^{3+}(aq) + Cl^-(aq) \rightarrow Fe(OH)_3(s) + Na^+(aq) + Cl^-(aq)$

54 Now balance the equation in frame 53 so that there are equal numbers of elements (including ions) on each side. (Hint: You may find it easier to balance the molecular version first. The unbalanced molecular version is $NaOH + FeCl_3 \rightarrow Fe(OH)_3 + NaCl$.)

$$\_\_Na^+(aq) + \_\_OH^-(aq) + \_\_Fe^{3+}(aq) + \_\_Cl^-(aq)$$
$$\rightarrow Fe(OH)_3(s) + \_\_Na^+(aq) + \_\_Cl^-(aq)$$

Answer: The balanced molecular version is

$$3NaOH + FeCl_3 \rightarrow Fe(OH)_3 + 3NaCl$$

The balanced complete ionic equation is

$$3Na^+(aq) + 3OH^-(aq) + Fe^{3+}(aq) + 3Cl^-(aq) \rightarrow Fe(OH)_3(s) + 3Na^+(aq) + 3Cl^-(aq)$$

 55

Eliminate the spectator ions on both sides of the ionic equation in frame 54. The resulting equation is:

_____ $\rightarrow$ _____

The resulting equation is called a(n) _____ ionic equation.

Answer: $Fe^{3+}(aq) + 3OH^-(aq) \rightarrow Fe(OH)_3(s)$; net ionic

 56

An aqueous solution of Na_2CO_3 is added to dilute hydrochloric acid. The resulting carbonic acid, H_2CO_3, breaks down to form CO_2 gas and H_2O. Complete and balance the molecular equation for this reaction.

_____ + _____ $\rightarrow$ _____ $+ CO_2 + H_2O$

Answer: $Na_2CO_3 + 2HCl \rightarrow NaCl + CO_2 + H_2O$

57

Write the complete ionic equation representing the reaction in frame 56. CO_2 escapes as a gas; therefore we place (g) behind its formula. Both CO_2 and H_2O are considered to be covalent molecules; therefore CO_2 and H_2O remain in molecular form.

_____ $\rightarrow$ _____

Answer: $2Na^+(aq) + CO_3^{2-}(aq) + 2H^+(aq) + 2Cl^-(aq) \rightarrow 2Na^+(aq) + 2Cl^-(aq) + H_2O(l) + CO_2(g)$

58

Write the net ionic version of the equation in frame 57.

_____ $\rightarrow$ _____

Answer: $2H^+(aq) + CO_3^{2-}(aq) \rightarrow H_2O(l) + CO_2(g)$

 59

Complete the following double displacement reaction in molecular form and balance.

$$CuSO_4 + 2LiOH \rightarrow \text{_____} + \text{_____}$$

Answer: $CuSO_4 + 2LiOH \rightarrow Cu(OH)_2 + Li_2SO_4$

What ions are represented on the left (reactant) side of the equation in frame 59?

_____ + _____ + _____ + _____ →

Answer: $Cu^{2+}(aq) + SO_4^{2-}(aq) + 2Li^+(aq) + 2OH^-(aq) →$

 Do any pair of the four ions in frame 60 form an insoluble precipitate? (Use the table on page 143.) _____ If yes, what pair of ions? _____

Answer: yes; Cu^{2+} and $2OH^-$ form an insoluble precipitate: $Cu(OH)_2$.

 Finish the complete ionic equation below. The precipitate should be in molecular form.

$$Cu^{2+}(aq) + SO_4^{2-}(aq) + 2Li^+(aq) + 2OH^-(aq)$$

→ _____ + _____ + _____

Answer: $Cu^{2+}(aq) + SO_4^{2-}(aq) + 2Li^+(aq) + 2OH^-(aq) → Cu(OH)_2(s) + 2Li^+(aq) + SO_4^{2-}(aq)$

63 Write the net ionic equation for the equation in frame 62.

_____ → _____

Answer: $Cu^{2+}(aq) + 2OH^-(aq) → Cu(OH)_2(s)$

64 Complete the following for the reaction of silver sulfate and sodium iodide.

(a) Complete and balance the following molecular equation.

$$Ag_2SO_4 + NaI →$$ _____ + _____

(b) What ions are represented on the reactant side of the above equation?

_____ + _____ + _____ + _____ →

(c) No gases, weak electrolytes, or covalent compounds are products. Will the reaction occur? Why or why not? _____

(d) Write the complete ionic equation.

_____

(e) Write the net ionic equation.

_____

Answers:

(a) $Ag_2SO_4 + 2NaI \rightarrow 2AgI + Na_2SO_4$

(b) $2Ag^+(aq) + SO_4{}^{2-}(aq) + 2Na^+(aq) + 2I^-(aq) \rightarrow$

(c) Yes, Ag^+ and I^- form a precipitate (AgI).

(d) $2Ag^+(aq) + SO_4{}^{2-}(aq) + 2Na^+(aq) + 2I^-(aq) \rightarrow 2AgI(s) + 2Na^+(aq) + SO_4{}^{2-}(aq)$

(e) $2Ag^+(aq) + 2I^-(aq) \rightarrow 2AgI(s)$

We can simplify this net ionic equation by dividing by a factor of 2. Thus: $Ag^+(aq) + I^-(aq) \rightarrow AgI(s)$. Either answer is acceptable.

 Let's try this again with another reaction.

(a) Complete and balance the following double displacement reaction in molecular form.

$$BaCl_2 + Na_2SO_4 \rightarrow \underline{\hspace{3cm}} + \underline{\hspace{3cm}}$$

(b) Does the reaction occur even though there are no gases, weak electrolytes, or covalent compounds formed? Why or why not?

_____

(c) Write the complete ionic equation for the above.

_____

(d) Write the net ionic equation.

_____

Answers:

(a) $BaCl_2 + Na_2SO_4 \rightarrow BaSO_4 + 2NaCl$

(b) Yes, $BaSO_4$ precipitates.

(c) $Ba^{2+}(aq) + 2Cl^-(aq) + 2Na^+(aq) + SO_4{}^{2-}(aq) \rightarrow BaSO_4(s) + 2Na^+(aq) + 2Cl^-(aq)$

(d) $Ba^{2+}(aq) + SO_4{}^{2-}(aq) \rightarrow BaSO_4(s)$

66 Here is another reaction to figure out.

(a) Complete and balance the following molecular equation. $AgNO_3 + KCl \rightarrow$

_____

(b) No product is a gas, a weak electrolyte, or a covalent compound. Does this reaction occur? Why or why not? _____

(c) Write the complete ionic equation for the above.

_____

(d) Write the net ionic equation.

_____

Answers:

(a) $AgNO_3 + KCl \rightarrow AgCl + KNO_3$

(b) Yes, AgCl precipitates.

(c) $Ag^+(aq) + NO_3^-(aq) + K^+(aq) + Cl^-(aq) \rightarrow AgCl(s) + K^+(aq) + NO_3^-(aq)$

(d) $Ag^+(aq) + Cl^-(aq) \rightarrow AgCl(s)$

67 One more time.

(a) Complete and balance the following molecular equation. One of the products of this reaction is H_2CO_3, which immediately breaks down to form H_2O and CO_2.

$$K_2CO_3 + H_2SO_4 \rightarrow \underline{\hspace{5cm}}$$

(b) Write the complete ionic equation for this reaction.

_____

(c) Write the net ionic equation.

_____

Answers:

(a) $K_2CO_3 + H_2SO_4 \rightarrow K_2SO_4 + H_2CO_3$
     (Since the H_2CO_3 breaks down into CO_2 and H_2O immediately, it should be written as follows: $K_2CO_3 + H_2SO_4 \rightarrow K_2SO_4 + H_2O + CO_2$.)

(b) $2K^+(aq) + CO_3^{2-}(aq) + 2H^+(aq) + SO_4^{2-}(aq) \rightarrow CO_2(g) + H_2O(l) + 2K^+(aq) + SO_4^{2-}(aq)$

(c) $CO_3^{2-}(aq) + 2H^+(aq) \rightarrow CO_2(g) + H_2O(l)$

Chemical formulas and equations are extremely useful tools for chemists. They save a lot of time and words when we try to communicate. You will encounter chemical equations throughout the rest of this book, so it is important that you know the material in this chapter.

In the next chapter you will use balanced chemical equations to examine weight relationships between reactants and products. (Ions and ionic equations will appear again in Chapters 9 and 11 through 13, so you can see how important they are to your understanding of chemistry.)

SELF-TEST

This self-test is designed to show how well you have mastered this chapter's objectives. Correct answers and review instructions follow the test.

1. What coefficients provide a balanced reaction for the neutralization of sodium hydroxide with sulfuric acid? $NaOH + H_2SO_4 \rightarrow H_2O + Na_2SO_4$

2. What coefficients provide a balanced reaction for the neutralization of magnesium hydroxide with phosphoric acid? $Mg(OH)_2 + H_3PO_4 \rightarrow H_2O + Mg_3(PO_4)_2$

3. Circle the reaction that is correct for the decomposition of ozone (O_3) to oxygen gas.

$$2O_3 \rightarrow 3O_2$$
$$4O_3 \rightarrow 6O_2$$
$$8O_3 \rightarrow 12O_2$$

4. Circle the reaction that is correct for the oxidation of glucose ($C_6H_{12}O_6$) to water and carbon dioxide gas.

$$C_6H_{12}O_6 + 6O_2 \rightarrow 6H_2O + 6CO_2$$
$$2C_6H_{12}O_6 + 12O_2 \rightarrow 12H_2O + 12CO_2$$
$$4C_6H_{12}O_6 + 24O_2 \rightarrow 24H_2O + 24CO_2$$

5. Identify the ions that are always soluble in solution. Li^+, NO_3^-, SO_4^{2-}, S^{2-}

6. Identify the ions that are sometimes insoluble in solution, depending on the metal ion found. NH_4^+, $C_2H_3O_2^-$, OH^-, Cl^-, K^+, NO_3^-, HCO_3^-

7. Complete and balance the following reactions as molecular equations, complete ionic equations, and net ionic equations.

   (a) ammonium hydroxide + sulfuric acid →

   _____

   _____

   _____

   (b) lead(II) nitrate + sodium chloride →

   _____

   _____

   _____

   (c) $Fe_2(SO_4)_3 + Ba(OH)_2 \rightarrow$

   _____

   _____

   _____

8. Complete and balance the following reactions as molecular equations, complete ionic equations, and net ionic equations.

    (a) lithium hydroxide + magnesium nitrate→

    _____

    _____

    _____

    (b) lead(II) nitrate + potassium iodide→

    _____

    _____

    _____

9. Are the reactions that produce a precipitate a type of single displacement or double displacement? _____

10. Indicate whether each of the following reactions is combination, decomposition, single displacement, or double displacement.

    (a) $2HBr \rightarrow H_2 + Br_2$_____

    (b) $4Ag + O_2 \rightarrow 2Ag_2O$_____

    (c) $2Mg + CO_2 \rightarrow 2MgO + C$_____

11. Indicate whether each of the following reactions is combination, decomposition, single displacement, or double displacement.

    (a) $NaCl + KBr \rightarrow NaBr + KCl$_____

    (b) $Zn + 2HCl \rightarrow ZnCl_2 + H_2$_____

    (c) $CO + Cl_2 \rightarrow COCl_2$_____

12. Using the table on page 143, underline the soluble compounds in the list below.

    $PbBr_2, AgNO_3, NaHCO_3, CuS, KOH, NH_4Br$

13. Using the table on page 143, underline the insoluble compounds in the list below.

    $KBr, NH_4Cl, AgCl, LiC_2H_3O_2, NaHCO_3, CuS$

14. Do all the equations in question 7 occur? Why or why not?

    (a) _____

    (b) _____

    (c) _____

15. A general chemistry student mixed several different ionic solutions together in a large beaker. The mixture contained S^{2-}, OH^-, PO_4^{3-}, and SO_4^{2-}. However, she noticed that nothing had happened upon mixing (i.e., no solids formed). What can the student infer about the ions from these results?

Compare your answers to the self-test with those given below. If you answer all questions correctly, you are ready to proceed to the next chapter. If you miss any, review the frames indicated in parentheses following the answers. If you miss several questions, you should probably reread the chapter carefully.

1. $2NaOH + H_2SO_4 \rightarrow 2H_2O + Na_2SO_4$ (frames 8–35)

2. $3Mg(OH)_2 + 2H_3PO_4 \rightarrow 6H_2O + Mg_3(PO_4)_2$ (frames 8–35)

3. $2O_3 \rightarrow 3O_2$ (frame 11)

4. $C_6H_{12}O_6 + 6O_2 \rightarrow 6H_2O + 6CO_2$ (frame 11)

5. Li^+ and NO_3^- (frame 41)

6. OH^-, Cl^-, HCO_3^- (frame 41)

7. (a) $2NH_4OH + H_2SO_4 \rightarrow (NH_4)_2SO_4 + 2H_2O$ (molecular)
   $2NH_4^+(aq) + 2OH^-(aq) + 2H^+(aq) + SO_4^{2-}(aq) \rightarrow 2NH_4^+(aq) + SO_4^{2-}(aq) + 2H_2O(l)$
   (complete ionic)
   $2OH^-(aq) + 2H^+(aq) \rightarrow 2H_2O(l)$ (net ionic)

   (b) $Pb(NO_3)_2 + 2NaCl \rightarrow PbCl_2(s) + 2NaNO_3$ (molecular)
   $Pb^{2+}(aq) + 2NO_3^-(aq) + 2Na^+(aq) + 2Cl^-(aq) \rightarrow PbCl_2(s) + 2Na^+(aq) + 2NO_3^-(aq)$
   (complete ionic)
   $Pb^{2+}(aq) + 2Cl^-(aq) \rightarrow PbCl_2(s)$ (net ionic)

   (c) $Fe_2(SO_4)_3 + 3Ba(OH)_2 \rightarrow 2Fe(OH)_3(s) + 3BaSO_4(s)$ (molecular)
   $2Fe^{3+}(aq) + 3SO_4^{2-}(aq) + 3Ba^{2+}(aq) + 6OH^-(aq) \rightarrow 2Fe(OH)_3(s) + 3BaSO_4(s)$
   (complete and net ionic because *both* products are precipitates)
   (frames 6–39, 49–67)

8. (a) $2LiOH + Mg(NO_3)_2 \rightarrow LiNO_3 + Mg(OH)_2$ (molecular)
   $2Li^+(aq) + 2OH^-(aq) + Mg^{2+}(aq) + 2NO_3^-(aq) \rightarrow 2Li^+(aq) + 2NO_3^-(aq) + Mg(OH)_2(s)$
   (complete ionic)
   $2OH^-(aq) + Mg^{2+}(aq) \rightarrow Mg(OH)_2(s)$ (net ionic)

   (b) $Pb(NO_3)_2 + 2KI \rightarrow PbI_2 + 2KNO_3$ (molecular)
   $Pb^{2+}(aq) + 2NO_3^-(aq) + 2K^+(aq) + 2I^-(aq) \rightarrow 2K^+(aq) + 2NO_3^-(aq) + Pb(I)_2(s)$
   (complete ionic)
   $Pb^{2+}(aq) + 2I^-(aq) \rightarrow PbI_2(s)$ (net ionic)
   (frames 6–39, 49–67)

9. Double displacement (frames 1–5)

10. (a) decomposition; (b) combination; (c) single displacement (frames 1–5)

11. (a) double displacement; (b) single displacement; (c) combination (frames 1–5)

12. $NaHCO_3$, KOH, NH_4Br, $AgNO_3$ (frames 41–43)

13. $AgCl$, CuS (frames 41–43)

14. (a) yes, because H_2O is produced

    (b) yes, because the precipitate $PbCl_2$ is formed

    (c) yes, because both products are precipitates (frames 40–48)

15. Since no solids formed the student must have had some mixture of alkali metals and/or ammonium ions in the solution (frames 6–39, 49–67, and solubility table on page 143)

7 Mole Concept

Just as a dozen is 12 and a gross is 144, a mole of any kind of particle is 6.022×10^{23} of the particles. The weight of a dozen eggs is not the same as the weight of a dozen oranges. Likewise, the weight of a mole of atoms of one element is not the same as the weight of a mole of atoms of another element. The weight of a mole of any kind of particle or unit, expressed in grams, is numerically the same as the weight of one of the individual particles or units, expressed in atomic mass units (amu).

The following are three examples of the mole concept.

1. From the table of atomic weights, the atomic weight of hydrogen is given as 1.008. This means that one H atom has a weight equal to 1.008 amu. It also means that the weight of 1 mole of H atoms is 1.008 grams. The weight of a single H atom is 1.674×10^{-24} grams. That is, 1.008 grams is the total weight of 6.022×10^{23} H atoms; therefore, the weight of one H atom is 1.008 grams divided by 6.022×10^{23}, which gives 1.674×10^{-24} grams.

2. The formula H_2 for elementary hydrogen means that each hydrogen molecule consists of two H atoms. The weight of two H atoms is 2×1.008 amu $= 2.016$ amu. The weight of 1 mole of H_2 molecules is, therefore, 2.016 grams. The weight of a single H_2 molecule is 3.348×10^{-24} grams.

3. The formula NO_3^- for a nitrate ion means that the ion consists of one nitrogen atom and three oxygen atoms (together with an additional electron that gives the ion its charge but does not contribute significantly to the weight within the range of our precision limits here). The weight of a nitrate ion is, therefore, the weight of one nitrogen atom plus the weight of three oxygen atoms, or 62.01 amu (to the nearest hundredth). The weight of 1 mole of nitrate ions is, therefore, 62.01 grams, and the weight of an individual nitrate ion is 1.03×10^{-22} grams.

In the past, problems involving chemical reactions have been called mass–mass, mass–volume, or volume–volume problems and have been solved using algebraic ratios or proportions. We feel the solution of these problems using the mole concept is superior and fosters a better understanding of what goes on in a reaction than ratio/proportion solving.

The problem solving presented here uses solely the mole concept when we are dealing with chemical reactions. A chemist, chemical engineer, or metallurgist uses this technique to answer very important questions, such as how much raw material will be required to produce a specified quantity of product. Management then uses the cost of those materials and labor, along with other production costs, to determine the price of the product.

OBJECTIVES

After completing this chapter you will be able to calculate the weight or number of moles of any reactant or product that will be used up (if a reactant) or produced (if a product) given a completely balanced chemical equation, a table of atomic weights, and a specified quantity (weight or moles) of any one reactant or product.

1 Remember, a mole is 6.022×10^{23} (Avogadro's number) particles such as atoms, ions, or molecules. The weight of this number of particles is expressed in grams. A mole of hydrogen ions (H^+) contains how many hydrogen ions? _____ How much do they weigh? _____ (Calculate answers in this chapter to the nearest hundredth unless otherwise indicated.)

Answer: 6.022×10^{23}; 1.01 grams

2 Unless otherwise stated, the term *mole* means the same as the term *gram-mole*. A **gram-mole** is the weight of 6.022×10^{23} particles expressed in grams. A gram-mole of atoms is equivalent to the atomic weight expressed in grams.

One mole of carbon atoms (C) weighs how many grams? _____. The atomic weight of carbon is how many grams? _____

Answer: 12.01; 12.01

3 One mole of the element lead (Pb) contains _____ atoms and weighs _____.

Answer: 6.022×10^{23}; 207.2 grams

4 How many atoms of carbon and sulfur are needed to make one molecule of carbon disulfide (CS_2)? _____

Answer: one atom of C and two atoms of S

5 In calculations, the abbreviation *mol* means mole or moles. To make 1 mol of carbon disulfide (CS_2) molecules requires how many moles of carbon atoms and sulfur atoms? _____

Answer: 1 mol of C and 2 mol of S

6 One mole of carbon tetrachloride (CCl_4) requires how many moles of C atoms and how many moles of Cl atoms? _____

Answer: 1 mol of C and 4 mol of Cl

7 Since 4 moles of Cl atoms and 1 mole of C atoms make up 1 mole of CCl_4, what does 1 mole of CCl_4 molecules weigh? _____

Answer: 153.81 grams (Remember to express the answer in grams.)

8 Half a mole of H_2SO_4 weighs how many grams?

Answer: 49.04 grams $\left(\text{weight of } H_2SO_4 = 0.50 \text{ mol } H_2SO_4 \times \dfrac{98.08 \text{ grams } H_2SO_4}{1 \text{ mol } H_2SO_4} = \right.$
$\left. 49.04 \text{ grams } H_2SO_4 \right)$

9 Thirty-six grams of H_2O represent how many moles? _____

Answer: 2 mol $\left(\text{mol } H_2O = 36 \text{ g } H_2O \times \frac{1 \text{ mol } H_2O}{18.02 \text{ g } H_2O} = 2.00 \text{ mol } H_2O\right)$

10 Nine grams of H_2O represent how many moles? _____

Answer: 0.50 mol $\left(\text{mol } H_2O = 9 \text{ g } H_2O \times \frac{1 \text{ mol } H_2O}{18.02 \text{ g } H_2O} = 0.50 \text{ mol}\right)$

11 One formula weight of methyl alcohol (CH_3OH) is 32.05 grams. Half a gram of CH_3OH represents how many moles (to the nearest thousandth)? _____

Answer: 0.016 mol $\left(0.50 \text{ g } CH_3OH \times \frac{1 \text{ mol } CH_3OH}{32.05 \text{ g } CH_3OH} = 0.016 \text{ mol } CH_3OH\right)$

12 A mole of Na atoms is 22.99 grams. A mole of Cl atoms is 35.45 grams. A mole of NaCl molecules is 58.44 grams. Which weighs more, a mole of Cl atoms or a mole of Na atoms? _____

Answer: a mole of Cl atoms

Now that you understand what a mole of a substance is, let's apply the concept to some reactions.

13 In the balanced reaction $Zn + 2HCl \rightarrow H_2 + ZnCl_2$, when zinc metal reacts with hydrochloric acid, hydrogen gas and zinc chloride are the products. How many moles of HCl are needed to produce 1 mole of $ZnCl_2$? _____

Answer: 2

14 Suppose that in the previous reaction, only half a mole of Zn was available. How many moles of $ZnCl_2$ could be produced? _____ How many moles of HCl would be used up? _____

Answer: ½; 1

15 Suppose that only ⅓ mole of Zn was available. How many moles of $ZnCl_2$ would be produced? _____ How many moles of HCl would be used up? _____

Answer: ⅓; ⅔

16 The reaction $2KClO_3 \rightarrow 2KCl + 3O_2$ is often used in the laboratory to produce oxygen gas. In this reaction, 2 moles of $KClO_3$ are required to produce 2 moles of KCl and 3 moles of O_2. If 2 moles of $KClO_3$ are reacted until completely consumed, how much oxygen (O_2) by weight would be produced? (Remember that oxygen gas exists as a diatomic molecule and that 3 moles of O_2 molecules weighs the same as 6 moles of O atoms.) _____

Answer: weight of O_2 = 3 mol $O_2 \times \frac{32.00 \text{ g } O_2}{1 \text{ mol } O_2}$ = 96.00 g O_2

17 Express the following weights to the nearest tenth.

(a) 2 moles of $KClO_3$ weigh _____ grams.

(b) 2 moles of KCl weigh _____ grams.

(c) 3 moles of O_2 weigh _____ grams.

Answer: (a) 245.1; (b) 149.1; (c) 96.0

18 If only 122.6 grams of $KClO_3$ are available, how many moles of KCl and O_2 would be produced? (First, convert grams of $KClO_3$ to moles of $KClO_3$.)

moles of KCl = _____

moles of O_2 = _____

Answer: mol KClO$_3$ = 122.6 g KClO$_3$ × $\frac{1 \text{ mol KClO}_3}{122.55 \text{ g KClO}_3}$ = 1 mol KClO$_3$

2 mol KClO$_3$ produced 2 mol KCl and 3 mol O$_2$ (from frame 16)

∴ 1 mol KClO$_3$ will produce 1 mol KCl and 1.5 mol O$_2$ (Throughout this book, we will use the symbol ∴ for the word "therefore.")

19   If only 122.6 grams of KClO$_3$ are available, how many grams of KCl and O$_2$ would be produced (to the nearest tenth)?

grams of KCl = _____

grams of O$_2$ = _____

Answer: From frame 18, 1 mol KCl and 1.5 mol O$_2$ are produced.

$$\therefore \text{g KCl} = 1 \text{ mol KCl} \times \frac{74.55 \text{ g KCl}}{1 \text{ mol KCl}} = 74.6 \text{ g KCl}$$

$$\therefore \text{g O}_2 = 1.5 \text{ mol O}_2 \times \frac{32.0 \text{ g O}_2}{1 \text{ mol O}_2} = 48.0 \text{ g O}_2$$

20   If only 24.5 grams of KClO$_3$ are available, how many moles of KClO$_3$ are available?

Answer: mol KClO$_3$ = 24.5 g KClO$_3$ × $\frac{1 \text{ mol KClO}_3}{122.55 \text{ g KClO}_3}$ = 0.20 mol KClO$_3$

21   If only 24.5 grams of KClO$_3$ are available, how many moles of KCl and O$_2$ would be produced? (First convert grams of KClO$_3$ to moles of KClO$_3$ as you did in frame 18.)

mol KCl _____

mol O$_2$ _____

How many grams of O$_2$ would be produced (to the nearest tenth)?

_____

Answer:

2 mol KClO$_3$ produces 2 mol KCl and 3 mol O$_2$

∴0.20 mol KClO$_3$ produces 0.20 mol KCl

∴0.20 mol KClO$_3$ produces 0.30 mol O$_2$

$$\therefore \text{g O}_2 = 0.30 \text{ mol O}_2 \times \frac{32.0 \text{ g O}_2}{1 \text{ mol O}_2} = 9.6 \text{ g O}_2$$

22 $2Fe + 3Cl_2 \rightarrow 2FeCl_3$

(a) 2 moles of of Fe weighs _____ grams.

(b) 3 moles of of Cl_2 weighs _____ grams.

(c) 2 moles of of $FeCl_3$ weighs _____ grams.

Answer: (a) 111.70; (b) 212.70; (c) 324.40

23 How many moles of iron (Fe) and chlorine gas (Cl_2) are required to produce 32.44 grams of $FeCl_3$? (First convert grams of $FeCl_3$ to moles of $FeCl_3$.)

moles of Fe _____
moles of Cl_2 _____

Answer: mol $FeCl_3$ = 32.44 g $FeCl_3$ $\times$ $\dfrac{1 \text{ mol } FeCl_3}{162.20 \text{ g } FeCl_3}$ = 0.20 mol $FeCl_3$

1 mol Fe and 1.5 mol Cl_2 will produce 1 mol $FeCl_3$

$\therefore$ 0.20 mol $FeCl_3$ will require 0.20 mol Fe and 0.30 mol Cl_2

24 In the reaction $2H_2 + O_2 \rightarrow 2H_2O$, if 360.4 grams of water are produced, how many moles of H_2 and of O_2 are required? How many grams of O_2 are required?

(a) moles of H_2 _____

(b) moles of O_2 _____

(c) grams of O_2 _____

Answer:

mol H_2O = 360.4 g H_2O $\times$ $\dfrac{1 \text{ mol } H_2O}{18.02 \text{ g } H_2O}$ = 20 mol H_2O

1 mol H_2 plus $\dfrac{1}{2}$ mol O_2 will produce 1 mol H_2O

(a) 20 mol H_2; (b) 10 mol O_2; (c) 320 g O_2 $\left(10 \text{ mol } O_2 \times \dfrac{32.0 \text{ g } O_2}{1.0 \text{ mol } O_2} = 320 \text{ g } O_2 \right)$

LIMITING REACTANT

For many real–life chemical reactions, one or more of the reactants is limited in quantity. For example, when a car burns 1 gallon of gasoline, the gasoline reacts with oxygen from the air and, while the engine runs, produces chemical reaction products, such as CO (carbon monoxide) and H_2O, as well as some others. After the gallon of gasoline is all used up, there is still plenty of oxygen in the air. In fact, after a whole tank full of gasoline is burned, there is still plenty of oxygen in the air. The limiting reactant in this case is the gasoline. It is completely consumed in the reaction.

25 For the reaction $C + O_2 \rightarrow CO_2$, only 3 grams of carbon (C) are available, but there is plenty of oxygen.

(a) Which reactant is the limiting reactant, C or O_2? _____

(b) How many grams of CO_2 could be produced? _____

Answer:

(a) C

(b) $\text{mol C} = 3 \text{ g C} \times \dfrac{1 \text{ mol C}}{12.01 \text{ g C}} = 0.25 \text{ mol C}$

1 mol will produce 1 mol CO_2

∴0.25 mol C will produce 0.25 mol CO_2

∴g CO_2 = 0.25 mol $CO_2 \times \dfrac{44.01 \text{ g CO}_2}{1 \text{ mol CO}_2} = 11.00 \text{ g CO}_2$

26 Only 4.01 grams of methane (CH_4) are available for the reaction $CH_4 + 2O_2 \rightarrow CO_2 + 2H_2O$. There is plenty of oxygen.

(a) Which reactant is the limiting reactant? _____

(b) How many grams of H_2O will be formed? _____

Answer:

(a) CH_4

(b) 9.01 g H_2O

The weight of 1 mol CH_4 = 12.01 g + (4 × 1.01 g) = 16.05 g

$$4.01 \text{ g CH}_4 \times \dfrac{1 \text{ mol CH}_4}{16.05 \text{ g}} = 0.25 \text{ mol CH}_4$$

1 mol CH_4 will produce 2 mol H_2O.∴0.25 mol CH_4 will produce 0.5 mol H_2O

$$0.5 \text{ mol H}_2O \times \dfrac{18.02 \text{ g}}{1 \text{ mol H}_2O} = 9.01 \text{ g H}_2O$$

27 For the following reaction, 18.02 grams of water are produced and all of the silane (SiH_4) is used up, although oxygen is still available after the reaction: $SiH_4 + 2O_2 \rightarrow SiO_2 + 2H_2O$.

(a) How many moles of silane (SiH_4) are used up? _____

(b) What is the limiting reactant? _____

Answer: (a) 0.50 mol SiH_4; (b) silane (SiH_4)

28 After the following reaction, all of the oxygen is used up in the reaction vessel and 60.09 grams of SiO_2 are produced. The reaction is $SiH_4 + 2O_2 \rightarrow SiO_2 + 2H_2O$.

(a) What weight of silane (SiH_4) was used up? _____

(b) Which do you think is the limiting reactant? _____

Answer: (a) 32.13 g SiH_4; (b) Since the oxygen was all used up, it is the limiting reactant. We can assume that some silane still exists in the reaction vessel.

You have just learned how to use the mole concept to solve problems that deal with chemical reactions. You would be able to tell a producer of vinyl chloride how much chlorine would be needed to prepare 1 million pounds of vinyl chloride or a headache remedy producer how much salicylic acid is needed to produce 10 million pounds of acetylsalicylic acid (aspirin), provided you know the formulas of the reactants and the products and their combining ratios as indicated in a balanced chemical equation.

Understanding the mole concept is of utmost importance. You will encounter its use frequently in any chemistry course and in the chapters that follow in this book.

SELF-TEST

This self-test is designed to show how well you have mastered this chapter's objectives. Correct answers and review instructions follow the test. Round answers to the nearest hundredth.

1. How many moles of each atom are in
    (a) 1.00 moles of CH_4. _____
    (b) 1.00 moles of H_3PO_4. _____
    (c) 2.00 moles of glucose ($C_6H_{12}O_6$). _____

2. How many moles of each atom are in
    (a) 1.50 moles of NH_3. _____
    (b) 2.00 moles of H_2CO_3. _____
    (c) 1.00 moles of sucrose ($C_{12}H_{22}O_{11}$). _____

3. Express the following weights to the nearest tenth.

   (a) 4.0 moles of C_2H_6 weigh _____ grams.

   (b) 4.0 moles of C_3H_8 weigh _____ grams.

   (c) 4.0 moles of C_4H_{10} weigh _____ grams.

4. Express the following weights to the nearest tenth.

   (a) 1.8 moles of O_3 weigh _____ grams.

   (b) 3.2 moles of NO_2 weigh _____ grams.

   (c) 5.6 moles of HNO_3 weigh _____ grams.

5. Express the following moles to the nearest hundredth.

   (a) 44.2 grams of Cl_2 is equal to _____ moles of Cl_2.

   (b) 541.0 grams of S_8 is equal to _____ moles of S_8.

   (c) 65.7 grams of H_2O is equal to _____ moles of H_2O

6. Express the following moles to the nearest hundredth.

   (a) 11.3 grams of H_2 is equal to _____ moles of H_2.

   (b) 755.0 grams of P_4 is equal to _____ moles of P_4.

   (c) 89.9 grams of CH_2O is equal to _____ moles of CH_2O

7. Consider the following balanced reaction for the burning of glucose ($C_6H_{12}O_6$):

$$C_6H_{12}O_6 + 6O_2 \rightarrow 6H_2O + 6CO_2$$

Determine the mole-to-mole ratio that is incorrect:

$$\frac{1\ mol\ C_6H_{12}O_6}{6\ mol\ O_2} ; \frac{1\ mol\ C_6H_{12}O_6}{6\ mol\ CO_2} ; \frac{1\ mol\ O_2}{6\ mol\ CO_2} ; \frac{6\ mol\ H_2O}{6\ mol\ CO_2}$$

8. Consider the following balanced reaction for the decomposition of ammonia (NH_3) to nitrogen and hydrogen gas: $2NH_3 \rightarrow N_2 + 3H_2$
   Determine the mole-to-mole ratio that is incorrect:

$$\frac{2\ mol\ NH_3}{1\ mol\ N_2} ; \frac{1\ mol\ N_2}{3\ mol\ H_2} ; \frac{3\ mol\ H_2}{1\ mol\ NH_3}$$

9. Using the balanced equation for the reaction of hydrochloric acid with magnesium hydroxide, determine the amount of water produced if 4.5 moles of HCl are completely reacted.

10. Using the balanced equation for the combustion of glucose in question 7, determine the amount (number of moles) of $C_6H_{12}O_6$ required to produce 18.0 moles of CO_2.

11. Iron metal reacts with oxygen to produce iron(III) oxide, commonly known as rust. If 2.40 moles of iron are completely reacted with excess oxygen, how much iron(III) oxide, in grams, is produced?

12. Consider the balanced equation in question 11. How many grams of iron are required to produce 10.7 grams of Fe_2O_3?

13. In the following reaction, 4 moles of oxygen (O_2) molecules are completely consumed to produce water. There is still some hydrogen left in the reaction vessel after the reaction.

$$2H_2 + O_2 \rightarrow 2H_2O$$

   (a) How many moles of water (H_2O) are formed?_____

   (b) What weight of water is formed?_____

   (c) Which reactant is the limiting reactant?_____

14. In the following reaction, 16.02 grams of methyl alcohol (CH_3OH) are burned in open air:

$$2CH_3OH + 3O_2 \rightarrow 2CO_2 + 4H_2O$$

   (a) How many moles of oxygen (O_2) are used up?_____

   (b) How many grams of water (H_2O) are formed?_____

   (c) Which reactant is the limiting reactant?_____

ANSWERS

Compare your answers to the self-test with those given below. If you answer all questions correctly, you are ready to proceed to the next chapter. If you miss any, review the frames indicated in parentheses following the answers. If you miss several questions, you should probably reread the chapter carefully.

1. (a) 1.00 mol C and 4.00 mol H

   (b) 3.00 mol H, 1.00 mol P, and 4.00 mol O

   (c) 12.00 mol C, 24.00 mol H, and 12.00 mol O (frames 4–6)

2. (a) 1.50 mol N and 4.50 mol H

   (b) 4.00 mol H, 2.00 mol C, and 6.00 mol O

   (c) 12.00 mol C, 22.00 mol H, and 11.00 mol O (frames 4–6)

3. In order to get the mass in grams, multiply the amount of moles by the formula weight of the substance: (grams of substance = moles of substance × formula weight of substance)

   (a) 120.3 g C_2H_6

(b) 176.4 g C_3H_8

(c) 232.5 g C_4H_{10} (frames 8–24)

4. In order to get the mass in grams, multiply the amount of moles by the formula weight of the substance: (grams of substance = moles of substance × formula weight of substance)

(a) 86.4 g O_3

(b) 147.2 g NO_2

(c) 352.9 g HNO_3 (frames 8–24)

5. In order to get the moles of substance, divide the grams of substance by the formula weight of the substance: (moles of substance = grams of substance/formula weight of substance)

(a) 0.62 mol Cl_2

(b) 2.11 mol S_8

(c) 3.65 mol H_2O (frames 8–24)

6. In order to get the moles of substance divide the grams of substance by the formula weight of the substance: (moles of substance = grams of substance/formula weight of substance)

(a) 5.61 mol H_2

(b) 6.09 mol P_4

(c) 2.99 mol CH_2O (frames 8–24)

7. $\dfrac{1 \text{ mol } O_2}{6 \text{ mol } CO_2}$ (frames 13–15)

8. $\dfrac{3 \text{ mol } H_2}{1 \text{ mol } NH_3}$ (frames 13–15)

9. The balanced reaction $2HCl + Mg(OH)_2 \rightarrow 2H_2O + MgCl_2$ shows that for every 2 moles of HCl used, there are 2 moles of water produced. Therefore, 4.5 moles of HCl yields 4.5 moles of water. (frames 13–15)

10. The balanced reaction $C_6H_{12}O_6 + 6O_2 \rightarrow 6H_2O + 6CO_2$ shows that for every 1 mole of glucose used, 6 moles of carbon dioxide are produced. Therefore, producing 18.0 moles of CO_2 requires 3.0 moles of glucose. (frames 13–15)

11. $4Fe + 3O_2 \rightarrow 2Fe_2O_3$

$$2.40 \; \cancel{\text{mol Fe}} \times \frac{2 \; \cancel{\text{mol Fe}_2\text{O}_3}}{4 \; \cancel{\text{mol Fe}}} \times \frac{159.69 \text{ g Fe}_2\text{O}_3}{1 \; \cancel{\text{mol Fe}_2\text{O}_3}} = 191.63 \text{ g Fe}_2\text{O}_3 \text{ (frames 18–24)}$$

12. $10.7 \text{ g Fe}_2\text{O}_3 \times \dfrac{1 \text{ mol Fe}_2\text{O}_3}{159.69 \text{ g Fe}_2\text{O}_3} \times \dfrac{4 \text{ mol Fe}}{2 \text{ mol Fe}_2\text{O}_3} \times \dfrac{55.845 \text{ g Fe}}{1 \text{ mol Fe}} = 7.5 \text{ g Fe}$

(frame 18–24)

13. 1 mol O_2 produces 2 mol H_2O.

(a) ∴ 4 mol O_2 produces 8 mol H_2O.

(b) g H_2O = 8 mol $H_2O \times \dfrac{18.02 \text{ g H}_2\text{O}}{1 \text{ mol H}_2\text{O}} = 144.16 \text{ g H}_2\text{O}$

(c) O_2 (frames 13–16, 28)

14.
$\text{mol CH}_3\text{OH} = 16.02 \text{ g CH}_3\text{OH} \times \dfrac{1 \text{ mol CH}_3\text{OH}}{32.05 \text{ g CH}_3\text{OH}} = 0.50 \text{ mol CH}_3\text{OH}$
2 mol CH_3OH uses 3 mol O_2 and produces 4 mol H_2O.

(a) ∴ 0.50 mol uses 0.75 mol O_2 and produces 1 mol H_2O.

(b) ∴ g H_2O = 1 mol $H_2O \times \dfrac{18.02 \text{ g H}_2\text{O}}{1 \text{ mol H}_2\text{O}} = 18.02 \text{ g H}_2\text{O}$.

(c) CH_3OH is the limiting reactant, since there is plenty of oxygen left after the reaction (the reaction takes place in open air). (frames 18–21, 25–27)

8 Gases

Up to this point you have dealt with the states of matter (gases, liquids, and solids) when they appear as reactants or products in chemical reactions, but we have not discussed their properties.

In this chapter we discuss the properties of gases, how they behave, and how they are affected by changes in pressure and temperature. Several new laws will be introduced that deal with the behavior and properties of gases. You will also see how the molecular weights of gases are obtained from experimental measurements of the weights and volumes of samples of gaseous substances. It is possible to determine the molecular weight of any gas using the laws that are presented because all gases respond in much the same manner to changes in temperature and pressure.

As you work through this chapter we suggest that you think about encounters you have had with gases, such as a bicycle pump, a flat tire, a balloon (filled with air or helium), and the odors of ammonia, perfumes, or skunks. When dealing with problems involving gases, remember to use your experience and common sense.

OBJECTIVES

After completing this chapter, you will be able to

- recognize and apply or illustrate: pressure, temperature, partial pressure, kinetic molecular theory, diffusion, effusion, standard temperature and pressure (STP), Boyle's Law, Charles's Law, Dalton's Law, Graham's Law, combined gas law equation, ideal gas law, absolute temperature, absolute zero, and compressibility;

- calculate the fourth term in Boyle's or Charles's Laws when given any three of the four terms in the equations;

- calculate the sixth term in the combined gas law equation when given any five of the six terms in the equation;

- calculate the molecular weight of any gas when given

1. its density (two ways: using the volume per mole of a gas at STP or Graham's Law);

2. the rate of effusion of the gas compared to the rate of another gas (Graham's Law);

3. P, V, R, T, and the weight of a sample of the gas (using the ideal gas law);

- calculate the weight of a liter of a gas or the volume occupied by 1 gram of a gas when given its molecular formula;

- state the assumptions of the kinetic molecular theory of gases and how they relate to ideal and real gases;

- calculate, using Dalton's Law,

  1. the partial pressure exerted by each gas within a mixture of gases when given the total pressure of the mixture and the mole fraction of each gas in the mixture (or the data necessary to calculate the mole fractions) and vice versa;

  2. the volume percent of each gas in a mixture when given the total volume of gases and the mole fraction of each gas (or the data needed to calculate the mole fraction) and vice versa.

KINETIC MOLECULAR THEORY

1 The next three chapters will deal with the three states of matter: *gases*, *liquids*, and *solids*. These three states are related. If a solid is heated sufficiently, it will melt and become a liquid. If a liquid is heated sufficiently, it will become a gas or vapor. If a gas is cooled sufficiently, it will liquefy, and if a liquid is cooled sufficiently, it will solidify.

Gases are defined on the basis of the kinetic molecular theory. This theory makes several assumptions about gases. The first assumption is that gases are made up of individual molecules that are in constant rapid motion. The molecules possess kinetic energy (energy of motion) and bounce into each other and the walls of the gas container. Molecules bump into the wall at the top of a container as often as they bump into the bottom. This contrasts with the molecules of a liquid, which can flow and glide over one another but are, as you know, kept toward the bottom of the container. For example, the liquid molecules of a glass of water stay at the bottom of the glass. In a solid, the molecules can vibrate and bend, but they cannot flow or glide over one another or bounce around freely.

The kinetic molecular theory also assumes that the molecules of a gas are much farther apart than molecules of a liquid or solid.

(a) In which state do molecules have the most freedom to move about? _____

(b) Which do you expect would contain more molecules, a liter of a gas such as oxygen or a liter of a liquid such as water? _____

Answer:

(a) gas (Gas molecules can move rapidly in any direction. Molecules of a liquid or solid are more restricted.)

(b) a liter of liquid (In a gas, the molecules are farther apart than in a liquid. Therefore, there would be fewer molecules in a given volume of a gas such as a liter.)

2 The constantly moving gas molecules possess kinetic energy (energy of motion). All gases at the same temperature are assumed to have the same average kinetic energy. In fact, temperature is simply a measure of the average kinetic energy. Heating a gas results in an increase in the temperature of the gas and an increase in the average kinetic energy of gas molecules.

The average kinetic energy of sample *A* of a gas is greater than that of sample *B* of that gas. Which gas sample (*A* or *B*) could be expected to have a higher temperature? (All other conditions are equal.) _____

Answer: sample **A** (At constant temperature, all gases are assumed to have the same kinetic energy. Therefore the gas with a higher kinetic energy would also be of higher temperature.)

3 The kinetic molecular theory assumes that the rapidly moving gas molecules collide with each other and the walls of the gas container without any loss of kinetic energy. The collision of the gas molecules with the walls of the gas container results in what we call **pressure**.

A balloon is kept inflated because of pressure caused by _____.

Answer: the collision of gas molecules with the inside wall of the balloon

4 Another assumption of the kinetic molecular theory is that the gas molecules themselves occupy no volume (thus leaving only the space between molecules as the volume occupied by a gas) and have no attraction for each other. Note that a gas that behaves completely according to the kinetic molecular theory assumptions is a theoretically ideal gas. Most actual gases approximate an ideal gas very well at moderate temperatures and pressures because of the large amount of space between the molecules. At low temperatures and high pressures, normal gases deviate from a theoretically ideal gas. We will cover some of these deviations later. Calculations involving pressure, volume, and temperature of gases are much simpler if we assume that gases behave as theoretically ideal. According to the kinetic molecular theory, all of the volume of an ideal gas can be attributed to _____.

Answer: the space between the molecules

5 Gases can be more readily compressed than liquids or solids. Since gas molecules are relatively far apart, an external force or pressure can readily push the molecules closer together.

If a certain volume (such as a liter) of gas is compressed to half that volume, what happens to the number of molecules of gas during compression? (increases, decreases, stays the same) _____

Answer: stays the same (Just the space between the molecules is decreased.)

6 Besides greater compressibility than liquids or solids, a characteristic of gases is greater **diffusibility** than liquids or solids. One gas rapidly diffuses into (mixes completely with) another gas because the space between gas molecules allows ready access to molecules of another gas.

Why could you expect gases to diffuse more rapidly than liquids under roughly comparable conditions? _____

Answer: The space between molecules of a gas is greater than the space between molecules of a comparable liquid.

7 When a gas is compressed, the same number of gas molecules occupies a smaller volume than before compression. That is, the gas container is smaller but contains the same number of molecules. We have defined pressure as the collision of gas molecules with the wall of the gas container. When a gas is compressed would you expect an increase, decrease, or no change in the number of collisions of gas molecules per square unit (such as a square centimeter) of container wall? _____ Why? _____

Answer: an increase (There are just as many gas molecules now striking the wall of the smaller container as were striking the wall of the larger container. However, the container is now smaller; therefore, the number of collisions per square unit of container wall surface has increased.)

8 Pressure is often measured as the relative number of collisions on a given area (square unit) of the gas container. Would you expect the pressure of a compressed gas to be greater than, less than, or the same as that of the gas before compression? _____

Answer: greater (Under compression, more gas molecules collide on a given area of the gas container.)

9 Answer these questions about the effect of pressure on a gas.

(a) At constant temperature, when the volume of a gas (with a specific and unchanging number of molecules) is decreased by compression, the pressure on a given area of the inside wall of the container is (decreased, increased, unchanged) _____

(b) At constant temperature (with a specific number of gas molecules), when the pressure of a gas increases, the volume of the gas is probably being (increased, decreased) _____

Answer: (a) increased; (b) decreased

BOYLE'S LAW

10 A scientist named Boyle formalized the relationship between pressure and volume that you have just seen in the previous frame. **Boyle's Law** states that at constant temperature, the volume of a confined dry gas is inversely proportional to the pressure. (Dry indicates no water present.) Mathematically, Boyle's Law is stated as:

$$PV = \text{a constant at constant temperature}$$

Pressure and volume are inversely proportional. Mathematically this means that if we multiply pressure by some number, we must divide the volume by that same number, or vice versa. For example, assume the constant is equal to the number 4. Assume that the pressure and the volume each are equal to 2:

$$PV = 2 \times 2 = 4$$

If we divide the pressure by 2, we must multiply the volume by 2 in order for the constant to remain unchanged:

$$PV = 1 \times 4 = 4$$

If we multiply the pressure by 10, we must divide the volume by 10 in order for the constant to remain unchanged:

$$PV = 20 \times 0.2 = 4$$

At constant temperature with a confined specific amount of gas molecules: (a) if the volume is decreased, the pressure is _____ and (b) if the pressure is increased, the volume is _____.

Answer: (a) increased; (b) decreased

11 The pressure (P) multiplied by the volume (V) is a constant (unchanging) number if the temperature remains constant. Note that Boyle's Law is based on a theoretically ideal gas. Real gases behave in a similar fashion to an ideal gas at moderate temperatures and pressures. Unless otherwise stated, we will assume that all gases behave exactly like an ideal gas. Since all gases are assumed to behave ideally, one gas will behave exactly like any other gas.

According to the equation for Boyle's Law at constant temperature:

(a) increasing P causes a corresponding _____ in V.

(b) increasing V causes a corresponding _____ in P.

Answer: (a) decrease; (b) decrease

12 Boyle's Law can be applied in practical situations involving the pressure and volume of a gas at constant temperature. A gas originally occupies a volume, which can be called V_1, at the pressure of P_1. By Boyle's Law, we know that multiplying P_1 times V_1 gives a number that is a constant. This same gas undergoes an increase in pressure to P_2 and a corresponding decrease in volume to V_2. There is no change in temperature. How does the numerical constant obtained by multiplying P_1 by V_1 compare with the numerical constant obtained by multiplying P_2 by V_2? _____

Answer: The numerical constants are exactly the same. (A constant does not change.)

13 $P_1 V_1$ is equal to a constant and if $P_2 V_2$ is equal to that same constant, then $P_1 V_1$ must equal $P_2 V_2$. This is a more practical expression of Boyle's Law. If the new pressure (P_2) of a quantity of gas is greater than the earlier pressure (P_1) of that gas, and the temperature is assumed to be constant, then the new volume (V_2) must be (greater, less) _____ than the earlier volume (V_1).

Answer: less (Pressure and volume are inversely proportional. As one increases, the other decreases.)

14 The equation $P_1 V_1 = P_2 V_2$ has four unknowns. If any three quantities are known, then the fourth can be determined. If P_1 is the unknown, the expression can be written as

$$P_1 = \frac{P_2 V_2}{V_1}$$

by simply dividing both sides of the equation by V_1. Suppose that V_2 is unknown and P_1, P_2, and V_1 are known. Rewrite the equation $P_1 V_1 = P_2 V_2$ so that only V_2 remains on one side of the equation.

Answer: $\dfrac{P_1 V_1}{P_2}$ (Divide both sides of the equation by P_2.)

15 V_1 is the original volume and P_1 is the original pressure. V_2 is the new volume and I_2 is the new pressure. The pressure of a gas is measured in the metric system as millimeters of mercury. In honor of Torricelli, who did pioneering work with pressure, a new unit was accepted. A pressure that will support a column of mercury 1 millimeter high is called a **torr**. Millimeters of mercury and torr are equivalent and interchangeable as measures of pressure. Some textbooks use millimeters of mercury; we will use torr. The volume of a gas is usually designated in liters or milliliters.

Five liters of a gas exist at a pressure of 700 torr. If the pressure is increased to 1400 torr, the volume of the gas is decreased. Let V_2 represent the unknown new volume. Identify V_1, P_1, and P_2 in this example.

$V1 =$ _____

$P1 =$ _____

$P_2 =$ _____

Answer: 5 liters; 700 torr; 1400 torr

16 Using the information from frame 15, determine the new volume (V_2). (Temperature is constant.)

$V_2 =$ _____

Answer: $\dfrac{P_1 V_1}{P_2} = \dfrac{700 \text{ torr} \times 5 \text{ liters}}{1400 \text{ torr}} = \dfrac{5}{2} = 2.5 \text{ liters}$

(Remember to express the volume in liters.)

17 Five liters of oxygen gas exist at a pressure of 700 torr. If the pressure is decreased to 600 torr and temperature is constant, what will be the new volume? (If rounding is required, round to the nearest hundredth for this answer and others unless otherwise indicated.)

$V_2 = $ _____

Answer: $\dfrac{P_1 V_1}{P_2} = \dfrac{700 \text{ torr} \times 5 \text{ liters}}{600 \text{ torr}} = \dfrac{700 \times 5}{600} = \dfrac{3500 \text{ liters}}{600} = 5.83 \text{ liters}$

18 Five liters of oxygen gas exist at a pressure of 700 torr. The pressure of the gas is reduced to a quarter of the original pressure. (A quarter of 700 torr pressure is $1/4 \times 700 = 175$ torr.) The temperature is constant. What is the new volume of oxygen?

$V_2 = $ _____

Answer: $\dfrac{P_1 V_1}{P_2} = \dfrac{700 \text{ torr} \times 5 \text{ liters}}{175 \text{ torr}} = \dfrac{700 \times 5}{175} = 20 \text{ liters}$

19 Assume the temperature does not change. Pressure and volume are inversely proportional. If a quantity of gas is compressed to half its volume, the new pressure will be doubled. If a quantity of gas is compressed to $\frac{1}{10}$ of its volume, its new pressure must be how many times as great as that of the original pressure? (If you wish to do so, try this problem on the 5 liters of gas at the pressure of 700 torr. The new volume would be 0.5 liters.) What is the new pressure in relation to the original pressure?

$P_2 = $ _____

Answer: $\dfrac{P_1 V_1}{V_2} = \dfrac{700 \text{ torr} \times 5 \text{ liters}}{0.5 \text{ liters}} = \dfrac{700 \times 5}{0.5} = 7000 \text{ torr}$

The new pressure is 10 times the original pressure.

20 A pressure of 760 torr or millimeters of mercury is equivalent to **1 atm (atmosphere)** of pressure. One atmosphere is the average pressure produced by the earth's atmosphere at sea level. A pressure of 1 atm supports a column of mercury that is 760 millimeters high. Ten liters of hydrogen gas at a pressure of 1 atm is compressed to a new pressure of 850 torr. What is the new volume of the gas?

$V_2 = $ _____

Answer: $\dfrac{P_1 V_1}{P_2} = \dfrac{760 \text{ torr} \times 10 \text{ liters}}{850 \text{ torr}} = \dfrac{760 \times 10}{850} = 8.94 \text{ liters}$

21 Another unit for pressure is the **kilopascal**. One atmosphere is equal to 760 torr, which is equal to 101.325 kilopascals (kPa). A quantity of gas with an original pressure of 100 kPa is compressed to 150 kPa. The new volume of gas is 2 liters. What was the original volume of gas?

$V_1 =$ _____

Answer: $\dfrac{P_2 V_2}{P_1} = \dfrac{150 \text{ kPa} \times 2 \text{ liters}}{100 \text{ kPa}} = 3 \text{ liters}$

22 An aerosol spray can has a volume of 0.15 liters. It is designed so that it can deliver 1000 times its original volume to a new pressure assumed to be equal to one atmosphere. What is the original pressure inside the can? (An answer in terms of either torr or atmospheres (atm) is acceptable.)

$P_1 =$ _____

Answer: There are several ways to solve this problem. The first two methods use the fact that 1000 times the original 0.15 liter volume is $1000 \times 0.15 = 150$ liters.

Method 1: $P_1 = \dfrac{P_2 V_2}{V_1} = \dfrac{1 \text{ atm} \times 150 \text{ liters}}{0.15 \text{ liters}} = \dfrac{150}{0.15} = 1000 \text{ atm}$

Method 2: $P_1 = \dfrac{P_2 V_2}{V_1} = \dfrac{760 \text{ torr} \times 150 \text{ liters}}{0.15 \text{ liters}} = \dfrac{760 \times 150}{0.15}$
$= 760{,}000 \text{ torr} = 1000 \text{ atm}$

Method 3: Merely knowing that the new volume is 1000 times greater than the original volume is useful. The new volume (V_2) equals $1000 \times V_1$. Substitute ($1000 V_1$) for V_2.

$P_1 = \dfrac{P_2 V_2}{V_1} = \dfrac{1 \text{ atm} \times 1000 V_1}{V_1} = 1000 \text{ atm}$

23 According to Boyle's Law, the pressure times the volume of a gas is a constant at constant temperature. Remember that all gases at a constant temperature are assumed to have the same average kinetic energy. Temperature is simply a measurement of the average kinetic energy of the gas.

Based on these statements, would you expect the average kinetic energy of a gas to remain the same if the temperature was changed? _____

Answer: No, a change in temperature indicates a change in the average kinetic energy.

24 The **absolute temperature scale** is a measure of the average kinetic energy of gases. Doubling the *absolute* temperature doubles the average kinetic energy. Doubling the kinetic energy results in doubling the pressure (twice as many collisions of gas molecules with the walls) if the volume is constant, or a doubling of the volume if the pressure is constant. What happens to the average kinetic energy of a gas as the absolute temperature increases? _____

Answer: The average kinetic energy increases.

CHARLES'S LAW

With the pressure constant, the volume of a gas is directly proportional to the temperature. (That is, if the volume is multiplied by a number, the temperature must be multiplied by that same number. If the volume is divided by some number, the temperature must also be divided by that number.)

A scientist named Charles first stated this relationship formally. Charles's Law can be stated mathematically as:

$$\frac{V}{T} = \text{a constant at constant pressure}$$

V represents the volume of a gas and T is the absolute temperature of a gas.

(a) In Boyle's Law, at constant temperature, the pressure and volume are (directly, inversely) _____ proportional.

(b) In Charles's Law, at constant pressure, the volume and absolute temperature are (directly, inversely) _____ proportional.

Answer: (a) inversely; (b) directly

26 Charles's Law can be rewritten as:

$$\frac{V_1}{T_1} = \frac{V_2}{T_2} \text{ (at constant pressure)}$$

V_1 is the original volume and V_2 is the new volume. T_1 is the original temperature and T_2 is the new temperature.

A quantity of gas at 200 K (absolute temperature scale) is heated to 400 K. The original volume is 3 liters. What is the new volume (with no change in pressure)?_____

Answer: Charles's Law applies here. Modify the equation so that just V_2 is on the left side of the equation by multiplying both sides by T_2.

$$\frac{V_1}{T_1} = \frac{V_2}{T_2}$$

$$V_2 = V_1 \times \frac{T_1}{T_2} = 3 \text{ liters} \times \frac{400 \text{ K}}{200 \text{ K}} = 6 \text{ liters}$$

27 Charles's Law is based on absolute temperature. The absolute (Kelvin) temperature scale is comparable to the Celsius scale in size of the degree, but the zero point is different. The zero point for the Kelvin scale is known as **absolute zero**. The zero point for the Celsius scale is the freezing point of water. The Kelvin scale zero point (0 K) is equivalent to −273°C. The freezing point of water is 0°C, or 273 K. To change from Celsius to Kelvin, just add 273 to the Celsius temperature. To change from Kelvin to Celsius, just subtract 273 from the Kelvin temperature. Fill in the blanks in the following conversions.

$0°C = 273\,K$
$10°C = 283\,K$
$50°C =$ _____ K
_____ $= 300\,K$

Answer: 323; 27°C

28 At absolute zero, an ideal gas has theoretically no volume at all. A gas loses volume at the rate of 1/273 for every degree of temperature drop at constant pressure. The temperature at which a theoretically ideal gas has no volume is _____ °Celsius.

Answer: −273

29 A quantity of gas at 10°C and 2 liters volume is heated to 50°C. Determine the new volume if the pressure remains constant. (Hint: To make calculations simpler when using Charles's Law, change all temperatures to the absolute, or Kelvin, scale first.)

Answer: Charles's Law applies here.

$$\frac{V_1}{T_1} = \frac{V_2}{T_2}$$

$T_2 = 50°C = 323\,K$

$V_1 = 2\,liters$

$T_1 = 10°C = 283\,K$

$V_2 = V_1 \times \frac{T_2}{T_1} = 2\,liters \times \frac{323\,K}{283\,K} = 2.28\,liters$

30 A sample of neon gas occupies 100 milliliters (mL) volume at 100°C. The sample is heated to 200°C. What is the new volume (to the nearest tenth) of neon (pressure unchanged)? _____

Answer:

$$V_1 = 100\,\text{mL}$$

$$T_1 = 100°C = 373\,\text{K}$$

$$T_2 = 200°C = 473\,\text{K}$$

$$V_2 = V_1 \times \frac{T_2}{T_1} = 100\,\text{mL} \times \frac{473\,\text{K}}{373\,\text{K}} = 126.8\,\text{mL}$$

 31   A sample of chlorine gas occupies $100\,\text{mL}$ at $-11°C$. Heating (with no increase in pressure) changes its volume to $150\,\text{mL}$.

(a) What is the new temperature as expressed on the absolute temperature scale? _____

(b) Express T_2 in Celsius degrees. _____

Answer:

(a) $V_1 = 100\,\text{mL}$
$\;\; V_2 = 150\,\text{mL}$
$\;\; T_1 = -11°C = 262\,\text{K}$
$\;\; T_2 = \dfrac{V_2}{V_1} \times T_1 = \dfrac{150\,\text{mL}}{100\,\text{mL}} \times 262\,\text{K} = 393\,\text{K}$

(b) 120°C

PARTIAL PRESSURE

32   When two or more different gases are mixed together, each gas in the mixture contributes to the total pressure of the mixture. The pressure contributed by each gas is in direct proportion to the number of molecules of the gas in the mixture. In a particular mixture of nitrogen gas and oxygen gas, there are twice as many nitrogen molecules as oxygen molecules. Each gas molecule contributes equal amounts to the total pressure regardless of what kind of molecule it is. Since there are twice as many nitrogen gas molecules as oxygen molecules in this mixture, we could expect the nitrogen molecules to contribute (half as much, twice as much, an equal amount) _____ to the total pressure as the oxygen molecules.

Answer: twice as much (The pressure contribution of a particular gas to the total pressure of a gas mixture is directly proportional to the relative number of molecules of the gas in the mixture. Twice as many molecules cause twice as much pressure contribution.)

33 In a gas mixture of nitrogen and oxygen, there are twice as many nitrogen molecules as oxygen molecules. One-third of the total pressure of the mixture is contributed by the oxygen molecules; two-thirds is contributed by the nitrogen molecules. The total pressure of the gas mixture is 900 torr.

(a) Nitrogen contributes how much to the total pressure? _____ torr

(b) Oxygen contributes how much to the total pressure? _____ torr

Answer: (a) 600 (Twice as much, or two-thirds of the total, is contributed by nitrogen: $^2/_3 \times 900 = 600$ torr); (b) 300 ($^1/_3 \times 900 = 300$ torr)

34 The total pressure of a mixture of gas is the sum of all **partial pressures**. The partial pressures are the pressures contributed by each of the gases. A partial pressure of a particular gas is proportional to the amount (relative number of molecules) of the gas within the mixture. In the previous example, 600 torr is the partial pressure exerted by nitrogen gas, 300 torr is the partial pressure of the oxygen gas, and 900 torr is the total pressure, which is the sum of all partial pressures in the mixture.

Dalton's Law of Partial Pressures (named after its discoverer) states that the total pressure of a mixture of gases is the sum of all of the partial pressures of the individual gases. The molecules of each gas exert the same pressure within the mixture as they would if they were not in the mixture. Dalton's Law can be written mathematically as:

$$P_T = p_1 + p_2 + p_3 + \ldots$$

The P_T (in capital letters) represents the total pressure. What do p_1, p_2, p_3, and so on (in lowercase letters) represent? _____

Answer: partial pressures

35 Determining partial pressures of gases is useful when analyzing gas mixtures. In order to determine the partial pressures of the individual gases in a mixture, we must first know the relative number of molecules of each gas in the mixture. Since a molecule is a very small unit, it is easier to use the mole. Partial pressures are proportional to the number of moles of the individual gases in a mixture. Partial pressure can be determined by dividing the moles of each particular gas by the total gas moles in the mixture and multiplying the result by the total pressure. A mixture of argon and xenon gases involves 2 mol of argon and 5 mol of xenon. The total number of moles of the mixture is $2 + 5 = 7$ mol. The total pressure exerted by the mixture is 700 torr. The total pressure exerted by argon is $2/7$ mol $\times$ 700 torr $= 200$ torr. What is the partial pressure exerted by xenon?

(Remember that according to Dalton's Law the total pressure is equal to the sum of the partial pressures.) _____

Answer: 700 torr – 200 torr = 500 torr

36 We could have solved the problem in frame 35 by using the following formula for partial pressure:

$$p_{gas} = \frac{n_{gas}}{n_T} \times P_T$$

The formula symbols represent the following quantities.

p_{gas} = partial pressure for the gas in question

n_{gas} = moles of the gas in question

n_T = total moles of all gases in the mixture

P_T = total pressure

A mixture of gases exerting a pressure of 840 torr consists of 0.2 mol of O_2 gas, 0.3 mol of H_2 gas, and 0.7 mol of CO_2 gas. The pressure contributed by the H_2 gas is _____ torr.

Answer: H_2, the gas in question, has 0.3 mol in the mixture. The total moles of gas is 0.2 + 0.3 + 0.7 = 1.2 mol.

$$p_{(H_2\ gas)} = \frac{n_{(H_2\ gas)}}{n_T} \times P_T = \frac{0.3\ mol}{1.2\ mol} \times 840\ torr = 210\ torr$$

(Because H_2 gas accounts for one-quarter of the total moles of gas in the mixture, it accounts for one-quarter of the total pressure.)

37 In the same gas mixture, the pressure contributed by the CO_2 gas is _____ torr.

Answer: $p_{CO_2} = \frac{n_{(CO_2\ gas)}}{n_T} \times P_T = \frac{0.7\ mol}{1.2\ mol} \times 840\ torr = 490\ torr$

38 The formula for partial pressure involves multiplying the **mole fraction** by the total pressure. The mole fraction is simply the number of moles of a particular gas divided by the total number of moles of gas in a mixture:

$$p_{gas} = \frac{n_{gas}}{n_T} \times P_T$$

Which part of the above formula represents the mole fraction? _____

Answer: $\frac{n_{gas}}{n_T}$

39 The gas mixture in frame 36 consisted of 0.2 mol of O_2, 0.3 mol of H_2, and 0.7 mol of CO_2. The total gas moles were $0.2 + 0.3 + 0.7 = 1.2$ mol. The mole fractions of each of the gas components are as follows.

$$CO_2 = \frac{n_{(CO_2\ gas)}}{n_T} = \frac{0.7}{1.2} = \frac{7}{12} = 0.583$$

$$H_2 = \frac{n_{(H_2\ gas)}}{n_T} = \frac{0.3}{1.2} = \frac{1}{4} = 0.250$$

$$O_2 = \frac{n_{(O_2\ gas)}}{n_T} = \frac{0.2}{1.2} = \frac{1}{6} = 0.167$$

Mole fractions can be expressed in fractional or decimal form. The mole fraction of a gas is the number of moles of a particular gas divided by the total number of moles of gas in a mixture.

(a) The mole fraction for CO_2 in the above gas mixture is _____ .

(b) The mole fractions of all gases in the mixture add up to _____.

Answer:

(a) $\frac{7}{12}$ or 0.583

(b) $1 \left(\frac{7}{12} + \frac{1}{4} + \frac{1}{6} = 1, \text{or } 0.583 + 0.250 + 0.167 = 1.00 \right)$

40 Mole fractions are very useful for determining the percentage volume for each gas in a gas mixture. All mole fractions in a gas mixture add up to 1. If we multiply the individual mole fractions by 100, the results are the percentages by volume of the various gases in the mixture. The previous mixture was made up of $0.250 \times 100 = 25.0\%$ hydrogen, $0.167 \times 100 = 16.7\%$ oxygen, and $0.583 \times 100 =$ _____% carbon dioxide.

Answer: 58.3

41 A gas mixture is made up of 25.0% H_2 gas, 16.7% O_2 gas, and 58.3% CO_2 gas. If the total volume of the mixture is 100.0 liters, the mixture could be expected to be made up of how many liters (to the nearest tenth) of H_2 gas, CO_2 gas, and O_2 gas? _____

Answer: 25.0, 16.7, 58.3

42 A corollary to Dalton's Law of Partial Pressures is the following:

$$V_T = v_1 + v_2 + v_3 + \ldots$$

The total volume (V_T) of a mixture of gases is the sum of all partial volumes (v_1, v_2, v_3, and so on).

A mixture is made up of 25.0 liters of H_2, 16.7 liters of O_2, and 58.3 liters of CO_2. What is V_T for the mixture?

Answer: 100 liters (25.0 + 16.7 + 58.3 = 100 liters)

43 Mole fractions are useful for determining the partial pressure contributed by a particular gas in a mixture or for determining the volume attributable to a gas in a mixture. Mole fractions are particularly useful in chemical analysis of a mixture of gases.

You have already determined mole fractions by dividing the moles of a gas by the total moles in the mixture. A mole fraction can also be determined by dividing the partial pressure of a gas by the total pressure of all gases in the mixture:

$$\frac{p_{gas}}{P_T} = \text{mole fraction of a gas}$$

A mole fraction can also be determined by dividing the partial volume of a gas by the total volume of all gases in the mixture:

$$\frac{v_{gas}}{V_T} = \text{mole fraction of a gas}$$

A mixture of gases is made up of 10 mL of neon, 20 mL of xenon, and 70 mL of nitrogen. What is the mole fraction of xenon in the mixture? _____

Answer: $V_T = v_1 + v_2 + v_3 = 10\text{ mL} + 20\text{ mL} + 70\text{ mL} = 100\text{ mL}$

$$\text{mole fraction (xenon)} = \frac{v_{xenon}}{V_T} = \frac{20\text{ mL}}{100\text{ mL}} = 0.2$$

44 A mixture of three gases (hydrogen, oxygen, and nitrogen) is chemically treated to absorb one gas at a time, using a process called selective absorption. Pressure is kept constant. The total original volume is 12 liters. After the oxygen is removed, the volume is 10 liters. After the hydrogen is removed, 6 liters of gas remain. The mixture contained:

(a) _____ liters of oxygen

(b) _____ liters of hydrogen

(c) _____ liters of nitrogen

Answer: $V_T = V_{oxygen} + V_{hydrogen} + V_{nitrogen}$

(a) 2 liters of oxygen $(12 - 10 = 2)$

(b) 4 liters of hydrogen $(10 - 6 = 4)$

(c) 6 liters of nitrogen (left after removal of oxygen and hydrogen)

45 Determine the mole fractions of each gas in the mixture in frame 44 and express them in decimal form. (Where necessary, round to the nearest thousandth.)

(a) Mole fraction of $O_2 =$ _____

(b) Mole fraction of $H_2 =$ _____

(c) Mole fraction of $N_2 =$ _____

Answer:

(a) $\dfrac{V_{O_2}}{V_T} = \dfrac{2 \text{ liters}}{12 \text{ liters}} = 0.167$

(b) $\dfrac{V_{H_2}}{V_T} = \dfrac{4 \text{ liters}}{12 \text{ liters}} = 0.333$

(c) $\dfrac{V_{N_2}}{V_T} = \dfrac{6 \text{ liters}}{12 \text{ liters}} = 0.5$

46 If the total pressure is 1 standard atmosphere (760 torr), what are the partial pressures of each gas in the mixture in frame 45?

(a) Partial pressure for O_2 gas = _____ torr

(b) Partial pressure for H_2 gas = _____ torr

(c) Partial pressure for N_2 gas = _____ torr

Answer: $P_{gas} = \text{mole fraction} \times P_T$

(a) 127 $(0.167 \times 760 = 127 \text{ torr})$

(b) 253 $(0.333 \times 760 = 253 \text{ torr})$

(c) 380 $(0.5 \times 760 = 380 \text{ torr})$

47   Let's review. A mixture of N_2 gas, NO gas, and NO_2 gas was analyzed by the selective absorption of nitrogen oxides. The initial volume of the sample is 300 mL. After treatment with water to remove NO_2, the volume is 220 mL. An iron(II) sulfate solution is used to remove NO, after which the volume left is 27 mL. Ignoring water vapor, determine the volume of each gas in the original mixture. The pressure remained constant at 1 standard atmosphere.

(a) $V_{NO2} = $ _____

(b) $V_{NO_2} = $ _____

(c) $V_{N_2} = $ _____

Answer: $VT = V_{NO_2} + V_{NO} + V_{N_2}$

(a) $V_{NO_2} = 80$ mL $(300 - 220 = 80$ mL$)$

(b) $V_{NO} = 193$ mL $(220 - 27 = 193$ mL$)$

(c) $V_{N_2} = 27$ mL (left over after removal of NO_2 and NO)

48   What is the mole fraction (to the nearest thousandth) of each gas in the mixture described in frame 47?

(a) Mole fraction of $NO_2 = $ _____

(b) Mole fraction of NO $= $ _____

(c) Mole fraction of $N_2 = $ _____

Answer:

(a) $\dfrac{V_{NO_2}}{V_T} = \dfrac{80\ mL}{300\ mL} = 0.267$

(b) $\dfrac{V_{NO}}{V_T} = \dfrac{193\ mL}{300\ mL} = 0.643$

(c) $\dfrac{V_{N_2}}{V_T} = \dfrac{27\ mL}{300\ mL} = 0.090$

49   Determine the volume percentage of the gas mixture in frames 47 and 48.

(a) Volume percentage of $NO_2 = $ _____

(b) Volume percentage of NO $= $ _____

(c) Volume percentage of $N_2 = $ _____

Answer: Volume percentage = mole fraction × 100%

(a) Volume percentage of NO_2 = 0.267 × 100 % = 26.7%

(b) Volume percentage of NO = 0.643 × 100 % = 64.3%

(c) Volume percentage of N_2 = 0.090 × 100 % = 9.0%

50 The total pressure of the mixture frames 47 and 48 prior to analysis was 1 standard atmosphere. Determine the partial pressure (to the nearest tenth) of each gas in the mixture.

(a) P_T = _____torr

(b) P_{NO_2} = _____torr

(c) P_{NO} = _____torr

(d) P_{N_2} = _____torr

Answer:

(a) P_T = 760 torr (1 standard atmosphere = 760 torr)

(b) P_{NO_2} = P_T × 0.267 = 760 torr × 0.267 = 202.9 torr

(c) P_{NO} = P_T × 0.643 = 760 torr × 0.643 = 488.7 torr

(d) P_{N_2} = P_T × 0.090 = 760 torr × 0.090 = 68.4 torr

You have just learned to determine the volumes, partial pressures, and mole fractions of individual gases within a gas mixture. In the next section you will learn how to determine the formula weight of an unknown gas.

GRAHAM'S LAW

 Earlier in this chapter, we mentioned that gases readily **diffuse** (mix spontaneously throughout each other). Even with no breeze blowing the smell of ammonia (a gas) or the scent of a surprised skunk (gaseous vapor) rapidly diffuses through the air. While diffusion denotes the mixing of gases, **effusion** is the process of a gas going through a small opening into a vacuum. (A vacuum is the absence of a solid, liquid, or gas.)

Diffusion indicates one gas mixing into another gas. Effusion indicates a gas moving into a vacuum. A scientist named Graham noted a relationship between the formula weight of a gas and the velocity or speed with which it effuses. Such

a relationship is very useful for determining the formula weights of unknown gases. It is also useful in the separation of two gases of different formula weights. Although Graham's Law predicts accurately the velocity of a gas during effusion, the same law can be used as a reasonable approximation of the velocity of a gas during diffusion.

According to **Graham's Law**, at a constant temperature, the rate (velocity) of effusion of a gas is inversely proportional to the square root of its formula weight. In the equation, a gas of known molecular weight is compared to an unknown gas. In all calculations with Graham's Law, it is assumed that the temperature and pressure are the same for both gases:

$$\frac{v_1}{v_2} = \sqrt{\frac{M_2}{M_1}}$$

v_1 = velocity or rate of effusion of the first gas

M_1 = formula weight of the first gas

v_2 = velocity or rate of effusion of the second gas

M_2 = formula weight of the second gas

Here is a practical example using Graham's Law. Let oxygen gas (O_2) be the known gas. Oxygen has an effusion rate v_{O_2} of 10 milliliters (mL) per second and a formula weight of 32 amu. An unknown gas has an effusion rate (v_X) of 5 mL per second. Determine the formula weight (M_X) of the unknown gas. (Note that the equation can be modified as follows.)

$$\frac{v_{O_2}}{v_X} = \sqrt{\frac{M_X}{M_{O_2}}} \quad \text{or} \quad \left(\frac{v_{O_2}}{v_X}\right)^2 = \frac{M_X}{M_{O_2}}$$

Answer:

$$\left(\frac{v_{O_2}}{v_X}\right)^2 = \frac{M_X}{M_{O_2}}$$

$$\left(\frac{10 \text{ mL per second}}{5 \text{ mL per second}}\right)^2 = \frac{M_X}{32 \text{ amu}}$$

$$(2)^2 = \frac{M_X}{32 \text{ amu}}$$

$M_X = 4 \times 32 \text{ amu} = 128 \text{ amu}$

52 In another situation involving the use of Graham's Law, helium gas (He) has a fairly high velocity of effusion, 50 mL per second, and has a formula weight of 4 amu. An unknown gas has a slower velocity of effusion, 10 mL per second.

Using the appropriate equation from the previous frame, determine the formula weight of the unknown gas.

Answer:

$$\left(\frac{v_{He}}{v_X}\right)^2 = \frac{M_X}{M_{He}}$$

$M_{He} = 4\,amu$

$v_{He} = 50\,mL\ per\ second$

$v_X = 10\,mL\ per\ second$

$$\left(\frac{50\,mL\ per\ second}{10\,mL\ per\ second}\right)^2 = \frac{M_X}{4\,amu}$$

$$(5)^2 = \frac{M_X}{4\,amu}$$

$M_X = 25(4\,amu) = 100\,amu$

 53

In a different situation, hydrogen (H_2) has an effusion rate of $20\,mL$ per second and its formula weight is $2\,amu$. Oxygen (O_2) has a formula weight of $32\,amu$.

(a) What is the expected effusion rate of O_2 in this situation? _____

(b) Oxygen, with a formula weight of $32\,amu$, is a heavier gas than hydrogen. Which gas has the greater effusion rate, the heavier oxygen or the lighter hydrogen? _____

Answer:

(a) Note that either of these equations is correct. We will use the equation on the right for convenience in the calculation.

$$\frac{v_{H_2}}{v_{O_2}} = \sqrt{\frac{M_{O_2}}{M_{H_2}}}\ \text{or}\ \frac{v_{O_2}}{v_{H_2}} = \sqrt{\frac{M_{H_2}}{M_{O_2}}}$$

$v_{H_2} = 20\,mL\ per\ second$

$M_{H_2} = 2\,amu$

$M_{O_2} = 32\,amu$

$$\frac{v_{O_2}}{v_{H_2}} = \sqrt{\frac{M_{H_2}}{M_{O_2}}} = \sqrt{\frac{2\,amu}{32\,amu}} = \sqrt{\frac{1}{16}} = \frac{1}{4}$$

$$v_{O_2} = \frac{1}{4} \times v_{H_2} = \frac{1}{4} \times 20\,mL\ per\ second = 5\,mL\ per\ second$$

(b) The hydrogen gas has an effusion rate of 20 mL per second and oxygen gas has an effusion rate of only 5 mL per second. The lighter hydrogen has the greater rate of effusion (four times greater than oxygen).

54 The heavier the formula weight of a gas, the slower the effusion rate, according to Graham's Law. Arrange the four gases listed below in order from the slowest to the fastest effusion rate.

CO_2 has a formula weight of 44.
N_2 has a formula weight of 28.
F_2 has a formula weight of 38.
O_2 has a formula weight of 32.

Slowest _____, _____, _____, _____, Fastest

Answer: CO_2, F_2, O_2, N_2

55 The equation for Graham's Law can be modified to deal with the density of gases. The relationship between density and effusion rates will be dealt with in a few frames. First, let's clarify the meaning of density as applied to gases. Density is a measure of weight per unit volume. For gases, we will use grams per liter as the measure of density.

You have already learned that the volume of a gas is dependent on pressure and temperature. To have an accurate measure of the density of a gas, we must specify a temperature and pressure. **Standard temperature and pressure** (abbreviated **STP**) have been set arbitrarily at 0°Celsius and 1 atm. At STP, 1 mol of *any* gas occupies 22.4 liters. Note that we are still treating every gas as a theoretically ideal gas. The volume of 22.4 liters per mole at STP is a reasonable approximation for real gases.

Assuming that O_2 gas behaves ideally, 1 mol of O_2 at STP occupies what volume? _____

Answer: 22.4 liters

56 Unless otherwise stated we will assume that all gases behave ideally. Since oxygen (O_2) has a volume of 22.4 liters per mole at STP, we can calculate its density. The formula weight (the weight of 1 mol) of O_2 is 32.0 grams. The density of O_2 at STP to the nearest hundredth of a gram per liter is:

$$\frac{32.0 \text{ grams}}{1 \text{ mol of } O_2} \times \frac{1 \text{ mol of } O_2}{22.4 \text{ liters}} = 1.43 \text{ grams/liter}$$

Calculate the density of N_2 at STP. _____

Answer: Using the periodic table, the formula weight of N_2 is 28.0 grams (2×14.0 grams)

$$\frac{28.0 \text{ grams}}{1 \text{ mol of } N_2} \times \frac{1 \text{ mol of } N_2}{22.4 \text{ liters}} = 1.25 \text{ grams/liter}$$

(Remember to express the density in grams per liter.)

 What is the density of Cl_2 at STP?_____

Answer: $\dfrac{70.9 \text{ grams}}{1 \text{ mol of } Cl_2} \times \dfrac{1 \text{ mol of } Cl_2}{22.4 \text{ liters}} = 3.17 \text{ grams/liter}$

 An unknown gas was found to have a formula weight of 128.0 grams. What is its density at STP?_____

Answer: $\dfrac{128.0 \text{ grams}}{1 \text{ mol}} \times \dfrac{1 \text{ mol}}{22.4 \text{ liters}} = 5.71 \text{ grams/liter}$

59 The equation for Graham's Law can be rewritten to include densities:

$$\frac{v_1}{v_2} = \sqrt{\frac{M_2}{M_1}} = \sqrt{\frac{d_2}{d_1}}$$

By eliminating the part that includes formula weights, we have:

$$\frac{v_1}{v_2} = \sqrt{\frac{d_2}{d_1}}$$

In the equation, d_1 indicates the density of gas 1 and d_2 indicates the density of gas 2. Here is an example using this formula. A known gas (oxygen) has a density (d_{O_2}) of 1.43 grams per liter and an effusion rate of 10.0 mL per second. An unknown gas (X) has an effusion rate of 5.0 mL per second. The density of the unknown gas can be determined as follows:

$$\left(\frac{v_{O_2}}{v_X}\right)^2 = \frac{d_X}{d_{O_2}}$$

$$\left(\frac{10.0 \text{ mL per second}}{5.0 \text{ mL per second}}\right)^2 = \frac{d_X}{1.43 \text{ grams per liter}}$$

$$d_X = 1.43 \text{ grams per liter} \times \left(\frac{10.0}{5.0}\right)^2$$

$$d_X = 1.43 \text{ grams per liter} \times 4$$

$$d_X = 5.72 \text{ grams per liter}$$

Another unknown gas has an effusion rate of 8.0 mL per second. Using oxygen gas as a reference (effusion rate and density are the same as in the previous example), determine the density of the unknown gas. _____

Answer:

$$\left(\frac{v_{O_2}}{v_X}\right)^2 = \frac{d_X}{d_{O_2}}$$

$$\left(\frac{10.0 \text{ mL per second}}{8.0 \text{ mL per second}}\right)^2 = \frac{d_X}{1.43 \text{ grams per liter}}$$

$$d_X = 1.43 \text{ grams per liter} \times \left(\frac{10.0}{8.0}\right)^2$$

$$d_X = 1.43 \text{ grams per liter} \times (1.25)^2$$

$$d_X = 2.23 \text{ grams per liter}$$

60 Carbon dioxide (CO_2) is used as a reference gas with a density of 1.96 grams per liter and an effusion rate of 12.0 mL per second. An unknown gas has an effusion rate of 10.0 mL per second. Determine the density of the unknown gas. _____

Answer:

$$\left(\frac{v_{CO_2}}{v_X}\right)^2 = \frac{d_X}{d_{CO_2}}$$

$$\left(\frac{12 \text{ mL per second}}{10 \text{ mL per second}}\right)^2 = \frac{d_X}{1.96 \text{ grams per liter}}$$

$$d_X = 1.96 \text{ grams per liter} \times \left(\frac{12}{10}\right)^2$$

$$d_X = 1.96 \text{ grams per liter} \times (1.2)^2$$

$$d_X = 2.82 \text{ grams per liter}$$

61 Once you have found the density of an unknown gas, it is a simple matter to determine its formula weight. Just reverse the procedure used in finding a density from formula weight. For example, the unknown gas in frame 59 had a density of 2.23 grams per liter. Assume STP.

$$\frac{2.23 \text{ grams}}{1 \text{ liter}} \times \frac{22.4 \text{ liters}}{1 \text{ mol}} = 50.0 \text{ grams per mol}$$

In frame 60, you determined that an unknown gas had a density of 2.82 grams per liter. Determine its formula weight to the nearest tenth of a gram per mole. Assume STP._____

Answer: $\dfrac{2.82 \text{ grams}}{1 \text{ liter}} \times \dfrac{22.4 \text{ liters}}{1 \text{ mol}} = 63.2 \text{ grams per mol}$

THE COMBINED GAS LAW

62 Earlier in this chapter, you learned about Boyle's Law and Charles's Law. We will now combine these two laws. To review, Boyle's Law states that $PV =$ constant (at constant temperature). A practical expression of Boyle's Law is $P_1 V_1 = P_2 V_2$ at constant temperature. Charles's Law states

$$\frac{V}{T} = \text{constant (at constant pressure)}.$$

A practical expression of Charles's Law is

$$\frac{V_1}{T_1} = \frac{V_2}{T_2} \text{ (at constant pressure)}.$$

Boyle's and Charles's Laws can be combined into

$$\frac{PV}{T} = \text{constant}.$$

A practical expression of the combined gas law is

$$\frac{P_1 V_1}{T_1} = \frac{P_2 V_2}{T_2}$$

In using the combined gas law or any gas law involving temperature, what temperature scale must be used?_____

Answer: Absolute or Kelvin or K (Any of these answers is acceptable.)

63 We can modify the combined gas law as follows.

$$\frac{P_1 V_1}{T_1} = \frac{P_2 V_2}{T_2}$$

Multiplying both sides by T_2 results in $\dfrac{P_1 V_1 T_2}{T_1} = P_2 V_2$.

Dividing both sides by P_2 results in $\dfrac{P_1 V_1 T_2}{T_1 P_2} = V_2$.

A sample of helium occupies 1000 liters (V_1) at 15°C (T_1) and 763 torr (P_1). The temperature is changed to -6°C and the pressure to 420 torr. Use the combined gas law from frame 62 to find the new volume of the gas (V_2) to the nearest liter. _____

Answer: Change all temperatures to Kelvin scale (add 273 to Celsius scale).

$$V_2 = \frac{P_1 V_1 T_2}{P_2 T_1} = \frac{1000 \text{ liters} \times 763 \text{ torr} \times 267 \text{ K}}{420 \text{ torr} \times 288 \text{ K}} = 1684 \text{ liters}$$

 A sample of oxygen gas occupies 400 mL at 20°C and 740 torr. If the conditions are changed so that the gas is warmed to 30°C and occupies 425 mL, what is its new pressure to the nearest torr? _____

Answer: $P_2 = \dfrac{P_1 V_1 T_2}{T_1 V_2} = \dfrac{740 \text{ torr} \times 400 \text{ mL} \times 303 \text{ K}}{425 \text{ mL} \times 293 \text{ K}} = 720 \text{ torr}$

THE IDEAL GAS LAW EQUATION

 In using the combined gas law, we have used an equation with six items, five of which are given and one of which is unknown. A simpler equation can be made from the combined gas law by including a constant to represent gases at STP:

$$PV = nRT$$

In this equation, P, V, and T describe the existing gas conditions. The symbol n represents the number of moles of the gas. The symbol R represents the universal gas constant. This equation is commonly known as the **ideal gas law equation**. Rewrite the ideal gas law equation so that only R remains on one side of the equation.

$R = $ _____

Answer: $\dfrac{PV}{nT}$ (divide both sides by nT)

66 The **universal gas constant** (R) is a numerical value made up of the combination of standard pressure (1 atm or 760 torr), standard temperature (273 K), and the fact that 1 mol of a gas occupies a volume of 22.4 liters at STP.

The following value of R is appropriate for calculations dealing with pressure expressed in *atmospheres*, volume expressed in *liters*, temperature expressed in *Kelvin* degrees, and the quantity of gas expressed in *moles*:

$$R = \frac{PV}{nT} = \frac{1\text{ atm} \times 22.4\text{ liters}}{1\text{ mol} \times 273\text{ K}} = 0.082 \frac{\text{atm} \times \text{liters}}{\text{mol} \times \text{K}}$$

We can also calculate a value of R for use in problems dealing with pressure expressed in *torr*, volume in *liters*, temperature in *Kelvin*, and quantity in *moles*:

$$R = \frac{PV}{nT} = \frac{760\text{ torr} \times 22.4\text{ liters}}{1\text{ mol} \times 273\text{ K}} = 62.4 \frac{\text{torr} \times \text{liters}}{\text{mol} \times \text{K}}$$

If volume is expressed in *milliliters*, pressure in *torr*, temperature in *Kelvin*, and quantity in *moles*, we have the following value of R:

$$R = \frac{PV}{nT} = \frac{760\text{ torr} \times 22{,}400\text{ mL}}{1\text{ mol} \times 273\text{ K}} = 62{,}400 \frac{\text{torr} \times \text{mL}}{\text{mol} \times \text{K}}$$

The appropriate value of R must be selected based on the units of P and V.

If V is expressed in liters and P in torr, then R is 62.4.

If V is expressed in liters and P in atmospheres, then R is 0.082.

If V is expressed in milliliters and P in torr, then R is 62,400.

It is assumed the temperature is expressed in Kelvin and n is expressed in moles. Here is an example using the ideal gas law equation. Half a mole of gas (n) has a pressure of 2 atm (P) and a temperature (T) of 341 K. We want to calculate its volume in liters using the ideal gas equation. Since V is expressed in liters and P in atmospheres, R must be

$$0.082 \frac{\text{atm} \times \text{liters}}{\text{mol} \times \text{K}}$$

Dividing both sides of the equation $PV = nRT$ by P results in

$$V = \frac{nRT}{P} = \frac{0.5\text{ mol} \times 0.082 \dfrac{\text{atm} \times \text{liters}}{\text{mol} \times \text{K}} \times 341\text{ K}}{2\text{ atm}}$$

$$= \frac{0.5 \times 0.082\text{ liters} \times 341}{2} = 7.0\text{ liters}$$

In another example, 2 mol of a gas occupies a volume of 4 liters when the temperature is 300 K. What is the pressure (expressed in torr) of the gas? _____

Answer: Since V is expressed in liters and P in torr, R must be $62.4 \dfrac{\text{torr} \times \text{liters}}{\text{mol} \times \text{K}}$.

$$P = \frac{nRT}{V} = \frac{2 \text{ mol} \times 62.4 \dfrac{\text{torr} \times \text{liters}}{\text{mol} \times \text{K}} \times 300 \text{ K}}{4 \text{ liters}} = 9360 \text{ torr}$$

67 Three moles of a gas are compressed to 10,000 torr. The temperature is 27°C. What is the volume of the gas in milliliters?_____

Answer:

Since V is expressed in milliliters and P in torr, R must be $62,400 \dfrac{\text{torr} \times \text{mL}}{\text{mol} \times \text{K}}$.

Celsius temperature must be changed to Kelvin ($27 + 273 = 300$ K).

$$V = \frac{nRT}{P} = \frac{3 \text{ mol} \times 62,400 \dfrac{\text{torr} \times \text{mL}}{\text{mol} \times \text{K}} \times 300 \text{ K}}{10,000 \text{ torr}} = 5616 \text{ mL}$$

68 If 1.5 mol of gas occupies 41 liters at a pressure of 3 atm, what is the temperature of the gas?_____

Answer: Since V is expressed in liters and P in atmospheres, R must be

$$0.082 \frac{\text{atm} \times \text{liters}}{\text{mol} \times \text{K}}$$

Divide both sides of the equation by nR:

$$T = \frac{PV}{nR} = \frac{3 \text{ atm} \times 41 \text{ liters}}{1.5 \text{ mol} \times 0.082 \dfrac{\text{atm} \times \text{liters}}{\text{mol} \times \text{K}}} = 1000 \text{ K}$$

69 We can also use the ideal gas equation to determine the formula weights of gases. The formula weight of neon gas (Ne) is 20.18 grams per mol. A sample of Ne gas weighing 201.8 grams represents the weight of how many moles of Ne gas?_____

Answer: $201.8 \text{ grams} \times \dfrac{1 \text{ mol of Ne}}{20.18 \text{ grams}} = 10 \text{ mol of Ne}$

70 In the equation $PV = nRT$, the symbol n represents the number of moles of a gas. The number of moles of a gas is equal to the following, where M is the formula weight of the gas in grams per mole:

$$\text{weight of gas} \times \frac{1}{M} = n$$

The equation $PV = nRT$ can be modified to include the above equation for the number of moles. Show this below.

$PV =$ _____

Answer: weight of gas $\times \dfrac{RT}{M}$

71 The equation you just wrote can be solved for formula weight:

$$M = \frac{\text{weight of gas} \times RT}{PV}$$

Determine the formula weight of a gas if 20 grams of the gas occupies a volume of 6 liters at a pressure of 2 atm and a temperature of 300 K. Express in grams per mole. _____

Answer: Since V is in liters and P is in atmospheres, R must be $0.082 \dfrac{\text{atm} \times \text{liters}}{\text{mol} \times \text{K}}$.

$$M = \frac{\text{weight of gas} \times RT}{PV} = \frac{20 \text{ grams} \times 0.082 \dfrac{\text{atm} \times \text{liters}}{\text{mol} \times \text{K}} \times 300 \text{ K}}{2 \text{ atm} \times 6 \text{ liters}} = 41 \text{ grams per mol}$$

72 A sample of gas weighing 1.00 grams has a volume of 500 mL at a temperature of 77 °C and a pressure of 546 torr. What is the formula weight of the gas? _____

Answer: Since V is in milliliters and P is in torr, R must be

$$62{,}400 \frac{\text{torr} \times \text{mL}}{\text{mol} \times \text{K}}$$

Change 77°C to Kelvin (77 + 273 = 350 K).

$$M = \frac{\text{weight of gas} \times RT}{PV} = \frac{1.00 \text{ grams} \times 62{,}400 \dfrac{\text{torr} \times \text{mL}}{\text{mol} \times \text{K}} \times 350 \text{ K}}{546 \text{ torr} \times 500 \text{ mL}} = 80 \text{ grams per mol}$$

73 All of the gas law equations thus far have dealt with theoretically ideal gases. The real gases in the world behave ideally at high temperatures and at low pressures. However, at very high pressures and low temperatures, real gases deviate from the ideal.

In real gases at high pressures, the volume of the gas molecules actually makes up a significant part of the gas volume. This causes the real gas to be less compressible than predicted by the ideal gas law.

At very low temperatures, the gas molecules attract one another and the real gas becomes more compressible than predicted by the ideal gas law. These attractive forces between the molecules are known as **van der Waals forces**.

(a) In the kinetic molecular theory, what assumption is made about the volume of the molecules of an ideal gas?_____

(b) The attraction of gas molecules for each other becomes significant at which range of temperature, very high or very low?_____

Answer: (a) that the molecules of an ideal gas occupy no volume; (b) very low

At sufficiently low temperatures, the van der Waals forces become strong enough to overcome the kinetic energy of the gas molecules. The result is that gas molecules stick together and form small droplets. These droplets combine to form a liquid. You will learn more about van der Waals forces in Chapter 10 on liquids.

Additional constants have been determined to add to the ideal gas law equation that make the equation more correct for the effects of gas molecule volume at high pressures and van der Waals forces at low temperatures. This equation, known as the van der Waals's equation of state for a real gas, is beyond the scope of this book.

Now you know that gases may be compressed (as in a bicycle pump), exert pressure on the walls of their containers (as in a tire or balloon), have different densities (air versus helium), and effuse or diffuse so you can smell them.

You also know that all gases behave in a similar manner and are affected by changes in temperature and pressure. The laws presented in this chapter apply only to the gaseous state. The remaining states of matter, liquids, and solids behave quite differently than gases. The next two chapters discuss the solid and liquid states. In contrast to gases, solids and liquids do not all behave in a similar manner.

SELF-TEST

This self-test is designed to show how well you have mastered this chapter's objectives. Correct answers and review instructions follow the test. Round answers to the nearest tenth.

1. Which of the following assumptions about the kinetic molecular theory of gases is incorrect?

    (a) Gases are made up of individual molecules that are in constant rapid motion.

    (b) Gas molecules are closer together than molecules of a liquid or solid.

(c) Gas molecules possess kinetic energy that varies with temperature.

(d) Gas molecules occupy no volume and therefore have no attraction for each other.

2. 15.5 Liters of neon gas at 1 atm of pressure is compressed to 4.4 liters of volume. What is the pressure at this new volume?

3. 2.75 Liters of nitrogen gas at 543 torr is compressed to a pressure of 965 torr. What is the resulting volume at this new pressure?

4. A sample of Cl_2 gas occupies a volume of 30.0 mL at 20°C. What volume would it occupy at 50°C? Assume the pressure remains constant and use Charles's Law._____

5. A sample of oxygen gas occupying a volume of 325.0 mL at 30°C is heated to 65°C. Calculate the final volume of this gas.

6. A sample of helium gas occupying a volume of 250.0 mL at constant pressure at 100K is heated and expands to a volume of 850.0 mL. Calculate the final temperature of the gas at these conditions.

7. 5.00 mol H_2 at 1.10 atm of pressure and 308 K occupies a volume of _____ L.

8. 8.55×10^{-2} mol N_2 at 740 torr of pressure and 298 K occupies a volume of _____ L (answer to the nearest hundredth).

9. A sample of CO_2 gas occupies 350 mL at 27°C and 790 torr. What pressure would it exert if it occupied 175 mL and the temperature was changed to 57°C? Use the combined gas law equation._____

10. What volume would 22.0 grams of N_2O occupy at STP? _____

11. What is the density of fluorine (F_2) at STP? _____

12. Arrange the following gases in the order of increasing effusion rates. CO, O_2, SO_2, NH_3, F_2 _____, _____, _____, _____, _____

13. A mixture of CO_2 and O_2 is used to "decompress" deep sea divers to prevent the "bends." If the mixture is 70% O_2 and 30% CO_2 by volume, what is:

(a) the volume of O_2 present in 50.0 liters of the mixture?_____

(b) the mole fraction of O_2 in the mixture?_____

(c) the pressure exerted by the O_2 if the pressure of the mixture is 1,000 torr?_____

14. Which of the following can account for the fact that you can smell a skunk even though you may not see it (more than one is correct)?

_____ (a) Gases are in rapid random motion.

_____ (b) Gases may be compressed.

_____ (c) Gas molecules are far apart.

_____ (d) Gas volumes are directly proportional to the absolute temperature.

_____ (e) Gas volumes are inversely proportional to the pressure.

15. Calculate the formula weight of 1.68 grams of a gas that occupies 1.35 liters at STP using the ideal gas law equation.

$$R = 0.082 \frac{atm \times liters}{mol \times K}$$

ANSWERS

Compare your answers to the self-test with those given below. If you answered all questions correctly, you are ready to proceed to the next chapter. If you got any wrong, review the frames indicated in parentheses following the answers. If you got several questions wrong, you should probably reread the chapter carefully.

1. (b.) is incorrect [frames 1-9]

2. $P_1V_1 = P_2V_2$, so $P_2 = \frac{P_1V_1}{V_2} = \frac{(1 \text{ atm})(15.5 \text{ L})}{4.4 \text{ L}} = 3.5 \text{ atm}$ [frames 10 – 24]

3. $P_1V_1 = P_2V_2$, so $V_2 = \frac{P_1V_1}{P_2} = \frac{(543 \text{ torr})(2.75 \text{ L})}{965 \text{ torr}} = 1.55 \text{ L}$ [frames 10 – 24]

4. $\frac{V_1}{T_1} = \frac{V_2}{T_2}$

$V_2 = \frac{V_1 T_2}{T_1} = \frac{30.0 \text{ mL} \times 323 \text{ K}}{293 \text{ K}} = 33.1 \text{ mL}$

(frames 25–31)

5. $\frac{V_1}{T_1} = \frac{V_2}{T_2}$, so $V_2 = \frac{V_1 T_2}{T_1} = \frac{(325.0 \text{ mL})(338 \text{ K})}{303 \text{ K}} = 362.5 \text{ mL}$ [frames 25 – 31]

6. $\frac{V_1}{T_1} = \frac{V_2}{T_2}$, so $T_2 = \frac{V_2 T_1}{V_1} = \frac{(850.0 \text{ mL})(100 \text{ K})}{250.0 \text{ mL}} = 340 \text{ K}$ [frames 25 – 31]

7. $PV = nRT$, so $V = \frac{nRT}{P} = \frac{(5.00 \text{ mol}) \left[0.082 \frac{L \times atm}{mol \times K}\right](308 \text{ K})}{1.10 \text{ atm}} = 115 \text{ L}$ [frames 65 – 73]

8. $PV = nRT$, so $V = \frac{nRT}{P} = \frac{(8.55 \times 10^{-2} \text{ mol}) \left[62,400 \frac{mL \times torr}{mol \times K}\right](298 \text{ K})}{740 \text{ torr}} = 2148.5 \text{ mL} = 2.15 \text{ L}$ [frames 65 – 73]

9. $$\frac{P_1V_1}{T_1} = \frac{P_2V_2}{T_2}$$

$$P_2 = \frac{P_1V_1T_2}{V_2T_1} = \frac{790 \text{ torr} \times 350 \text{ mL} \times 330 \text{ K}}{175 \text{ mL} \times 300 \text{ K}} = 1738 \text{ torr}$$

(frames 62–64)

10. $$V_{N_2O} = 22.0 \text{ grams} \times \frac{1 \text{ mol N}_2\text{O}}{44.0 \text{ grams N}_2\text{O}} \times \frac{22.4 \text{ liters}}{\text{mol}} = 11.2 \text{ liters N}_2\text{O}$$

(frame 55)

11. $$d_{F_2} = \frac{\text{weight F}_2}{\text{liters F}_2} = \frac{38.0 \text{ grams F}_2 \text{ per mol}}{22.4 \text{ liters per mol}} = 1.7 \text{ grams per liter}$$

(frames 56–58)

12. formula weight of CO = 28.01

   formula weight of O_2 = 32.00

   formula weight of SO_2 = 64.06

   formula weight of NH_3 = 17.04

   formula weight of F_2 = 38.00

   Therefore, increasing effusion rate from left to right is:
   SO_2, F_2, O_2, CO, NH_3 (frames 51–54)

13. (a) $v_{O_2} = V_T \times \%O_2 = 50.0 \text{ liters} \times 0.70 = 35.0 \text{ liters}$

   (frames 39, 40)

   (b) mole fraction $O_2 = \dfrac{V_{gas}}{V_T} = \dfrac{70\%}{100\%} = 0.70$

   (c) $p_{O_2} = \text{mole fraction } O_2 \times P_T = (0.70)(1000 \text{ torr}) = 700 \text{ torr}$

   (frames 36, 37)

14. (a) and (c) (frames 1, 6, 51)

15. $$M = \frac{\text{weight of gas} \times RT}{PV}$$

$$M = \frac{1.68 \text{ g} \times 0.082 \dfrac{\text{atm} \times \text{liters}}{\text{mol} \times \text{K}} \times 273 \text{ K}}{1 \text{ atm} \times 1.35 \text{ liters}}$$

$$M = 27.9 \text{ grams per mol}$$

(frames 65–72)

THE USES OF HYDROGEN GAS

The lightest gas we have is made up of two hydrogen atoms (H_2), known as elemental hydrogen gas or just hydrogen gas. Not only is it the lightest elemental gas we have, it is also extremely flammable, being able to sustain a fire after it is ignited. Oxygen (O_2) is a heavier diatomic gas but it is not as flammable since these substances require O_2 to burn.

When the reactants hydrogen gas and oxygen gas mix no, reaction occurs:

$$2H_2(g) + O_2(g) \longrightarrow \text{no reaction}$$

However, the mixture will undergo a chemical reaction with the help of a spark or flame. The results can prove disastrous and the amount of heat released is tremendous even when only moderate quantities of H_2 and O_2 are present for reaction:

$$2H_2(g) + O_2(g) \xrightarrow{\text{spark or flame}} 2H_2O(g) + \text{heat}$$

On 6 May 1937 in Manchester Township in New Jersey, a German passenger airship named the Hindenburg caught fire and was destroyed. This aircraft was filled with hydrogen gas, making use of this light gas for flight (Figure 8.1).

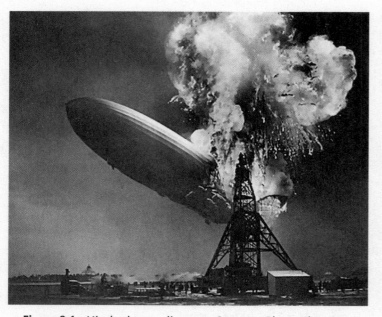

Figure 8.1 Hindenburg disaster. Source: Photo by Sam Shere, now in USA public domain.

At one point, the aircraft sprung a leak of hydrogen gas and it began to mix with the oxygen in the air. The mixture was able to ignite and the Hindenburg was quickly engulfed in flames. The aircraft crashed and 36 people died out of the 96 people onboard. One person on the ground also died.

This was a large-scale chemical reaction where water vapor was produced as the product.

Other reactions that involve hydrogen are the chemical reactions in fuel cells. A fuel cell is a device that converts chemical energy into electrical energy. These cells are similar to batteries but require a continuous source of fuel, most often hydrogen gas (H_2) (Figure 8.2).

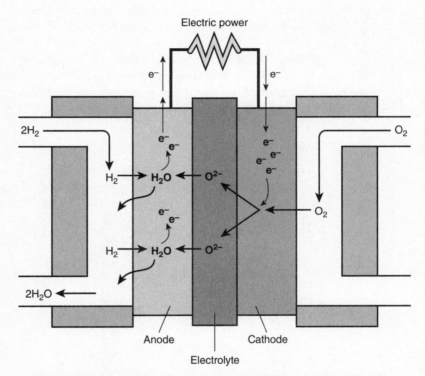

Figure 8.2 Schematic of a hydrogen fuel cell (https://en.wikipedia .org/wiki/File:Solid_oxide_fuel_cell.svg).

The chemical reactions that occur within the fuel cell are:

$$\text{Anode}\,(-): \quad 2H_2 + 2O^{2-} \;\rightarrow\; 2H_2O + 4e^-$$
$$\text{Cathode}\,(+): \quad O_2 + 4e^- \;\rightarrow\; 2O^{2-}$$
$$\text{Overall reaction} \quad 2H_2 + O_2 \;\rightarrow\; 2H_2O$$

You'll notice that the overall reaction within the fuel cell is exactly the same reaction as the tragedy that occurred in the Hindenburg disaster. Again, hydrogen gas and oxygen gas are the reactants and water is the product. A key difference in the fuel cell and the Hindenburg aircraft is that the fuel cell uses far less hydrogen gas, making them much safer.

Fuel cells are finding applications in diverse ways. For instance, they have been used for automobiles, satellites, and submarines. It's their ability to generate clean energy that clearly sets them apart from the typical automobile engine. In fact, fuel cells offer two advantages that typical automobiles do not possess. First, fuel cells are more efficient than the combustion engine. Their increased efficiency could range from 5% more efficient up to 35% more efficient. Second, their by-product is water, instead of the CO_2 and other gases that can be released from cars. In other words, the fuel cell does not contribute to the production of greenhouse gases!

Fuels cells have a lot of potential for the future; however, major hurdles need to be overcome before they are in the general public's automobiles. Their technology makes them quite expensive and they are prone to failure (i.e., they have a lower reliability than the combustion engine).

9 Solids

Look around you. What do you see? Lots of visible things, such as buildings, tables, chairs, wires, windows, and trees. This chapter is an introduction to the structure and properties of the solid state.

The chapter will let you compare solids and gases, and will explore how bonding between particles affects the properties of solids.

We will study the shapes of solids, that is, the different ways atoms, ions, and molecules arrange themselves in three dimensions. Apart from shape, four types of solids have distinctive properties dependent upon the type of bonding between particles. You already know that different substances have different thermal properties. We will show how a specific thermal property may be used to calculate approximate atomic weights of pure metals. Finally, we will discuss the relationship between the size of an ion and its parent atom, and the sizes of atoms and ions that contain the same number of electrons.

OBJECTIVES

After completing this chapter, you will be able to

- recognize and apply or illustrate: specific heat, crystal lattice, allotrope, polymorphic, heat of fusion, melting point, sublimation, diffraction, isoelectronic, amorphous;

- explain the differences between "glasses" and crystals;

- explain the differences between gases and solids;

- explain why an ion has a different radius than its parent atom;

- determine if a group of ions and atoms is isoelectronic;

- calculate the approximate atomic weight of a pure metal when given the specific heat of the metal;

- calculate the amount of heat required to melt a solid knowing the heat of fusion of the solid;

- locate the melting point and other properties of a solid on a graph that shows what happens when a solid is heated;

- describe the differences in properties, bonding, and structure among ionic, covalent, molecular, and metallic solids.

1 In Chapter 8, the kinetic molecular theory of gases described gases as molecules that move rapidly and readily to conform to the shape of the container. A solid, on the other hand, will not conform readily to the shape of its container unless it is forced to do so, for example if it is crushed to a powder. The atoms or molecules of a solid are fixed in place and do not move; they only vibrate within the confines of their positions in well-defined structures.

Ice cubes and coal lumps need to be crushed to conform to the walls of a container. Are ice cubes and coal lumps gases or solids? _____

Answer: solids

2 The forces attracting one gas molecule to another are negligible (except at low temperatures and high pressures). These are the van der Waals forces discussed in Chapter 8. In solids, the forces of attraction between atoms and molecules are relatively strong, and a great deal of energy is required to overcome these forces. In gases, most of the volume (theoretically all of the volume) is the space between the molecules. Gases are easily compressed. Changes in temperature and pressure have comparatively little effect on the volume of a solid. Even extreme pressure causes only a slight reduction in the volume of solids.

We know that there is a lot of space between molecules or atoms in a gas. Based on the above information, would you expect a lot or a little space between the molecules or atoms of a solid? _____

Answer: little

3 **Melting** is defined as changing from the solid state to the liquid state. The temperature at which both the liquid and solid state are present at the same time is known as the melting point of the solid. If a liquid is changing to a solid, this same melting point temperature is also called the freezing point. Each solid has a specific, unique temperature as its melting point.

(a) Melting is the change of state from which to which: solid to liquid or liquid to solid? _____

(b) Is the temperature at which a liquid changes to a solid its melting point or its freezing point? _____

Answer: (a) solid to liquid; (b) freezing point

4 Molecules on the surface of some solids escape from the solid and enter a gaseous state directly without entering the liquid state first. These solids exhibit a measurable vapor pressure. One of the best known examples of this is dry ice, which is solid carbon dioxide. The phenomenon of a solid changing directly to the gaseous state is known as **sublimation**. The liquid state is bypassed.

Solid mothballs become smaller over a period of time without the appearance of any liquid. The distinctive odor of mothballs in the air indicates that some of the solid has vaporized. Wet clothes hung out on a clothesline freeze if the outside temperature remains below the freezing point! Although the temperature remains below the freezing point, the clothes do dry. The ice crystals go directly to the gaseous state over a period of time. You may also have noticed that ice cubes shrink when left in the freezer for a long time. Have the mothballs, ice cubes, and ice crystals in the frozen clothes melted or sublimed? _____

Answer: sublimed

CRYSTALS AND THEIR SHAPES

5 Many naturally occurring solids have very distinct geometrical shapes. Table salt, NaCl, looks like cubes. Calcite, $CaCO_3$, looks like a rectangular solid. Alum, $KAl(SO_4)_2 \bullet 12H_2O$, looks like two pyramids base to base. A solid that exists in a definite three-dimensional geometrical shape is known as a **crystalline solid** or **crystal**. Crystalline solids have a characteristic shape regardless of the size of the piece. There are seven basic crystal shapes that have been identified in naturally occurring solids. They are shown on page 206, along with some variations of the basic shapes, giving a total of 14 arrangements, or crystal lattices, in which crystalline solids occur.

If a large crystal is cleaved (split), would the smaller pieces still be crystals?

_____

Answer: yes (Crystalline solids have a characteristic shape regardless of the size of the pieces. This is how a jeweler cuts a large diamond into several smaller ones.)

The seven crystal systems and their variations are determined by the arrangement of atoms, ions, and molecules within the crystals. The arrangement of atoms within the crystal is known as the **crystal lattice**. We can determine this arrangement by a process called **X-ray diffraction**. Planes of atoms and ions that make up the crystal diffract X-rays similar to the way a mirror reflects light. X-ray diffraction analysis of crystals is possible because the wavelengths of X-rays are very close in size to the diameters of the atoms. Visible light has much longer wavelengths than X-rays; therefore, the individual atoms within the crystal lattice will not diffract normal light. When X-rays are used, the resulting diffraction may be photographed or sent directly to a computer for analysis and mathematical determination of the structure.

We now discuss some more properties of crystalline and noncrystalline solids.

6   Some substances can assume more than one crystalline shape (as shown in the figures below). Sulfur, for example, can be found as rhombic crystals or monoclinic crystals. A substance is **polymorphic** if it exists in more than one crystalline form. Calcium carbonate, $CaCO_3$, occurs in nature with either the hexagonal structure or the rhombic structure. Is $CaCO_3$ an example of a polymorphic solid? _____ Explain why or why not. _____

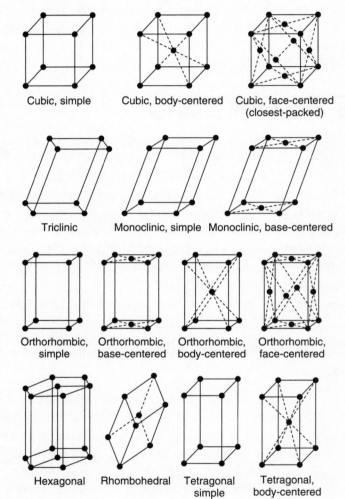

Cubic, simple   Cubic, body-centered   Cubic, face-centered
                                              (closest-packed)

Triclinic   Monoclinic, simple   Monoclinic, base-centered

Orthorhombic,   Orthorhombic,   Orthorhombic,   Orthorhombic,
simple   base-centered   body-centered   face-centered

Hexagonal   Rhombohedral   Tetragonal   Tetragonal,
                           simple   body-centered

Answer: yes; it exists in more than one crystalline form.

7   When a chemical element is found to be polymorphic, the different crystalline forms of the element are known as **allotropes** (or **allotropic forms**) of the element. Carbon is a polymorphic element. Graphite and diamond are both allotropic forms of carbon.

(a) Sulfur is an element that occurs in two forms. Thus, rhombic sulfur and monoclinic sulfur are _____ of sulfur.

(b) White phosphorus, red phosphorus, and yellow phosphorus are crystalline forms of phosphorus; therefore, phosphorus is called a _____ substance.

Answer: (a) allotropes; (b) polymorphic

8 Not all solids are crystalline. A solid that does not have a definite crystalline shape is said to be **amorphous**. Glass is a common example of an amorphous solid. It is a solid, but it is not crystalline.

If a piece of glass were broken into smaller pieces, could you predict the shape of those pieces? Why or why not? _____

Answer: No, because glass is not crystalline, it has no definite geometrical form, such as one of the basic crystal shapes.

9 The entire category of solid substances that appear to be crystalline but are not is called **glasses**. Glasses have several characteristics that differ from crystals. Crystals have sharp melting points. Melting destroys the attractions that give solids their rigid structure. Below the sharp melting point, crystals are solids. Above the sharp melting point, crystals become liquids. Glasses have relatively large melting ranges instead of a sharp melting point. Glasses generally soften over several degrees of increasing temperature before finally becoming liquid.

Ice melts at 0°C. It does not become soft before melting. Is ice more likely to be a glass or a crystalline solid? _____

Answer: crystalline solid

10 Crystals cleave (split) along planar (flat) surfaces when struck. Even when broken, crystals retain their basic structural shape. Glasses, on the other hand, fracture to form uneven surfaces when struck. Breakage is not along any definite pattern.

A jeweler can split a gemstone with a jeweler's chisel. The surfaces along the split are flat and the cut is clean. Is the gemstone a crystal or a glass? _____

Answer: crystal

11 When heated, a solid substance becomes soft and pliable several degrees before actually melting. Is the substance likely to be crystalline? _____

Answer: no (Crystals have sharp melting points and do not soften before melting.)

In crystals, the strengths of the chemical bonds between the atoms, ions, or molecules are all the same. This accounts for their very sharp melting point. In glasses, the strengths of the bonds vary throughout the solid. This accounts for the wide melting point range of glasses. In which class of solids (glasses or crystals) would it be possible for some bonds to be broken at one temperature while other bonds would not be broken until a slightly higher temperature was reached?_____

Answer: glasses (With bonds of varying energies, some bonds would be broken at lower temperatures than other bonds.)

HEAT OF FUSION

 When crystalline solids melt, the crystal lattice is broken down. The solids lose their shape and become fluid. The energy required to change a solid to a liquid while remaining at the melting point temperature is called the **heat of fusion**. The energy required to change a block of ice at $0°C$ to water at $0°C$ is known as the _____ for H_2O.

Answer: heat of fusion

 Heat of fusion can be expressed as either *calories per gram* or *calories per mole*. The calorie (cal) is a unit of energy in the form of heat. The heat of fusion of H_2O is 80 calories per gram. The energy required to melt a 1 gram ice cube at $0°C$ to water at $0°C$ is _____

Answer: 80 calories (Since the heat of fusion for H_2O is 80 calories per gram, it takes 80 calories to melt 1 gram of ice.)

 The heat of fusion can be graphically shown by plotting temperature against the energy requirement.

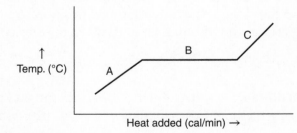

Remember that the heat of fusion is the energy required to liquefy a solid at the melting point. When a solid is heated, its temperature rises *until* the melting point. At the melting point the temperature remains constant until all the solid has melted. After melting is completed, the temperature of the substance (now totally liquid) rises as more heat is applied. Which part of the graph shows the heat of fusion? (A, B, C) _____

Answer: B (because the temperature remains constant at the heat of fusion)

16 If the graph below represents H_2O, at what point on the temperature scale should the melting point (0°C) be placed? (A, B, C, D, E) _____

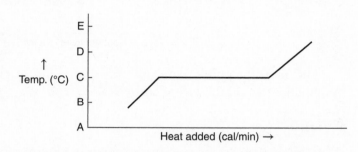

Answer: C (the temperature at which the heat of fusion is observed)

17 The graph below represents heat applied to a substance such as H_2O or some other crystalline substance.

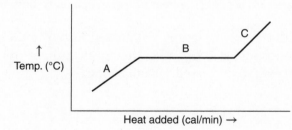

For each of the indicated points, say whether the substance is solid, liquid, or both solid *and* liquid.

(a) at point A _____

(b) at point B _____

(c) at point C _____

Answer: (a) solid; (b) both solid and liquid (energy as the heat of fusion is being used to melt the solid); (c) liquid

18 The next graph represents the reverse process. Heat is removed from a system and a liquid freezes into a solid. In this case, the heat of fusion energy must be removed for the liquid at the freezing point to become solid at the freezing point.

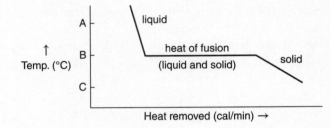

The freezing point (same temperature as the melting point) is indicated by which letter on the temperature scale? (A, B, C) _____

Answer: B

19 Remember that the heat of fusion for H_2O is 80 calories per gram, with the temperature remaining at the melting point. Based on this information, how many calories of heat do you think must be removed to freeze 1 gram of water if the temperature remains at the freezing point? _____

Answer: 80 calories (The heat of fusion applies to both melting and freezing.)

20 For the moment, ignore the heat of fusion in the graph below. Before the heat of fusion, what happens to the temperature of the solid when heat is added: is it increased, decreased, or unchanged? _____

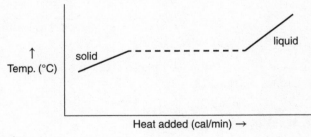

Answer: increased

SPECIFIC HEAT

21 Each solid substance as well as each liquid requires a different quantity of heat added to increase the temperature by a given amount. To raise the temperature of 1 gram of solid iron 1° on the Celsius scale requires 0.108 calories. To raise the temperature of 1 gram of solid lead 1° on the Celsius scale requires 0.0306 calories.

Which substance requires more heat, measured in calories, to raise its temperature 1°: solid lead or solid iron? _____

Answer: solid iron (Iron requires more than three times as much heat as lead to raise the temperature 1°C for each gram.)

22 The calories required to raise the temperature of 1 gram of a substance by 1°C is called the specific heat of the substance. The specific heat of copper is 0.093 calories per gram. The specific heat of aluminum is 0.219 calories per gram. Which has the lower specific heat, aluminum or copper? _____

Answer: copper

23 Specific heat is useful because it is related to atomic weight.

$$\text{atomic weight} = 6.2 \div \text{specific heat}$$

The atomic weight found by this formula is a reasonable approximation of the actual atomic weight of solid elements. The formula shows a relationship between _____ and _____.

Answer: atomic weight; specific heat

24 The formula is useful for determining approximate atomic weights and identifying unknown substances.

$$\text{atomic weight (g/mol)} = \frac{6.2 \, \text{cal/mol}}{\text{specific heat (cal/g)}}$$

For example, the specific heat for iron is $0.108 \, \text{cal/g}$.

$$\text{atomic weight of iron (g/mol)} = \frac{6.2 \, \text{cal}}{\text{mol}} \div \frac{0.108 \, \text{cal}}{\text{g}}$$

$$= \frac{6.2 \, \text{cal}}{\text{mol}} \times \frac{\text{g}}{0.108 \, \text{cal}} = \frac{6.2 \, \text{g}}{0.108 \, \text{mol}} = 57.4 \, \text{g/mol}$$

Since the actual atomic weight of iron is 55.8 grams/mol, the result is an approximate atomic weight.

What is the approximate atomic weight (to the nearest tenth) of nickel if it has a specific heat of 0.105 cal/g? _____

Answer: atomic weight of nickel = $\dfrac{6.2\,\text{cal/mol}}{0.105\,\text{cal/g}}$ = 59.0 g/mol

(The actual atomic weight of nickel is 58.71 g/mol.)

25 A solid element is found to have a specific heat of 0.0565 cal/g. Determine the approximate atomic weight of the element (to the nearest tenth). _____

Answer: atomic weight = $\dfrac{6.2\,\text{cal/mol}}{0.0565\,\text{cal/g}}$ = 109.7 g/mol

atomic weight = 109.7 g/mol

26 Look at the periodic table and determine which solid element is closest in atomic weight to 109.7 g/mol. _____

Answer: silver (The closest element to an atomic weight of 109.7 is silver, Ag, with an actual atomic weight of 107.87 g/mol. Another possibility is cadmium, Cd, with an atomic weight of 112.41 g/mol. Further chemical analysis is necessary to positively identify the element. This is a good approximation, however, and limits the choices.)

Now that you have learned some of the properties of solid substances, we will examine the properties of four general types of solids and cite commonly occurring examples of each.

TYPES OF SOLIDS

27 The melting points of different solids vary with the differences in interparticle attraction found among the various solids. Crystalline solids are classified in four categories based upon differences in interparticle attraction and in the nature of the particles (atoms, ions, or molecules) that make up the crystals. The four categories are ionic solids, covalent solids, molecular solids, and metallic solids. (You should remember the terms *ionic* and *covalent* as two types of chemical bonds discussed in Chapter 3.)

Ionic solids are made up of ions held together by ionic bonds. Negative and positive ions are bonded together in a regular arrangement of negative to positive to negative to positive and so on, throughout the neatly stacked arrangement within the crystal lattice. The strong electrostatic attraction between the ions is responsible for the hardness and generally high melting points of ionic solids. Examples of ionic solids are NaCl and KCl.

(a) In an ionic solid is the attraction between ions weak or strong? _____

(b) Would you expect an ionic solid to have a low or high melting point? _____

(c) Would you expect an ionic solid to be hard or soft? _____

Answer: (a) strong; (b) high; (c) hard

28 Ionic solids do not conduct electricity when solid, but when heated to the liquid state the molten ionic compounds conduct electricity well. In the liquid state, the crystal lattice is broken and the ions are free to move about and carry the electric current. Most ionic solids dissolve easily in water. The aqueous solutions of ionic solids also conduct electricity well.

A very hard solid with a high melting point dissolves in water, but the aqueous solution does not conduct electricity. Would you expect the solid to be ionic?_____

Answer: no (Ionic solids dissolve readily in water and conduct electricity.)

29 **Covalent solids** (also known as network solids) consist of atoms held together completely by covalent bonds extending from atom to atom throughout the crystal lattice. The entire crystal lattice is interlocked in a series of such covalent bonds. Covalent bonds are very strong, and such solids are very hard with very high melting points. A relatively large amount of energy is required to break the bonds.

A solid substance with a high melting point could be (an ionic solid, a covalent solid, either an ionic solid or a covalent solid) _____

Answer: either an ionic solid or a covalent solid (Covalent solids often have higher melting points than ionic solids, but both have relatively high melting points.)

30 Covalent solids are nonconductors of electricity in either the solid or the molten (liquid) state. Most covalent solids are not soluble in water. Typical examples of covalent solids are diamond (made of carbon), silicon carbide (SiC, better known as carborundum), and aluminum nitride (AlN). All of these examples are useful as industrial abrasives for grinding and polishing because of their hardness.

A solid is heated until it melts at a high melting point. The molten substance conducts electricity. This solid is likely to be (an ionic solid, a covalent solid, either an ionic or covalent solid) _____

Answer: an ionic solid (Ionic solids conduct electricity when molten, while covalent solids do not.)

31 **Molecular solids** are made up of either polar or nonpolar molecules. (See Chapter 3 if you need to review polar and nonpolar compounds.) Nonpolar molecules are held together by weak van der Waals forces. Note that the atoms within a molecule may be covalently bonded, but in a molecular solid the molecules themselves are not interconnected with covalent bonds. Both ionic and covalent solids are relatively hard and have high melting points because of the strong bonds within the solids. Try some predictions about the hardness and the melting points of nonpolar molecular solids. Would you expect nonpolar molecular solids with their weak interconnecting forces to:

(a) be relatively hard or relatively soft? _____

(b) have melting points that are high or low? _____

Answer: (a) relatively soft; (b) low melting points

32 Two examples of nonpolar molecular solids are dry ice (CO_2) and naphthalene (mothballs). Polar molecular solids such as ice (H_2O), solid ammonia (NH_3), and solid sulfuric acid (H_2SO_4) have melting points and degrees of hardness between those of nonpolar molecular solids and ionic solids. There is some electrostatic attraction between polar molecules. As a group, polar and nonpolar molecular solids are generally soft, have comparatively low melting points, and do not conduct electricity.

A solid is very hard but does not conduct electricity in the solid or liquid state. The solid is probably (molecular, ionic, covalent) _____

Answer: covalent

33 A fourth category of crystalline solids is metallic solids. Examples are elemental iron, copper, and silver. Solid metals have been described as made up of positive ions held together by highly mobile electrons or as positive ions in a "sea" of electrons. These electrons move freely, with the result that the metals are excellent conductors of electricity. The melting points and hardnesses of various metals vary over wide ranges. Which of the four types of crystalline solids conducts electricity as a solid: ionic, covalent, molecular, or metallic?

_____

Answer: metallic (Ionic solids must be melted or dissolved before they will conduct electricity.)

34 In metallic solids, each metal atom is surrounded by eight other atoms as its nearest neighbors in a cubic closest-packed crystalline structure, or 12 other atoms in a hexagonal closest-packed structure. The outer shell electrons are loosely held in each atom. The mobility of these electrons accounts for the fact that metals are excellent conductors of both electricity and heat.

Metallic solids generally assume two closest-packed crystalline structures. What are these geometric structures? _____

Answer: cubic and hexagonal

35 As you should recall, metallic solids also have certain characteristics such as **malleability** (can be rolled or beaten into fine sheets), **ductility** (can be drawn into fine wires), and **luster** (shine) in both the solid and molten states. Which two of the following characteristics of metallic solids best explain the use of some metallic solids in electrical wiring? (luster, malleability, ductility, cubic or hexagonal closest-packed crystal structure, mobility of outer shell electrons)

_____

Answer: ductility and mobility of outer shell electrons

36 You have just learned the properties of these types of solids: ionic, covalent, molecular, and metallic.

(a) Which are the hardest, so that some examples of these solids are used for industrial drilling and abrasives? _____

(b) Which solids are the softest? _____

(c) Which are the best conductors of electricity and heat? _____

(d) Which solids are held in a crystalline structure by strong electrostatic attractions?_____

Answer: (a) covalent; (b) molecular; (c) metallic; (d) ionic

In the next section we will discuss how X-ray diffraction has been used to determine the sizes of atoms and ions and how the sizes vary from group to group and period to period in the periodic table.

ATOMIC AND IONIC RADII

 An important use of X-ray diffraction analysis is the determination of atomic and ionic sizes. Atomic and ionic sizes are usually listed in angstrom units, Å (1 Ångstrom unit = 1×10^{-8} centimeters). The radii of some typical atoms and their ions are listed below.

Sizes of Positive Ions and Their Parent Atoms

Atom	Radius (Å)	Ion	Radius (Å)
Li	1.23	Li^+	0.60
Na	1.57	Na^+	0.95
K	2.03	K^+	1.33
Rb	2.16	Rb^+	1.48
Cs	2.35	Cs^+	1.69
Be	0.89	Be^{2+}	0.31
Mg	1.36	Mg^{2+}	0.65
Ca	1.74	Ca^{2+}	0.99
Sr	1.91	Sr^{2+}	1.13
Ba	1.98	Ba^{2+}	1.35

In this chart, locate the Na atom. What is the atomic radius of a Na atom?

_____

Answer: 1.57 Å (Remember to express atomic radii in angstrom units, Å.)

38 Using the same chart, find the radius of the sodium ion, Na^+. _____

Answer: 0.95 Å

39 The positive ion Na^+ has a smaller radius than its parent atom Na. Look at all the positive ions listed on the chart and compare the positive ions to their parent atoms. How do the radii of positive ions compare with the radii of their parent atoms? _____

Answer: Positive ions are all smaller than their parent atoms.

40 Positive ions are smaller than their parent atoms because a positive ion has *lost* one or more electrons. The protons now outnumber the electrons, so the protons (which are in the nucleus) exhibit more attraction for the remaining electrons. The protons thus draw the remaining electrons closer to the nucleus, resulting in a smaller radius than the original atom. In addition, electrons are lost from the outermost shell. Loss of these outermost shell electrons also makes the ion smaller than the original atom.

A negative ion has *gained* one or more electrons. Based upon what you have just learned about positive ions, how should the radius of a negative ion compare with the radius of its parent atom? _____

Answer: The negative ion radius should be larger than the radius of the parent atom.

Sizes of Negative Ions and Their Parent Atoms

Atom	Radius (Å)	Ion	Radius (Å)
F	0.72	F⁻	1.36
Cl	0.99	Cl⁻	1.81
Br	1.14	Br⁻	1.95
I	1.33	I⁻	2.16
O	0.74	O^{2-}	1.40
S	1.04	S^{2-}	1.84
Se	1.17	Se^{2-}	1.98
Te	1.37	Te^{2-}	2.21

41 Look at the chart of sizes of negative ions and their parent atoms.

(a) What is the radius of an O^{2-} ion? _____

(b) How many centimeters is it? _____

Answer:

(a) 1.40 Å

(b) $1.40 \, \text{Å} \times \left(\dfrac{1 \times 10^{-8} \text{cm}}{1 \, \text{Å}} \right) = 1.40 \times 10^{-8} \, \text{cm}$

42 Ions and atoms that have the same number of electrons are called **isoelectronic**. The ion Na^+ has the same number of electrons as an atom of Ne. All of the following have the same number of electrons and are isoelectronic:

$$N^{3-}, O^{2-}, F^-, Ne, Na^+, Mg^{2+}, Al^{3+}, Si^{4+}$$

Do any two of these isoelectronic ions or atoms have the same number of protons? _____

Answer: no (Remember that only the number of electrons is changed. The number of protons in the nucleus remains unchanged and is a fixed number for each element, its atomic number.)

43 An increase in the positive charge per electron results in a decrease in the radius. All of the following have 10 electrons, so they are isoelectronic. Their atomic numbers (shown below the isoelectronic ions) range from 7 to 14.

N^{3-}	O^{2-}	F^-	Ne	Na^+	Mg^{2+}	Al^{3+}	Si^{4+}
7	8	9	10	11	12	13	14

In the negative ions, the electrons outnumber the protons. In the positive ions, the protons outnumber the electrons. Would you expect the radii of these isoelectronic species to increase from left to right or to decrease from left to right? _____

Answer: decrease from left to right (The number of protons increases while the number of electrons remains constant, therefore increasing the attraction on the electrons and making each successive ion smaller.)

When an atom has more protons than electrons, the increase in the positive charge per electron results in a decrease in the radius. When an atom has fewer protons than electrons, the decrease in the positive charge per electron results in an increase in the radius.

Iron, Fe, has an atomic number of 26. A neutral atom of Fe has 26 electrons. An Fe^{2+} ion has 24 electrons. An Fe^{3+} ion has 23 electrons. List the Fe atom and its two ions in order of decreasing size (largest first). _____

Answer: Fe, Fe^{2+}, Fe^{3+}

You have just learned some of the properties of solids and how they compare with gases. Our knowledge of the structures of solids and their properties helps us to develop new uses for solids and new combinations of solids that are stronger, better conductors of heat and electricity, and more resistant to chemical attack.

In Chapter 10 we will discuss the liquid state and its properties. A comparison of liquids to the solid and gaseous states is also presented.

SELF-TEST

This self-test is designed to show how well you have mastered this chapter's objectives. Correct answers and review instructions follow the test.

1. The temperature at which a solid changes to a liquid is which, its melting point or freezing point?

2. Dry ice, the solid form of carbon dioxide gas, does not melt to liquid carbon dioxide. Instead solid CO_2 directly changes to gaseous CO_2. The process of changing from a solid directly to a gas is known as _____.

3. There are _____ known basic crystal systems.

4. The process known as _____ allows us to determine the arrangement of atoms, ions, and molecules within a crystal lattice.

5. Sulfur exists in more than one crystalline form. Is it amorphous or polymorphic? _____

6. Graphite and diamond are two forms of carbon. What are they called? (amorphous, allotropes, isotopes) _____

7. State at least two basic differences between crystals and glasses.

    _____

8. The graph below represents the heating of a solid. Indicate what is happening or what is represented by each of the labeled points.

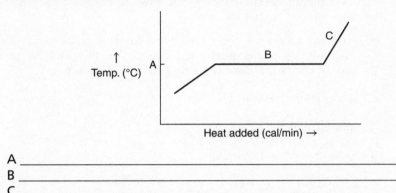

A _____
B _____
C _____

9. The heat required to raise the temperature of 1 gram of substance by 1°C is called the _____ of a substance.

10. What is the approximate atomic weight (to the nearest tenth) of aluminum if it has a specific heat of 0.215 cal/g?

11. What is the approximate atomic weight (to the nearest tenth) of titanium if it has a specific heat of 0.125 cal/g?

12. Which one of each of the following sets will have the larger radius?

    (a) Mg or Mg^{2+} _____

    (b) S or S^{2-} _____

    (c) F^- or Ne _____

    (d) Ca^{2+}, K^+, F^-, or O^{2-} _____

13. Which one of each of the following sets will have the smaller radius?

    (a) Ca or Ca^{2+} _____

    (b) P or P^{3-} _____

    (c) Li^+ or Br^- _____

    (d) F^-, Cl^-, Br^-, or I^- _____

14. Which of the following ions and atoms are isoelectronic with argon (Ar)? (Cl^-, S^{2-}, P, K, Ca^{2+}, Kr) _____

15. Which of the following are isoelectronic with krypton (Kr)? (Cl^-, Br^-, Rb^+, As^{3-}, Se, K^+)

ANSWERS

Compare your answers to the self-test with those given below. If you answer all questions correctly, you are ready to proceed to the next chapter. If you miss any, review the frames indicated in parentheses following the answers. If you miss several questions, you should probably reread the chapter carefully.

1. for many substances the melting and freezing point temperature is the same with both solid and liquid present at the same time but since the question noted "solid to liquid", the best answer is melting point [frames 1-3]

2. sublimation [frame 4]

3. seven [frame 5]

4. X-ray diffraction [frame 5]

5. polymorphic (frame 6)

6. allotropes (frame 7)

7. (a) Crystals have sharp melting points, while glasses have broad melting point ranges.

   (b) Crystals cleave along planar surfaces, while glasses fracture to form rounded or uneven surfaces.

   (c) Crystals have equal bond strengths in all directions, while glasses have variable bond strengths.

   (d) Crystals diffract X-rays, but glasses do not.

   (frames 9–12)

8. (a) melting point

   (b) heat of fusion (solid and liquid present while the solid is melting)

   (c) only liquid present, while temperature is increasing (frames 15–20)

9. specific heat [frame 22]

10.
$$\text{atomic weight} \left(\frac{g}{mol} \right) = \frac{6.2 \frac{cal}{mol}}{\text{specific heat} \left(\frac{cal}{g} \right)} = \frac{6.2 \text{ cal/mol}}{0.215 \left(\frac{cal}{g} \right)} = 28.8 \text{ g/mol}$$ *Due to this being an approximation it is relatively close to the atomic weight of aluminum which is 26.982 g/mol.* [frames 21-26]]

11.

$$\text{atomic weight} \left(\frac{g}{mol}\right) = \frac{6.2\frac{cal}{mol}}{\text{specific heat}\left(\frac{cal}{g}\right)} = \frac{6.2\ cal/mol}{0.125\left(\frac{cal}{g}\right)} = 49.6\ g/mol \quad Due \quad to$$

this being an approximation it is relatively close to the atomic weight of titanium which is 47.867 g/mol. [frames 21-26]]

12. (a) Mg, (b) S^{2-}, (c) F^-, (d) O^{2-} (frames 37–40, 43)

13. (a) Ca^{2+} (b) P (c) Li^+ (d) F^- [frames 37-40, 43]

14. Cl^-, S^{2-}, Ca^{2+} (frame 42)

15. Br^-, Rb^+, As^{3-} [frame 42]

EVERYDAY CHEMISTRY

Piezoelectric Crystals

In this chapter, you learned about some solids that are crystals with very specific shapes. Did you know that some crystals when struck or deformed just a little can produce a spike of electricity? If you have ever used an outdoor gas grill, many have a red button that when pressed actuates a small spring hammer that strikes a crystal which produces a momentary high voltage that is conducted by wire to a gap, resulting in a spark. That spark ignites the gas and air mixture in the gas grill. No battery is needed and the process can be repeated many times as long as the mechanical and electrical connections are intact. A crystal of this type is **piezoelectric**. Not only is the crystal capable of producing a spike of electricity but the reverse can also be true. A very accurate quartz watch relies on the effect of a crystal that is excited by a tiny electric current vibrates at a consistent rate depending on how it is cut and its thickness. That same type of crystal is used in computers and many other devices to provide accurate timing.

A crystal can be carefully cut in shape and size to produce a vibration at a specific frequency when subjected to electric current. During World War II, the United States relied on Brazil to mine quartz crystals which were cut and shaped so that radio transmitters for communication could be relied on at various selected radio frequencies. The shortage of naturally occurring quartz for cutting those crystals caused a major research effort to produce man-made crystals. Most of the crystals in use today are man-made. The piezoelectric effect is also relied on for tiny speakers and beepers as well as for larger devices used in ultrasound.

The reason that a crystal can produce electricity when mechanically stressed and deformed slightly or can vibrate when subjected to an alternating electric current is that the arrangement of atoms and electrical charges within the crystal, called **dipole moments**, are balanced and therefore neutral. When squeezed, an unbalance occurs causing the electrical charges to no longer be neutral and producing a momentary voltage on two faces of a crystal until the crystal returns to its balanced position. Feeding an alternating electric current to the crystal faces causes the crystal to deform and vibrate.

The development of piezoelectric devices and materials is a fruitful area for continuing research in chemistry.

10 Liquids

Liquids are easily recognized by comparison with the most widely known liquid, water. You know from experience that water flows when we pour it, boils when heated enough, freezes when cooled enough, disappears when left in the open, does not mix with oil or gasoline, supports "water bugs" on its surface, and cools when it evaporates, as when perspiration leaves your skin. These properties of water are characteristic of all liquids. Every liquid has its own boiling point, freezing point, flow, surface, and "heat" properties.

This chapter will acquaint you with the characteristics of the liquid state. Every gas becomes a liquid when it is cooled to a sufficiently low temperature. This is the result of intermolecular attractions. The properties of liquids are largely determined by the magnitude of these intermolecular forces. The forces most often encountered are van der Waals forces or hydrogen bonds, formed by polar molecules, and very weak forces called London forces, which are primarily of interest in the behavior of nonpolar molecules.

Every liquid becomes a solid when sufficiently cooled. This happens at the temperature at which the kinetic energy of the molecules becomes so small that intermolecular attraction brings about a complete loss of fluidity. The solid that results has a rigid structure that is characterized by its own unique type of crystal shape. Liquids fall between gases and solids in their "ordering." Remember, gases fill their container and take its shape, and are in rapid random motion with no order.

Solids, on the other hand, are very highly ordered and do not fill or take the shape of their container unless forced to do so.

OBJECTIVES

After you have completed this chapter, you will be able to

- recognize and apply or illustrate: boiling point, vapor pressure, viscosity, surface tension, miscible, immiscible, volatility, density, hydrogen bonding, polar compounds, heat of vaporization, equilibrium, and Le Chatelier's Principle;

- state the effect of temperature on the surface tension, viscosity, and volatility of a liquid;

- give an explanation of why the vapor pressure of a liquid changes with temperature and how this relates to boiling point and critical temperature;

- explain what properties are necessary for a good refrigerant;

- given a graph representing what happens to the temperature of a substance as heat is added to it, locate the melting point, the boiling point, the states of matter that are present, and the regions representing heats of fusion and vaporization;

- given the graph of the vapor pressure versus the temperature of a liquid, determine: the vapor pressure of the liquid at a specific temperature and the boiling point of the liquid at a specific atmospheric pressure;

- calculate the amount of heat that must be supplied to vaporize a certain amount of a liquid when given its heat of vaporization;

- predict whether or not two liquids will mix when put together.

1 We begin this chapter with a brief comparison of the states of matter based upon the **kinetic molecular theory.** The kinetic molecular theory was first encountered in explaining the behavior of gases. When cooled to a sufficiently low temperature, the average kinetic energy drops and every gas becomes a liquid. The molecules of liquid substances still move in random motion according to the kinetic molecular theory. However, the molecules in the liquid state move more slowly, and there is very little space between molecules.

When a gas is compressed, the space between the molecules of a gas is actually compressed. Since there is very little space between the molecules of a liquid, would you expect liquids to be readily compressed? _____

Answer: no (In reality, even large pressures have very little effect on the volume of a liquid.)

2 The molecules of a liquid are not as structured as those in the solid state. The liquid molecules are free to slide over one another and to flow and assume the shape of the container. The molecules of a solid can vibrate and bend but cannot flow and slide over one another. The three states of matter (gaseous, liquid, and solid) differ in terms of the freedom of movement of molecules.

(a) Molecules have the greatest freedom of movement in which state?

_____

(b) Molecules have the least freedom of movement in which state?_____

Answer: (a) gaseous; (b) solid

3 Many of the properties of liquids are determined by the attraction of molecules for each other. These **intermolecular attractive forces** cause the molecules to stick together and coalesce into a liquid. The intermolecular attractions affect the flow rate of a liquid. Heavy oils, syrup, and molasses flow more slowly than water or gasoline. Compared to liquids that flow rather quickly, liquids of slower flow rate would have (stronger, weaker) _____ intermolecular attractive forces.

Answer: stronger (The stronger the intermolecular attraction, the more the molecules stick to each other and the slower the flow rate of a liquid.)

4 The **viscosity** of a liquid refers to its resistance to flow. A liquid with a high resistance to flow such as heavy oils or molasses would exhibit high viscosity. A liquid with very little resistance to flow would exhibit low viscosity. High viscosity is associated with relatively (strong, weak) _____ intermolecular attractive forces.

Answer: strong

5 A liquid with relatively weak intermolecular forces would probably exhibit low _____ (use the term that refers to resistance to flow).

Answer: viscosity

6 Razor blades and certain insects that are heavier than water can float on top of the water due to surface tension. Remember that the molecules of a liquid attract each other through intermolecular attractive forces. **Surface tension** is caused by a difference in direction of intermolecular attractive forces between those molecules at the surface of a liquid and those in the body of the liquid.

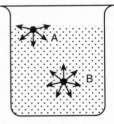

    A molecule in the body of the liquid (molecule B) is surrounded by attraction from other liquid molecules. There is equal attraction from all sides. A molecule at the surface of the liquid (molecule A) is attracted only toward the interior of the liquid. It is not balanced by an attraction from above. Which molecule (A or B) is involved in surface tension? _____

Answer: molecule A

7   Because molecule A and all other molecules making up the surface of the liquid are pulled inward by intermolecular attraction, the liquid tends to form as small a surface as possible. This inward pull of the surface molecules is known as surface tension.

A sphere is the three-dimensional figure with the smallest surface area. Water and other liquids tend to form droplets or small spheres because the inward pull on the surface molecules tends to produce as small a surface area as possible. A falling raindrop is actually a small sphere that is distended somewhat by the pull of gravity.

Liquids tend to form spherical droplets because of the inward pull on surface molecules, which is called _____

Answer: surface tension

8   The inward pull of surface molecules depends upon intermolecular attractive forces. Strong intermolecular attractive forces provide a strong inward pull. Compared to a liquid with relatively weak intermolecular attractive forces, a liquid with relatively strong intermolecular forces could be expected to show:

(a) (greater, lesser) _____ surface tension

(b) (higher, lower) _____ viscosity

Answer: (a) greater; (b) higher

VAPOR PRESSURE

9   **Vapor pressure** is the pressure exerted by the vapor (gaseous) phase of a liquid in a closed container. To understand vapor pressure, you must understand several other concepts. You may already be familiar with evaporation. **Evaporation** means passing from the liquid state to the gaseous (vapor) state. Remember that all of the molecules in a liquid together have an *average* kinetic energy that is a measure of the temperature of the liquid. However, not all individual molecules have the same kinetic energy. Some have above average and some have below average kinetic energy. Some molecules with greater kinetic energy will escape from the liquid if they are located near the surface. These molecules evaporate and enter the gaseous (vapor) state.

(a) The molecules of liquid that evaporate first are those of (higher, lower) _____ than average kinetic energy.

(b) If a saucer of water is uncovered and not touched for a day or two, some of the water "disappears." In terms of the kinetic molecular theory, explain what happened to the water that is missing. _____

Answer: (a) higher; (b) Some of the water has evaporated. Those molecules with greater than average kinetic energy that are located near the surface of the liquid have escaped and entered the vapor state.

10 The molecules of greatest kinetic energy evaporate first, while the molecules of lower kinetic energy remain. The result is a drop in temperature of the remaining molecules. (Temperature is simply a measurement of the average kinetic energy of the molecules. As the average kinetic energy drops, so does the temperature.)

A liquid in a well-insulated container has a rapid rate of evaporation. Its temperature is measured prior to evaporation. After evaporation of a portion of the liquid, would you expect the temperature of the remaining liquid to be higher or lower than the original temperature? _____

Answer: lower (because the average kinetic energy of the remaining liquid molecules is also lower)

11 To make up for the loss of kinetic energy through evaporation, a liquid undergoing evaporation will draw heat from its surroundings. That is why rubbing alcohol feels cold on your skin as it evaporates. Heat is removed from the skin during evaporation. Sweating is a means used by your body to control its temperature. A natural breeze or the breeze from a fan can increase the rate of evaporation. As a result, your skin temperature is (warmed, cooled) _____.

Answer: cooled (The evaporating sweat draws away heat and cools the skin.)

12 If a container of liquid has a cover, the liquid will continue to evaporate, but after a time, some of the vapor (gaseous) molecules will return to the liquid state in a process called **condensation**, which is essentially the opposite of evaporation. After a longer time, the rate of condensation will equal the rate of evaporation. When an equal rate of evaporation and condensation is reached, the pressure exerted by the vapor molecules is known as the **vapor pressure**.

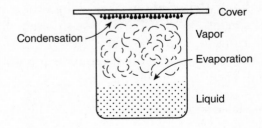

When the rate of evaporation is equal to the rate of condensation, the liquid and vapor are said to be in **equilibrium**. Liquid molecules continue to evaporate when equilibrium is reached, but just as many vaporized molecules condense into liquid. Both processes continue at equal rates. The processes are opposite but equal in rate.

$$\text{Vapor} \underset{\text{evaporation}}{\overset{\text{condensation}}{\rightleftharpoons}} \text{Liquid}$$

In order for equilibrium to be maintained with two opposite processes, the rates of the two opposite processes must be _____.

Answer: equal

13 An increase in temperature will increase the rate of evaporation so that the system is no longer in equilibrium. The system will achieve a new equilibrium at the new temperature, however, because the rate of condensation also increases until both rates are again equal.

$$\text{Vapor} \underset{\text{evaporation}}{\overset{\text{condensation}}{\rightleftharpoons}} \text{Liquid} \qquad \text{Vapor} \underset{\text{evaporation}}{\overset{\text{condensation}}{\rightleftharpoons}} \text{Liquid}$$

(temporarily unequal rates)   (new equilibrium reached)

An increase in the confining pressure will result in an increase in the rate of condensation. What could be expected to happen to the two rates after a period of time at the new pressure? _____

Answer: After a period of time, a new equilibrium is formed and the rates of evaporation and condensation would again be equal.

14 A change in temperature or pressure will cause a change in the rates of evaporation or condensation. After a time, a new equilibrium is set up, with the result that the rates of evaporation and condensation are again equal, although the rates may be somewhat faster or slower than in the previous equilibrium. **Le Chatelier's Principle** governs equilibrium systems. According to Le Chatelier's Principle, a "stress" on a system in equilibrium will cause a shift to a new equilibrium so as to relieve the stress.

The "stress" placed on the equilibrium system of evaporation and condensation rates can be a change in either _____ or _____

Answer: temperature; pressure

15 Le Chatelier's Principle governs the direction of change in a system that is at equilibrium and then undergoes a stress to attain a new equilibrium. An increase in temperature in a vapor-liquid equilibrium causes an increase in the evaporation rate. To achieve equilibrium, how must the condensation rate change? _____

Answer: The condensation rate must also increase.

16 At the new equilibrium, both evaporation and condensation rates are again equal, although both rates are increased when compared to the previous equilibrium. Remember that vapor pressure is the pressure that a vapor exerts when a liquid and its vapor (evaporation and condensation) are at equilibrium (at a specified temperature). According to Le Chatelier's Principle, an increase in temperature can shift the equilibrium to increase the rates of evaporation and condensation to form a new equilibrium. The average kinetic energy of the liquid molecules is also increased according to the temperature increase. Similarly, a decrease in temperature can shift the equilibrium to decrease the rates of evaporation and condensation (and decrease the average kinetic energy). The greater the rate of evaporation, the greater the vapor pressure.

Would you expect the vapor pressure of a liquid to increase or decrease if the temperature is increased? _____

Answer: increase (An increase in average kinetic energy allows more higher energy molecules to evaporate. The increased evaporation rate would increase the amount of vapor and the pressure it exerts.)

BOILING POINT

17 Water is the most widely known liquid. We have seen water freeze, flow when poured, or boil when the temperature is high enough. The boiling point of water is usually listed as 100°C. This is true only if the atmospheric pressure is 1 atmosphere (1 atm, which equals 760 torr or 101.325 kilopascals, kPa). At higher pressure, the boiling point of water is raised. At pressures lower than 1 atm, the boiling point of water is below 100°C.

When water is boiled in a pressure cooker, the pressure inside is greater than 1 atm. The boiling point of water could be expected to be (100°C, below 100°C, above 100°C) _____.

Answer: above 100°C

18 The boiling point of a liquid is the temperature at which the vapor pressure of a liquid equals the surrounding pressure. Based upon this definition, water would boil at 100°C at 1 atmosphere pressure because its vapor pressure was also equal to _____ at that temperature.

Answer: 1 atm

19 The surrounding pressure of an open pan of boiling water is usually 1 atm. In the pressure cooker, the surrounding pressure is greater than 1 atm. In order for water to boil in a pressure cooker, the vapor pressure must equal this greater surrounding pressure.

Suppose that instead of increasing the surrounding pressure as with a pressure cooker, we attempted to boil water at a pressure of less than 1 atm, for example, on a high mountain. Would you expect water to boil at 100°C, below 100°C, or above 100°C? _____

Answer: below 100°C (In fact, the boiling point of water on a high mountain, where the atmospheric pressure is less than 1 atm, *is* below 100°C.)

20 The higher the temperature, the greater the vapor pressure. When the vapor pressure of a liquid equals the surrounding pressure (pressure of the outside atmosphere or in some vessel such as a pressure cooker), the liquid will boil. The graph below plots the vapor pressure of water against temperature.

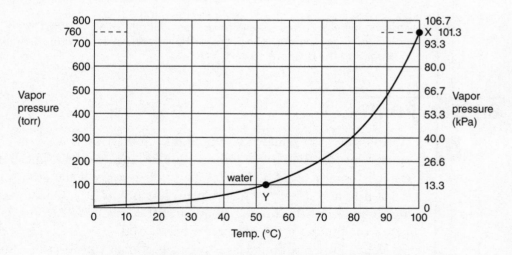

Change of vapor pressure with temperature for water

(a) At the normal boiling point of water (100°C), what is its vapor pressure? (See point X on the graph.) _____

(b) Water can be made to boil at various temperatures, provided that the vapor pressure is equal to the surrounding or confining pressure. If the surrounding pressure is only 350 torr (46.7 kPa), at what vapor pressure would the water boil? _____

(c) Using the graph, at what temperature (approximate) would water boil if the surrounding pressure was only 100 torr (13.3 kPa)? (See point Y on the graph.) _____

Answer: (a) 760 torr or 1 atm or 101.3 kPa; (b) 350 torr or 46.7 kPa (remember that, at boiling, the vapor pressure of the liquid equals the surrounding pressure); (c) slightly more than 50°C (At that temperature, the vapor pressure is 100 torr or 13.3 kPa. If the surrounding pressure and vapor pressure are equal, water boils.)

 21 The question of the boiling points of other liquids may have occurred to you. Do all liquids have the same boiling points? Look at the graph on the following page.

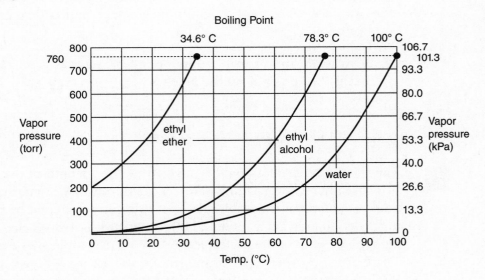

Vapor pressure and temperature for three common liquids

At a standard pressure of 1 atm (760 torr, 101.3 kPa), what is the boiling point of:

(a) water? _____

(b) ethyl alcohol? _____

(c) ethyl ether? _____

Answer: (a) 100°C; (b) 78.3°C; (c) 34.6°C

22 As a general rule, compounds of elements in the same family on the periodic table increase in boiling point with increasing molecular weight if all other factors are equal.

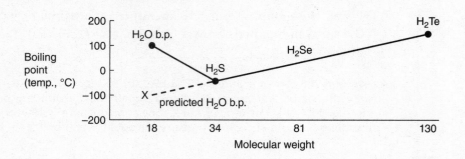

However, all factors are not always equal. Water (H_2O), for example, has a *predicted* boiling point of below 0°C when only molecular weights are considered. Another factor must account for the much higher than predicted boiling point of water.

(a) What is the approximate boiling point of water as predicted by the graph above? (See point X on the graph.) _____

(b) What is the actual boiling point of water at 1 atm? _____

Answer: (a) –100°C (approximately); (b) 100°C

23 The boiling point of a liquid is affected by the strength of the intermolecular attractive forces. Several kinds of forces can be classified as intermolecular attractions. One type of intermolecular attractive force is caused by polar molecules (first encountered in Chapter 3). Polar molecules have small electrostatic charges on opposite ends of the molecules. As a result, the molecules attract each other. There are also other types of intermolecular attractive forces such as van der Waals forces and London forces. You have already encountered van der Waals forces. **London forces** are much weaker intermolecular attractions in nonpolar compounds. They are caused by electrons from one molecule affecting the orbit of electrons in another molecule at very specific and constantly changing moments in time. Polar compounds have much stronger intermolecular attractions than nonpolar compounds. Stronger intermolecular attractions tend to keep molecules within the liquid form, with the result that the boiling point is higher than that predicted by molecular weight comparisons.

Compound A and compound B have approximately the same molecular weight. However, compound A has a much higher boiling point than compound B. Based upon what you have just learned, which compound (A or B) could be expected to have the stronger intermolecular attractive forces? _____

Answer: compound A

24 Water is a polar compound. The other hydrogen compounds of Group VIA elements such as H_2S, H_2Se, and H_2Te are compounds that are much less polar than H_2O. This explains why water has a much higher boiling point than predicted on the basis of molecular weight alone.

Besides water, certain other hydrogen compounds such as HF and NH_3 are also highly polar compounds with especially strong intermolecular attraction. Because elements such as O, F, and N are strongly electronegative, a high degree of polarity is found in these compounds. The result is that the positively charged hydrogen atom of one molecule may be strongly attracted to the negatively charged atom of another molecule. These intermolecular attractions occur throughout the liquid. Because the attractions are especially strong in certain compounds that are made up of hydrogen and electronegative elements, this type of intermolecular attraction is known as **hydrogen bonding**.

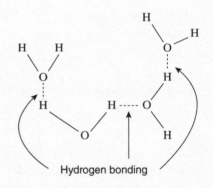

Hydrogen bonding

(a) Two compounds have similar molecular weights. Hydrogen bonding occurs in one compound but not in the other. Which compound would probably have the higher boiling point, the one with hydrogen bonding or the one without? _____

(b) Hydrogen bonding, polarity between molecules of a liquid, and van der Waals forces can all be classified as what kind of forces? _____

Answer: (a) the one with hydrogen bonding; (b) intermolecular attractive forces

25 Besides affecting the boiling point, intermolecular attractive forces also affect the evaporation of liquids. A liquid with relatively weak intermolecular forces will evaporate more readily than a liquid with relatively strong intermolecular forces. **Volatility** is a measure of the tendency of liquids to evaporate. A highly volatile liquid evaporates readily.

At room temperature, ether (diethyl ether) is highly volatile. Glycerin is very low in volatility. Assume that an equal quantity of each liquid is allowed to evaporate.

(a) Which liquid would you expect to evaporate first? _____

(b) Would you expect a liquid with strong intermolecular attractive forces to be highly volatile or low in volatility? _____

Answer: (a) Ether will evaporate first because of its high volatility. (b) low in volatility

26 Suppose we have two liquids with similar molecular weights. The liquid with stronger intermolecular attractive forces could be expected to have:

(a) (greater, lesser) _____ viscosity.

(b) (greater, lesser) _____ surface tension.

(c) (higher, lower) _____ boiling point.

(d) (higher, lower)_____ volatility.

(e) (higher, lower) _____ vapor pressure.

Answer: (a) greater; (b) greater; (c) higher; (d) lower; (e) lower

MISCIBILITY

27 The attraction of polar covalent molecules in liquids is one type of intermolecular attraction that affects characteristics such as boiling points, surface tension, volatility, vapor pressure, and viscosity. The polarity of molecules in liquids also affects **miscibility**, the ability of liquids to mix. Two liquids will mix if both are polar or both are nonpolar. If one liquid is polar and the other is nonpolar, the two liquids will not mix.

Oil and water do not mix. We have learned that water is a polar molecule. Therefore, would you expect oil to be polar or nonpolar? _____

Answer: nonpolar

28 Two polar liquids are miscible; two nonpolar liquids are miscible. However, a polar liquid and a nonpolar liquid are *not* miscible.

Water is a polar compound. Water and oil are not miscible. Gasoline and oil are miscible. Is gasoline polar or nonpolar? _____

Answer: nonpolar (Since water and oil are not miscible, oil must be nonpolar. Gasoline is nonpolar because it mixes with oil.)

29 Immiscible liquids (those not capable of being mixed) generally separate as two layers, with the more dense liquid on the bottom. Gasoline, jet fuel, or oil usually float on top of water because they are less dense than water and are not miscible with water. Based on this information, would you recommend water as a means for putting out gasoline, jet fuel, or oil fires? _____ Why or why not? _____

Answer: no; Since these liquids float on top of water, adding water might simply spread the fire. (A more appropriate means for putting out a fire of this type is foam or carbon dioxide, to cut off oxygen to the fire. Foam floats on top of oil or jet fuel.)

30 We can predict whether liquids are miscible or immiscible by knowing if the liquids are composed of polar or nonpolar molecules. The following table lists some common polar and nonpolar compounds. Read the table carefully and then answer the questions.

Compounds	Examples
Polar molecules	
H bonded to O, N, F, Br, Cl, and I	H_2O, NH_3, HF, HCl
O bonded to C, F, Cl, Br, I, and other elements	Alcohols (C_2H_5OH, CH_3OH), glycerine ($C_3H_5(OH)_3$), ethylene glycol ($C_2H_4(OH)_2$), HOCl, HOBr, HOI, acetone (CH_3COCH_3)
Nonpolar molecules	
Linear, planar, or symmetrical compounds with more than two atoms	CCl_4, CF_4, CO_2, BF_3, ether (C_2H_5—O—C_2H_5)
Containing only C and H	Gasoline (mixture of C_7H_{16} [heptane], C_8H_{18} [octane], C_6H_6 [benzene], and $C_6H_5CH_3$ [toluene])
Oils, fats (higher molecular weight containing C, H, and O).	Waxes, lipids, cooking oils

(a) Is benzene (C_6H_6) miscible with water (H_2O)? _____

(b) A liquid compound made of just C and H is added to a liquid binary compound with strong electronegativity differences. Will the two liquids mix?

_____

(c) Is ethylene glycol ($C_2H_4(OH)_2$) miscible with water (H_2O)? _____

(Note: Ethylene glycol has the structure shown here. The O atoms are bonded to C atoms.)

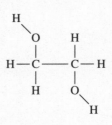

(d) Is the symmetrical compound CCl_4 miscible with an oil made up of carbon and hydrogen? _____

Answer:

(a) no (Benzene is a nonpolar compound because it is composed of only carbon and hydrogen. Water is polar since it is composed of H bonded to O.)

(b) no (Compounds made of just carbon and hydrogen are nonpolar. Binary compounds with strong electronegativity differences between the two atoms are polar.)

(c) yes (Both H_2O and $C_2H_4(OH)_2$ are polar molecules. A mixture of ethylene glycol and water is commonly found in automobile radiators to protect against freezing. These liquids must be miscible in order for the cooling system to be protected.)

(d) yes (Both liquids are nonpolar. The CCl_4 compound can be used to dissolve oil and grease stains in cloth, although it is now banned for cleaning purposes worldwide because of its effect on the ozone layer of the atmosphere.)

HEAT OF VAPORIZATION

31 Recall that the heat of fusion is the energy necessary to melt a solid completely to the liquid state (see Chapter 9). While a solid is melting at the melting point through addition of heat of fusion, its temperature does not change. At the boiling point of a liquid, a similar process occurs. In H_2O, the boiling point is 100°C at 1 atm pressure. When heated, the temperature of water increases until it reaches 100°C. After water reaches its boiling point, the temperature does not increase until all the liquid has vaporized to steam. The heat required to vaporize water at the boiling point is called the **heat of vaporization**. The following graph represents the heating of ice cubes until all of the H_2O is vaporized.

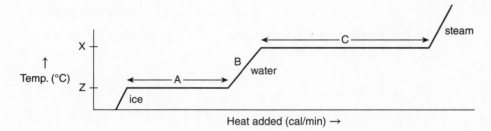

(a) What letter represents the heat of vaporization? _____

(b) What letter represents the boiling point? _____

(c) Does the water temperature change while the heat of vaporization is being applied? _____

Answer: (a) C; (b) X (remember the boiling point is a temperature); (c) no

32 The graph on the next page represents H_2O. At each of the indicated points, tell whether the H_2O is solid (ice), liquid (water), gaseous vapor (steam), a mixture of ice and water, or a mixture of water and steam.

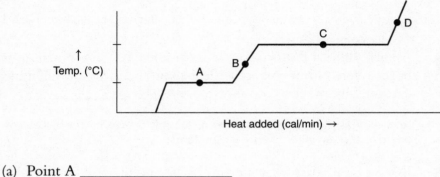

(a) Point A _____

(b) Point B _____

(c) Point C _____

(d) Point D _____

Answer: (a) ice and water (or solid and liquid); (b) water (or liquid); (c) water and steam (liquid and vapor); (d) steam (vapor)

33 The heat of vaporization of water is 540 cal/gram. It takes how many calories to convert 1 gram of liquid H_2O at 100°C to steam at 100°C? _____

Answer: 540 cal (This is simply a restatement of the definition of the heat of vaporization.)

34 The heat of vaporization for water is 540 cal for each gram of H_2O. Two grams of water would require twice as many calories.

Answer: $2 \text{ grams} \times \dfrac{540 \text{ cal}}{1 \text{ gram}} = 1080 \text{ cal}$

How many calories of heat would be needed to completely vaporize 10 grams of water at the boiling point (100°C)? _____

Answer: $10 \text{ grams} \times \dfrac{540 \text{ cal}}{1 \text{ gram}} = 5400 \text{ cal}$

35 The heat of vaporization is the energy needed to break the intermolecular attractive forces that keep a liquid in liquid form. It takes 540 cal of heat energy to break the intermolecular attractive forces and vaporize 1 gram of liquid H_2O at 100°C. It is important to remember that vaporization is a reversible process.

When H_2O condenses from steam at $100\,^{\circ}C$ to liquid at $100\,^{\circ}C$, it releases an amount of energy equal to the heat of vaporization.

When 1 gram of steam at $100\,^{\circ}C$ condenses to liquid H_2O at $100\,^{\circ}C$, it releases how many calories of heat? _____

Answer: 540 cal

36 A burn suffered from contact with steam at $100\,^{\circ}C$ is much more severe than a burn suffered from contact with the same quantity of boiling water. Can you explain why, in terms of vaporization? _____

Answer: This more severe burn from steam is caused by the release of 540 cal of heat energy per gram when the steam condenses.

37 Just as a liquid has a characteristic boiling point, each liquid also has a characteristic heat of vaporization. Certain gases that are readily condensed and have high heats of vaporization are useful in such refrigeration systems as air conditioners, refrigerators, or freezers. Refrigeration systems take advantage of the fact that a liquid which is vaporized at its boiling point requires heat. The resulting vapor (gas) releases heat when it condenses again at the boiling point. In a refrigeration system, a gas is compressed to the point where condensation takes place. The heat released by condensation is given off to the surroundings. The coils behind a refrigerator become warm to the touch because heat is being released after condensation. After the liquid releases its heat to the surroundings, it is piped inside the refrigerator and allowed to vaporize and expand very rapidly.

(a) Vaporization is a process that (requires, releases) _____ heat.

(b) Condensation is a process that (requires, releases) _____ heat.

Answer: (a) requires; (b) releases

38 Because the vaporization of the liquid inside the refrigerator piping is a process that requires heat, the heat inside the refrigerator is removed and the food and air inside the refrigerator are cooled. After the gas has removed heat from the inside of the refrigerator, the gas is piped outside and compressed again. The cycle of condensation and vaporization is continued. Refrigeration depends upon the requirement of heat for the process of _____ and the release of heat for the process of _____

Answer: vaporization; condensation

39 The condensation–vaporization cycle is important for refrigeration because heat is taken from one place (inside the refrigerator) through vaporization and released at another place (outside the refrigerator) through condensation. Condensation takes place because the gas is compressed through a mechanical compressor. The compressor forces the gas into less space so that the intermolecular attractive forces are strong enough to overcome the kinetic energy of the gas molecules. The molecules then coalesce and the gas is liquefied.

It is possible, however, for the kinetic energy of the molecules to be so great that the intermolecular attractive forces cannot overcome the kinetic energy and the gas will not liquefy no matter how great the pressure. If the coolant gas molecules in a refrigerator possessed so much kinetic energy that the inter-molecular attractive forces could not cause the gas to be liquefied no matter how great the pressure, could the refrigerator continue working through the vaporization–condensation cycle? _____ Explain. _____

Answer: no; The gas could not be liquefied under such conditions, and no condensation and therefore no vaporization could take place.

40 Remember that the temperature is a measure of the average kinetic energy of a gas. The temperature above which the intermolecular attractive forces can no longer overcome the kinetic energy in order to liquefy a gas is known as the **critical temperature**. At any temperature above the critical temperature, a gas will not liquefy no matter how much pressure is applied.

If you were selecting a coolant gas for a refrigerator, would you select one with a critical temperature that is well below room temperature or well above?

_____

Answer: well above room temperature (The refrigerator will no longer function at any temperature above critical temperature because condensation cannot take place at any temperature above critical temperature. The critical temperature should therefore be well above the normal temperature of the surroundings in which the refrigerator must operate.)

41 The critical temperature is different for each gas. For example, hydrogen has an extremely low critical temperature of −240°C. In order to liquefy hydrogen gas, the temperature must be (above, at, below) _____ −240°C.

Answer: at or below

SELF-TEST

This self-test is designed to show how well you have mastered this chapter's objectives. Correct answers and review instructions follow the test.

1. Consider two liquids in a chemistry lab, liquid #1 and liquid #2. Liquid #1 flows very quickly when poured out of its beaker while liquid #2 flows very

slowly. What conclusion can be made concerning the intermolecular attractive forces for the two liquids?

2. Water bugs walk on water, some razor blades can float on water, and water forms droplets. Are these examples of viscosity, surface tension, or vapor pressure? _____

3. The process of a liquid changing phases to the gaseous state is known as _____, while _____ describes the process of gaseous molecules changing phases to the liquid state.

4. What principle describes the shift in equilibrium of a chemical system to relieve stress?

5. If we increase the temperature of a liquid, what happens to its vapor pressure? _____

6. Below is a graph representing how the vapor pressure of a liquid changes as the temperature increases.

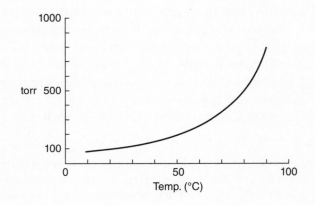

(a) What would be the approximate vapor pressure of the liquid at 50°C?

    _____

(b) If the pressure on the surface of the liquid is only 500 torr, at what approximate temperature would the liquid boil? _____

7. Consider the following two liquids, liquid A and liquid B. Liquid A and liquid B are pure liquids. Both liquids are isomers of one another. In other words the molecules that make up both liquids have the same molecular formula but different molecular structures. Liquid A's structure allows for hydrogen bonding while liquid B does not. Which liquid do you predict to have the higher boiling point and why?

8. Water has a normal boiling point of 100°C and ethyl alcohol (46 g/mol) has a normal boiling point of 78°C. Which has the lower intermolecular attractive forces? _____

9. Containers of three different liquids, A, B, and C, were opened to the air. All three had the same volume to begin with, but the contents of A disappeared very rapidly while the contents of C were the last to disappear. Which liquid has the lowest vapor pressure? _____

10. Suppose we have two liquids with similar molecular weights. The liquid that has the weaker intermolecular attractive forces could be expected to have:

    (a) (greater, lesser) _____viscosity.

    (b) (higher, lower) _____boiling point.

    (c) (greater, lesser) _____ surface tension.

11. Which of the following compounds would be miscible with water? C_3H_8, CH_3OH, HCl, NH_3, C_6H_6, $C_{12}H_{26}$.

12. Which would be more serious, a burn from water at 100°C or a burn from steam at 100°C? _____ Why? _____

13. Below is a graph representing a liquid being heated at a constant rate. What is happening at B and C? _____

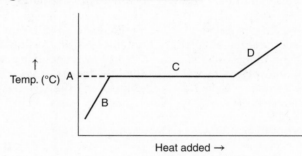

Heat added →

14. A liquid was vaporized in a closed container at a temperature of 200°C. The container was such that one end could be pushed in to increase the pressure on the vapor. While maintaining the temperature of the vapor at 200°C, the pressure on the vapor was increased to 1000 atm but none of the vapor condensed (liquefied). The 200°C temperature is most likely above the _____ temperature of the gas.

15. Neon gas has an extremely low critical temperature of -229°C. In order to liquefy neon gas, the temperature must be (above, at, below) -229°C.

ANSWERS

Compare your answers to the self-test with those given below. If you answer all questions correctly, you are ready to proceed to the next chapter. If you miss any, review the frames indicated in parentheses following the answers. If you miss several questions, you should probably reread the chapter carefully.

1. The more intermolecular forces the molecules of liquid possess, the slower the flow rate. Conversely, the less intermolecular forces molecules have the faster the flow rate. Liquid #1 must have few intermolecular attractive forces than those found in liquid #2. This would explain why liquid #1 flows faster than liquid #2. [frames 3-5]

2. surface tension (frames 6, 7)

3. evaporation, condensation [frames 9, 12]

4. Le Chatelier's Principle [frame 14]

5. Vapor pressure will increase. (frames 16, 20)

6. (a) approximately 200 torr; (b) approximately 80°C (frame 20)

7. Liquid A is expected to have the higher boiling point since it has hydrogen bonding interactions. [frames 18-24]

8. alcohol (It has the lower boiling point.) (frame 23)

9. C. The containers held three liquids with different boiling points, vapor pressures, and degree of volatility. Container A held the liquid that was most volatile and had the highest vapor pressure and lowest boiling point. Container C held the liquid that was least volatile and had the lowest vapor pressure and highest boiling point. Container B held a liquid with properties somewhere between those of A and C. (frame 25)

10. (a) lesser, (b) lower, (c) lesser [frames 3-22]

11. CH_3OH, HCl, NH_3 [frames 27-30]

12. steam; Steam has 540 cal more heat per gram at 100°C (heat of vaporization) than water does at 100°C. (frames 31, 36)

13. At B the temperature of the liquid is increasing as heat is being added. At C the liquid is boiling while the temperature remains constant. The flat portion represents heat of vaporization. (frames 31, 32)

14. critical (frames 40, 41)

15. at or below [frame 41]

THE LIQUID STATE

Liquids are one of the three common states of matter. Molecules in a liquid are held loosely together, and not nearly as closely as those found in solids. In fact, molecules in the liquid state are able to have reasonable contact with one another but the distances between each molecule are farther apart than those found in solids. However, the freedom of their movement allows them to take the shape of the container they fill.

Although there is more distance between the molecules in the liquid state that distance does not even come close to that found between gas particles and molecules. In the gaseous state, the third state of matter, particles are often several atoms or molecules away from one another.

Liquids are very important for life processes. For instance, our cellular fluid, gastric juices, and saliva all work within a liquid medium. The liquid that makes up these biological solutions is water, and water is essential for all life. Without water, life as we know it would not exist. Water is a very special liquid.

One of the amazing properties of water is that it is a liquid on Earth. This is incredible since water has a molecular mass of 18.02 g/mol. Why is this so important? Smaller molecules have fewer dispersion forces than larger ones. Dispersion forces are one of several intermolecular forces that help attract molecules to one another. The larger the molecule (i.e., the more dispersion forces it has) the more likely it will be a liquid or solid at room temperature.

Water is a very small molecule yet it can exist as a liquid on our planet where the atmospheric pressure at sea level is one atmosphere (1 atm). Even more remarkable is the wide range of temperatures on our planet. Earth can be rather hot, especially around the equator. Having liquid water is even more remarkable when considering how hot it can get here! Let's look into this further to help understand why having water on Earth is so amazing.

Many compounds that have a greater molar mass (i.e., more dispersion forces) are actually gases at 1 atm of pressure at room temperature. For instance, carbon monoxide and carbon dioxide are both gases at 1 atm. Carbon monoxide has a molar mass of 28.01 g/mol whereas the molar mass of carbon dioxide is 44.01 g/mol.

Other compounds that exist as gases at 1 atm and room temperature include chlorine (Cl_2, molar mass 70.90 g/mol), propane (C_3H_8, molar mass 44.10 g/mol), ozone (O_3, molar mass 48.00 g/mol), and phosgene ($COCl_2$, molar mass 98.91 g/mol). As you can see, especially in phosgene's case, water's molar mass is significantly less, yet it is a liquid and not a gas at room temperature and 1 atm.

What makes water a liquid under our normal conditions? The answer is that it has an important additional intermolecular force known as hydrogen bonding. The oxygen atom is electronegative and it is single bonded to two hydrogen atoms. The oxygen atom of one water molecule is attracted to the hydrogen atom of another water molecule.

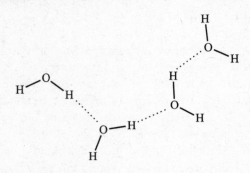

The number of hydrogen bonds to water varies with temperature. It can be as high as four hydrogen bonds per water molecule or a low as one. At lower temperatures water experiences more hydrogen bonding per molecule, while at higher temperatures there is a decrease in the number of hydrogen bonding interactions.

Hydrogen bonding is a rather weak intermolecular force by itself, but when the number of hydrogen bonding interactions adds up over enough water molecules the effects are significant. In fact, the sum of hydrogen bonding interactions is the main reason why we can have liquid water on our planet! Instead of boiling at temperatures lower than $0°C$, water has a boiling point of $100°C$, all thanks to hydrogen bonding interactions!

The unique nature of water does not end there. In a 2019 *Nature* publication researchers discovered *superionic water*, aka ice XVIII. Ice XVIII is a unique phase of water that exists at extremely high temperatures and pressures. In this phase water molecules are broken down into oxide ions (O^{2-}) that crystallize into an oxygen lattice where the hydrogen ions move around freely within it. As interesting as this sounds, these freely moving hydrogen ions make superionic water as conductive as metals!

11 Solutions and Their Properties

You have just been introduced to the three states of matter – solids, liquids, and gases – and their "observable" properties. Most substances that you encounter in your daily existence are mixtures of these states of matter. The mixtures may be solids in liquids, liquids in liquids, gases in liquids, gases in gases, or any combination of the three states. The most common mixtures are solids in liquids.

This chapter deals with the most common of all liquids, water, and the properties of mixtures of water with other states of matter, particularly solids. This chapter also is concerned with the ways of expressing the quantitative relationships between mixtures of states of matter.

OBJECTIVES

After you have completed this chapter, you will be able to

- recognize and apply or illustrate: solute, solvent, dilute, concentrated, molarity, molality, mole fraction, weight percent, solubility, hydrate, hydration, salvation, saturated, unsaturated, supersaturated, electrolyte, nonelectrolyte, hygroscopic, Raoult's Law, acid anhydride, and basic anhydride;

- given all but one of the following data, calculate the remaining unknown data: molecular weight of the solute, number of grams of the solute, number of grams of solvent (or density and volume), and molality of the solution;

- given all but one of the following data, calculate the remaining unknown data: the molality of a solution (or the data needed to find it), the molal freezing point depression constant, and the lowering of the freezing point of the solution (or the data needed to find it);

- given all but one of the following data, but with the molal boiling point elevation constant, calculate: the rise in the boiling point of the solution, the molality of a solution (or the data needed to find it), the molal freezing point depression constant, and the lowering of the freezing point of the solution (or the data needed to find it);

- given all but one of the following data, calculate the remaining unknown data: number of moles of solute, molarity of the solution, and volume of the solution;

- be able to write balanced chemical equations for the reaction of water with oxides of the alkali metals and oxides of the alkaline earth metals.

SALTS AND OXIDES

1 A solution is a **homogeneous mixture** of two or more substances. The solution looks the same throughout, as though it consists of only one substance. No different layers are visible in a true solution. Lemonade, tea, and soda pop are some common solutions made of water and other chemical substances.

A solution is composed of one or more solutes and one solvent. A **solute** is a substance that is dissolved and is usually the substance of lesser quantity in a solution. The **solvent** is the substance doing the dissolving and of greater quantity in a solution.

Suppose a teaspoon of table salt is dissolved in a glass of water.

(a) Which is the solute, water or salt? _____

(b) Which is the solvent? _____

(c) From your experience, is the resulting mixture a solution? _____

Answer: (a) salt; (b) water; (c) yes (it appears the same throughout and has no visible layers)

2 In a sugar and water solution, sugar is the _____ and water is the _____.

Answer: solute; solvent

3 Various types of solutions can include gases dissolved in liquids, liquids in liquids, or solids in liquids. A salt-water solution is an example of a solution involving a solid dissolved in a liquid.

In a fish aquarium, an electric pump is often used to bubble air in the aquarium water. Some of the oxygen remains dissolved in the water. This solution is an example where a _____ is the solute and a _____ is the solvent.

Answer: gas; liquid (The solution consists of a gas dissolved in a liquid.)

4 The most common solvent is water. The most common solutes dissolved in water are the soluble solids encountered in Chapter 6.

A soluble ionic solid in aqueous solution is called an **electrolyte** because such a solution conducts electricity. In aqueous (water) solution, the ions of an ionic compound **dissociate** and are free to move about to conduct electricity. A nonelectrolyte may also dissolve, but the solute remains in molecular form and will not conduct electricity.

(a) An aqueous solution of NaCl conducts electricity. Is the solute NaCl an electrolyte or nonelectrolyte? _____

(b) A solute made up of dissociated positive and negative ions is a(n) (electrolyte, nonelectrolyte) _____

(c) Does a solute that remains in molecular form conduct electricity? _____

Answer: (a) electrolyte; (b) electrolyte; (c) no

5 Instead of a solute just dissolving in water, it may react with water to form a different compound. Oxides of alkali metals (Na_2O, K_2O, Li_2O) and of alkaline earth metals (MgO, CaO, BaO) react with water to form metal hydroxide solutions. For example:

$$Na_2O + H_2O \rightarrow 2NaOH$$

$$CaO + H_2O \rightarrow Ca(OH)_2$$

Sodium hydroxide, NaOH, ionizes to yield Na^+ and OH^-. Calcium hydroxide, $Ca(OH)_2$, ionizes to yield Ca^{2+} and $2OH^-$.

A compound that ionizes to form OH^- in aqueous solution is called a **base**. After the metal oxide has reacted with the water, there is actually a new solute in the solution, the metal hydroxide. In the equations above, what compounds are the solutes once the reaction has occurred? _____

Answer: NaOH (or Na^+ and OH^-) and $Ca(OH)_2$ (or Ca^{2+} and $2OH^-$)

6 The general equation for the reaction of an alkali metal oxide with water is as follows. (Substitute any alkali metal for "M.")

$$M_2O + H_2O \rightarrow 2MOH$$

(MOH is a base since it dissociates to form OH^- ions.)

When K_2O is mixed with an excess of water, the resulting aqueous solution contains what base as the solute? _____

Answer: KOH (K^+, OH^-)

7 The general equation for the reaction of any alkaline earth metal oxide with water is as follows. (Substitute any alkaline earth metal for "M.")

$$MO + H_2O \rightarrow M(OH)_2$$

When MgO is mixed in an excess of water, the resulting aqueous solution contains _____ as the solute.

Answer: $Mg(OH)_2$ (Mg^{2+}, $2OH^-$). Note that the oxidation number of the metal ion does not change in going from metal oxide to metal hydroxide.

8 Many nonmetal oxides react with water to form acids that are ionized in aqueous solutions, yielding H^+ ions and a negative ion. A new solute is the result.

$$SO_2 + H_2O \rightarrow H_2SO_3$$
$$H_2SO_3 \rightarrow 2H^+ + SO_3^{2-}$$

In the above reaction the resulting solution contains _____ as the solute.

Answer: H_2SO_3 ($2H^+$, SO_3^{2-})

9 The oxides of a number of nonmetals such as carbon, sulfur, nitrogen, and phosphorus react with water to form **acids**. When the nonmetal oxides are mixed in water, the resulting solute is an acid because of the reaction of the oxide with the water.

$$N_2O_5 + H_2O \rightarrow 2HNO_3$$

What is the new solute in this reaction? _____

Answer: HNO_3 (H^+ and NO_3^- ions)

10 What is the resulting solute in the following mixture of carbon dioxide and water?

$$CO_2 + H_2O \rightarrow H_2CO_3$$

Answer: H_2CO_3 ($2H^+$, CO_3^{2-})

11 Nonmetal oxides (such as N_2O_5, SO_2, SO_3, P_2O_3, P_2O_5, CO_2, and so on) react with water to form acids. Note that the nonmetal does not change oxidation number in going from the oxide to the acid. The formulas of some of the acids are obtained simply by "adding" the atoms together. Here are some obvious examples.

$$SO_2 + H_2O \rightarrow H_2SO_3$$
$$CO_2 + H_2O \rightarrow H_2CO_3$$

Others are not as easy to see.

$$N_2O_5 + H_2O \rightarrow 2HNO_3$$

We must rely upon the formulas and names of the common acids you learned in Chapter 5.

What acid is formed when SO_3 reacts with H_2O? _____

Answer: sulfuric acid, H_2SO_4 ($H_2O + SO_3 \rightarrow H_2SO_4$)

 An unexpected product is sometimes formed when an automobile is equipped with a catalytic converter. The catalytic converter allows sulfur dioxide and sulfur trioxide and water vapor to mix together and react. The reaction of water and sulfur trioxide produces a pollutant, which is then sprayed out of the automobile exhaust system. Identify the pollutant. _____

Answer: H_2SO_4 (sulfuric acid)

13 Nonmetal oxides are known as **acid anhydrides** (acids without water) because they form acids when added to water. Metal oxides are known as **basic anhydrides** (bases without water) because they form bases when added to water.

Identify the following as either acid anhydrides or basic anhydrides.

(a) MgO _____

(b) P_2O_5 _____

(c) N_2O_3 _____

(d) BaO _____

Answer: (a) basic anhydride; (b) acid anhydride; (c) acid anhydride; (d) basic anhydride

 Acid anhydride $+ H_2O \rightarrow$ _____. Basic anhydride $+ H_2O \rightarrow$ _____

Answer: acid; base

HYDRATES AND WATER OF HYDRATION

15 Certain compounds actually contain water within the structure of their crystal lattices. Those compounds are called **hydrates**. The water they contain is known as the **water of hydration**. Probably the best known hydrate is copper sulfate pentahydrate, which has a characteristic deep blue color and is symbolized as $CuSO_4 \cdot 5H_2O$. The dot between $CuSO_4$ and $5H_2O$ means that water is actually a part of the hydrate molecule, and the formula weight of the hydrate must include the water.

A hydrate is a compound containing what as a part of its crystal lattice? _____

Answer: water (H_2O)

16 Determine the formula weight in grams (to the nearest tenth) of copper sulfate pentahydrate. (You might want to make a table like the ones we used in Chapter 4 to determine formula weights. If so, use a separate sheet of paper to set up the table and calculate.)

Formula weight of $CuSO_4 \cdot 5H_2O$ = _____

Answer:

Element	Number of atoms	Gram atomic weight	Atoms × weight
Cu	1 Cu	63.5	$1 \times 63.5 = 63.5$
S	1 S	32.1	$1 \times 32.1 = 32.1$
O	9 O	16.0	$9 \times 16.0 = 144.0$
H	10 H	1.0	$10 \times 1.0 = \underline{10.0}$
			Formula weight is 249.6 grams

17 It is possible to determine the percentage that the water contributes to the total weight of a hydrate. Just divide the weight of the water molecules by the entire formula weight of the hydrate (from the previous frame) and multiply the result by 100. For copper sulfate pentahydrate, divide the weight of 5 mol of water by the formula weight of the entire hydrate and multiply by 100.

$$\frac{\text{weight of } 5H_2O}{\text{formula weight of } CuSO_4 \cdot 5H_2O} \times 100 = \frac{\text{percentage of weight}}{\text{contributed by water}}$$

The percentage of weight contributed by water in $CuSO4 \cdot 5H_2O$ is _____ (Round this and all other answers in this chapter to the nearest tenth unless otherwise indicated.)

Answer: The weight of $5H_2O$ is:

Element	Number of atoms	Gram atomic weight	Atoms × weight
H	10 H	1.0	$10 \times 1.0 = 10.0$
O	5 O	16.0	$5 \times 16.0 = \underline{80.0}$
			Formula weight is 90.0 grams

$$\text{percentage weight of water} = \frac{\text{weight of } 5H_2O}{\text{formula weight of } CuSO_4 \cdot 5H_2O} = \frac{90.0\,g}{249.6\,g} \times 100 = 36.1\%$$

18 Another hydrate is magnesium sulfate heptahydrate, more commonly known as Epsom salts. Determine the formula weight in grams of this hydrate: $MgSO_4 \cdot 7H_2O$. _____

Answer: 246.4 grams

19 The percentage by weight of water in Epsom salts ($MgSO_4 \cdot 7H_2O$) is _____%. (The weight of 7 mol of H_2O is $7 \times 18.0 = 126.0$ grams).

Answer: % $H_2O = \dfrac{126.0\,g}{246.4\,g} \times 100 = 51.1\%$

20 When heated sufficiently, a hydrate will release all of its water molecules. Copper sulfate pentahydrate ($CuSO_4 \cdot 5H_2O$) is a deep blue color. Upon heating (symbolized by Δ), it loses its water molecules and becomes white.

$$CuSO_4 \cdot 5H_2O \quad \xrightarrow{\Delta} \quad CuSO_4 + 5H_2O$$

(blue hydrate)   (white anhydrous salt)

Upon heating, 1 mol of CuSO4 · $5H_2O$ yields 1 mol of CuSO4 · How many mole(s) of H_2O result? _____

Answer: 5

21 The white $CuSO_4$ is the anhydrous (waterless) salt of the $CuSO_4 \cdot 5H_2O$ hydrate. In the presence of water, the white anhydrous $CuSO_4$ absorbs water into its crystal lattice and returns to the blue hydrate ($CuSO_4 \cdot 5H_2O$). Every mole of $CuSO_4$ absorbs 5 mol of H_2O to form 1 mol of $CuSO_4 \cdot 5H_2O$ hydrate. One mole of water weighs 18.0 grams. One mole of $CuSO_4$ absorbs how many grams of H_2O to form 1 mol of $CuSO_4 \cdot 5H_2O$ hydrate? _____

Answer: 90.0 (5×18.0)

22 Suppose you had 3 mol of $CuSO_4 \cdot 5H_2O$ and heated it. Three moles of $CuSO_4 \cdot 5H_2O$ hydrate produces how many moles of H_2O when heated? _____ How many grams of H_2O? _____

Answer: 15; 270.0 (15×18.0)

23 If 1 kilogram of $CuSO_4 \cdot 5H_2O$ hydrate is heated, how many grams of water are produced? (Hint: Convert the kilogram of $CuSO_4 \cdot 5H_2O$ to moles of $CuSO_4 \cdot 5H_2O$.)

Answer:

$$\text{mol of CuSO}_4 \cdot 5H_2O = 1000\,g \times \frac{1\,mol}{249.6\,g} = 4.0\,mol$$

$$4.0\,mol\ CuSO_4 \cdot 5H_2O\ \text{yield}\ (4.0 \times 5) = 20.0\,mol\ H_2O$$

$$\text{grams of }H_2O = 20.0\,mol \times \frac{18\,g}{1\,mol} = 360.0\,g$$

24 Suppose we know that Epsom salt is a hydrate, but we don't know the number of molecules of water that are part of the hydrate formula. In the following equation, a known weight of Epsom salt hydrate is heated. The resulting anhydrous $MgSO_4$ is weighed.

$$MgSO_4 \cdot \text{?}H_2O \xrightarrow{\Delta} MgSO_4 + \text{?}H_2O$$

Upon heating, the hydrate loses its water. To determine the weight of water lost, subtract the weight of _____ from the weight of _____

Answer: $MgSO_4$; $MgSO_4 \cdot \text{?}\,H_2O$

25 In our equation from frame 24, if the weight of H_2O is known and the weight of anhydrous $MgSO_4$ is known, both weights can be converted to moles. The resulting mole fractions can be converted to simple whole numbers to determine the proper formula. Suppose that 3.5 mol of H_2O and 0.5 mol of $MgSO_4$ were produced when $MgSO_4 \cdot \text{?}H_2O$ was heated. The appropriate formula for the hydrate would be $MgSO_4 \cdot$ ___H_2O. (Divide each mole result by the smallest mole result to convert both to simple whole numbers.)

Answer: $\left(\dfrac{0.5}{0.5}\right) = 1$ and $\left(\dfrac{3.5}{0.5}\right) = 7$

The formula is $MgSO_4 \cdot 7H_2O$.

26 The hydrate $BaCl_2 \cdot \text{?}H_2O$ is heated to give:

$$BaCl_2 \cdot \text{?}H_2O \xrightarrow{\Delta} BaCl_2 + \text{?}H_2O$$

A sample of the hydrate weighed 488.4 grams before heating. After heating, the anhydrous $BaCl_2$ weighed 416.4 grams. Determine:

(a) the weight of the water formed. _____

(b) the moles of water formed. _____

(c) the moles of $BaCl_2$ formed. _____

(d) the true hydrate formula. _____

Answer:

(a) weight of $H_2O = 488.4\,g - 416.4\,g = 72.0\,g\ H_2O$

(b) mol $H_2O = 72.0\,g \times \left(\dfrac{1\,mol}{18.0\,g}\right) = 4\,mol\ H_2O$

(c) mol $BaCl_2 = 416.4\,g \times \left(\dfrac{1\,mol}{208.2\,g}\right) = 2\,mol\ BaCl_2$

(d) $\left(\dfrac{mol\ H_2O}{mol\ BaCl_2}\right) = \dfrac{4}{2} = 2$

The formula is $BaCl_2 \cdot 2H_2O$.

27 What does the dot (·) between $BaCl_2$ and $2H_2O$ in the $BaCl_2 \cdot 2H_2O$ hydrate indicate? _____

Answer: that water is an integral part of the crystal lattice of the hydrate (See frame 15 for review.)

28 Some chemical substances have the ability to absorb moisture from the atmosphere. Such substances are said to be **hygroscopic**. Flour and table salt are hygroscopic. White anhydrous $CuSO_4$ absorbs water to form the blue hydrate CuSO4 · $5H_2O$. Because of this very distinct color change, solid anhydrous CuSO4 is used to detect the presence of water in other liquids. If the solid turns from white to blue when added to a liquid, water is present.

Some anhydrous salts absorb enough water to dissolve themselves in their own water of hydration. They are called **deliquescent**. Anhydrous $CaCl_2$ is such a substance and is often used on dusty dirt roads or race tracks to control the dust by keeping it moist.

Substances that absorb moisture from the atmosphere are _____.

Answer: hygroscopic (deliquescent if they absorb enough to dissolve in it)

KINDS OF SOLUTIONS

29 There is an upper limit to the amount of substance that a quantity of solvent can dissolve depending upon the type of solvent, the substance to be dissolved, and the temperature. A solution containing the maximum amount of solute at a given temperature is said to be **saturated**. Additional solute material will not dissolve in a saturated solution. An **unsaturated** solution contains less than the maximum amount of solute at a given temperature. Additional solute material will dissolve in an unsaturated solution.

Half a gram of NaCl, table salt, is added to an aqueous solution of NaCl. The added salt dissolves. What kind of solution is it, saturated or unsaturated? _____

Answer: unsaturated

30 A solution can also be **supersaturated,** that is, made to hold more dissolved solute than normally expected at a given temperature. This can occur when a saturated solution is slowly cooled. The cooler solution then holds more dissolved solute than normal. If an additional crystal of solute is added to a supersaturated solution, the excess dissolved solute will quickly crystallize and come out of solution. After the excess solute comes out of solution, the remaining solution is saturated at the new temperature.

A crystal of NaCl is added to an aqueous solution of NaCl. The crystal will not dissolve but has no effect on the solution. What kind of solution is it— unsaturated, saturated, or supersaturated? _____

Answer: saturated

31 A crystal of sodium thiosulfate is added to an aqueous sodium thiosulfate solution. Immediately, some of the solute crystallizes out of solution.

(a) What kind of solution was it before the crystal was added? _____

(b) What kind of solution is it after the solute crystallizes? _____

Answer: (a) supersaturated; (b) saturated

32 A solution with a relatively large amount of solute in a given amount of solvent is considered to be a **concentrated** solution. Concentrated is not a very precise term. It simply means that a relatively large proportion of the solution is made up of solute.

A solution with a relatively small amount of solute in a given amount of solvent is considered to be a **dilute** solution. Dilute means that a relatively small proportion of the solution is made up of solute.

Would a saturated solution of NaCl be considered concentrated or dilute? _____

Answer: concentrated (because it contains the maximum amount of NaCl)

33 A small amount of concentrated aqueous sodium chloride solution is added to a very large amount of solvent (water). The resulting aqueous solution is probably (concentrated, dilute) _____

Answer: dilute

From this point on in this chapter, we do a great deal of manipulation of mathematical formulas and continue to use dimensional analysis in our problem solving. We suggest you obtain assistance if you have difficulty.

PROPORTIONS IN SOLUTIONS

34 It is useful to know precisely how dilute or concentrated a solution is. There are several methods of showing the proportion of solute to the solvent. One method of showing precise proportions is to give percent by weight. For example, a 10% solution of NaCl in water indicates that 10% of the solution weight is attributed to NaCl and 90% of the solution weight is attributed to the solvent.

In 100 grams of solution that is 10% NaCl by weight, there are how many grams of NaCl? _____ How many grams of water? _____

Answer:

$$\text{grams of NaCl} = 10\% \text{ NaCl} \times \frac{100 \text{ grams of solution}}{100\%} = 10 \text{ g NaCl}$$

$$\text{grams of H}_2\text{O} = \text{grams of solution} - \text{grams of NaCl} = 100 - 10 = 90 \text{ g H}_2\text{O}$$

35 Our NaCl solution from frame 34 gives us the following:

$$\text{solution percent by weight} = \frac{\text{weight of solute}}{\text{weight of solution}} \times 100\%$$

$$10\% \text{ NaCl solution} = \frac{10 \text{ grams of NaCl}}{100 \text{ grams of solution}} \times 100\%$$

Determine the percent by weight of a solution of NaCl if there are 30 grams of NaCl dissolved in water to yield a total solution weight of 200 grams. Remember, in this case, the "part" is the solute weight and the "whole" is the total weight of the solution: percent = (part/whole) × 100

Answer: % NaCl = $\dfrac{30 \text{ g NaCl}}{200 \text{ g solution}} \times 100\% = 15\%$ NaCl

36 Twenty-five grams of NaOH (solute) are added to 175 grams of water (solvent). What is the % NaOH in the solution? _____ (Note: The total solution weight includes the weight of both solvent and solute.)

Answer: total solution weight = 25 grams of solute + 175 grams of solvent = 200 grams

$$\% \text{ NaOH} = \frac{25 \text{ g NaOH}}{200 \text{ g solution}} \times 100\% = 12.5\% \text{ NaOH}$$

37 In 100 grams of an aqueous solution that is 15% NaCl by weight, there are how many grams of NaCl? _____ How many grams of water? _____

Answer:

$$g\,NaCl = 15\%\,NaCl \times \frac{100\,g\,solution}{100\%\,solution} = 15\,g\,NaCl$$

$$g\,H_2O = g\,solution - g\,NaCl = 100 - 15 = 85\,g\,H_2O$$

38 Concentrated HCl is actually an aqueous solution. Concentrated HCl is 37.9% by weight HCl. How many grams of concentrated 37.9% HCl solution does it take to yield 10 grams of HCl (solute)? (The grams of solution is the unknown.)

Answer: $g\,solution = 10\,g\,HCl \times \dfrac{100\,g\,solution}{37.9\,g\,HCl} \times 26.4\,g\,solution$

39 Let's try another one. What weight of dilute nitric acid solution contains 20 grams of HNO_3? The solution is 19% HNO_3 by weight. The total weight of the solution is _____ grams.

Answer: $g\,solution = 20\,g\,HNO_3 \times \dfrac{100\,g\,solution}{19\,g\,HNO_3} = 105.3\,g\,solution$

40 It is also possible to determine the volume of the solution if both the weight and the density of the solution are known.

$$density = \frac{weight}{volume}$$

Weight is usually expressed in grams and volume in milliliters (mL). If a solution has a density of 1.5 g/mL, how much volume would 3 grams of the solution occupy? _____

Answer:

$$density = \frac{weight}{volume}$$

$$volume = \frac{weight}{density} = \frac{3\,g}{1.5\,g/mL} = 2\,mL$$

41 In frame 39, it was determined that a weight of solution of 105.3 grams contained 20 grams of HNO_3 in a 19% aqueous HNO_3 solution. The total solution weighs 105.3 grams. The density of 19% HNO_3 solution is 1.11 g/mL. The volume occupied by 105.3 g at 1.11 g/mL is _____ mL.

Answer: $volume = \dfrac{weight}{density} = 105.3\,g \times \dfrac{1\,mL}{1.11\,g} = 94.9\,mL$

42 A liter of milk weighs 946 grams. What is the density of the milk in g/mL (to the nearest thousandth)? _____

Answer: density $= \dfrac{946\,g}{liter} \times \dfrac{1\,liter}{1000\,mL} = 0.946\,g/mL$

43 Instead of using weight percentage of solute in the solution and converting to volume, it is possible to determine **volume percentage** of the solute in the solution. Volume percentage is most often used when dealing with a solution involving two liquids, such as alcohol and water. Finding the volume percentage of the solute involves the familiar percentage formula.

$$\% = \frac{part}{whole} \times 100\%$$

$$\text{solute percent by volume} = \frac{\text{solute volume}}{\text{total volume}} \times 100\%$$

A certain brand of homogenized milk contains 35 mL of butterfat in a total volume of 1 liter. What is the percentage by volume of butterfat? _____

Answer: solute percent by volume

$$= \frac{\text{solute volume}}{\text{total volume}} \times 100\% = \frac{35\,mL}{1000\,mL} \times 100\% = 3.5\% \text{ butterfat}$$

44 The drugstore sells a solution of 3% hydrogen peroxide (H_2O_2) by volume. In a 150 mL quantity of this solution, there are how many milliliters of H_2O_2? (Hint: The volume of the solute is the unknown.) _____

Answer: volume $H_2O_2 = 150\,mL$ solution $\times \dfrac{3\,mL\,H_2O_2}{100\,mL\,solution} = 4.5\,mL\,H_2O_2$

MOLARITY

45 Instead of the percentage by weight or volume of a solute within the solution, we may wish to deal with moles of solute within the solution. **Molarity** describes the moles of a solute within a total solution of 1 liter. A solution that is 0.3 molar means that 0.3 mol of solute are contained in 1 liter of solution. One liter of 0.5 molar NaCl solution contains how many moles of NaCl? _____

Answer: 0.5

46 Molarity (abbreviated by the italicized capital letter M) indicates the moles of solute contained within 1 liter of solution. If 3 mol of sugar were dissolved in enough water to make a total solution of 1 liter, the solution would be _____ M sugar.

Answer: 3

47 Again, molarity indicates moles of solute per liter of solution. The molarity specifies the concentration of solute within the solution. All of the 3 *M* sugar solution from frame 46 is a 3 *M* sugar solution, even the last few drops, because molarity is the concentration of the solute. If you took 1 liter of 3 *M* sugar solution and poured half of it (500 mL) into another container, what would be the molarity (concentration) of sugar solution in each container? _____

Answer: 3 M (The concentration of sugar solute remains the same—3 mol/liter.)

48 We could also say that each container had 1 ½ mol per half liter concentration. It is easier to use a standard such as moles per liter, however. An analogy is miles per hour as a designation for speed. If you were driving at a steady 50 mph for half an hour, you would expect to have driven 25 miles. You were also driving at a speed of 25 miles per half hour, but it is more convenient to use miles per hour as a standard for speed rather than miles per half hour. If you took 1 ½ mol of sugar and added enough water to make a total of 500 mL of solution, the concentration of sugar solute would be _____ *M*. (Remember molarity = moles per liter.)

Answer: $M = \dfrac{\text{mol}}{1 \text{ liter}} = \dfrac{1.5 \text{ mol}}{0.5 \text{ liter}} = 3 \text{ } M$

49 If 0.1 mol of NaCl was dissolved in water to make 100 mL of solution, the concentration of NaCl in the solution will be _____ *M*.

Answer: $M = \dfrac{\text{mol}}{\text{liter}} = \dfrac{0.1 \text{ mol}}{0.1 \text{ liter}} = 1 \text{ } M$

50 The weight of a solute can be changed to moles to determine the molarity of a solution.

$$\text{mol of solute} = \frac{\text{weight of solute}}{\text{formula weight}}$$

$$\text{molarity} = M = \frac{\text{mol of solute}}{\text{liters}}$$

The two formulas can be combined as follows:

$$M = \frac{\text{weight of solute} \times \left(\dfrac{1 \text{ mol solute}}{\text{formula weight of solute}} \right)}{\text{liters of solution}}$$

The formula weight for NaCl is 58.4 grams. (To review, see Chapter 4.) If 87.8 grams of NaCl solute is dissolved in enough water to make 3 liters of solution, what molarity would the resulting NaCl solution have?

$M =$ _____

$$\text{Answer: } M = \frac{87.8 \text{ g NaCl} \times \left(\dfrac{1 \text{ mol NaCl}}{58.4 \text{ g Nacl}} \right)}{3.0 \text{ liters solution}} = \frac{1.5 \text{ mol NaCl}}{3.0 \text{ liters solution}} = 0.5 \, M$$

51 The molarity of a 1 liter solution containing 37.5 grams of $Ba(MnO_4)_2$ is _____ M.

$$\text{Answer: } M = \frac{37.5 \text{ g Ba(MnO}_4)_2 \times \left(\dfrac{1 \text{ mol Ba(MnO}_4)_2}{375.1 \text{ g Ba(MnO}_4)_2} \right)}{1 \text{ liter solution}} = 0.1 \, M \, Ba(MnO_4)_2$$

52 The formula weight of NaCl is 58.4 g/mol. How many grams of NaCl are needed to make 2 liters of a 0.3 M NaCl solution? _____

$$\text{Answer: g NaCl} = 2 \text{ liters solution} \times \frac{0.3 \text{ mol NaCl}}{1 \text{ liter solution}} \times \frac{58.4 \text{ g NaCl}}{1 \text{ mol Nacl}} = 35.0 \text{ g NaCl}$$

53 In aqueous solution, a strong electrolyte such as NaCl dissociates completely (separates into Na^+ and Cl^- ions). A solution that is 0.3 M NaCl solution is actually made up of 0.3 M Na^+ ion and 0.3 M Cl^- ion.

A 0.2 M solution of HCl (which is a strong electrolyte) would be made up of _____ M H^+ ion and _____ M Cl^- ion.

Answer: 0.2; 0.2

54 H_2SO_4 is also a strong electrolyte that dissociates into two H^+ ions and one SO_4^{2-} ion.

$$H_2SO_4 \rightarrow 2H^+ + SO_4^{2-}$$

That is, one molecule of H_2SO_4 yields two H^+ ions and one SO_4^{2-} ion in aqueous solution. One mole of H_2SO_4 molecules would yield 2 mol of H^+ ions and 1 mol of SO_4^{2-} ions. Two moles of H_2SO_4 would yield 4 mol H^+ ions, and so on. So a 0.1 M solution of H_2SO_4 results in 0.2 M H^+ ions and 0.1 M SO_4^{2-} ions.

Assume that H_3PO_4 is a strong electrolyte and dissociates completely.

$$H_3PO_4 \rightarrow 3H^+ + PO_4^{3-}$$

A solution of H_3PO_4 that is $0.1\ M$ results in _____ $M\ H^+$ ions and _____ $M\ PO_4^{3-}$ ions.

Answer: 0.3; 0.1

55 A liter of $0.1\ M$ solution of H_3PO_4 contains how many times the concentration of hydrogen ions as does a liter of $0.1\ M$ solution of HCl? _____

$$H_3PO_4 \rightarrow 3H^+ + PO_4^{3-}$$

$$HCl \rightarrow H^+ + Cl^-$$

Answer: three times as much

56 A liter of $0.1\ M$ solution of NaOH, which is a strong electrolyte, produces a _____ M concentration of OH^- ions and a _____ M concentration of Na^+ ions.

Answer: 0.1; 0.1

57 What would be the OH^- ion concentration in a $0.1\ M\ Ba(OH)_2$ solution? Assume $Ba(OH)_2$ is a strong electrolyte. _____

Answer: 0.2 *M* OH⁻

We will use H^+ ion and OH^- ion molar concentrations later in Chapter 13, when we further discuss acids and bases.

COLLIGATIVE PROPERTIES OF SOLUTIONS AND MOLALITY

58 The physical properties of a solution are somewhat different from the properties of the pure solvent. The amount and kind of solute in the solution affects such properties as boiling point, freezing point, and vapor pressure. These properties that vary according to the ratio of the weights of the solute and solvent are known as **colligative properties**.

In general, as solute is added the resulting solution will have a lower freezing point, lower vapor pressure, and higher boiling point than the pure solvent.

What happens to the freezing point of the water in an automobile cooling system when radiator antifreeze (the solute) is added to water (the solvent)?

_____

Answer: The freezing point decreases or lowers (The typical automobile radiator antifreeze solution is made up of half water as the solvent and half ethylene glycol as the solute. The freezing point is typically about −37°Celsius. The boiling point of the solution is increased.)

59   When dealing with solutions for chemical reactions such as neutralization, it is practical to use molarity (M) as the standard for the concentration of the solute in the solution. Molarity indicates moles of solute within each liter of solution. When working with colligative properties, however, it is more useful to deal with a different standard of the concentration of solute in the solution. This standard is **molality**. Molality, abbreviated by the small letter m, indicates the moles of solute in each *kilogram* of solvent.

When dealing with physical properties such as the boiling point elevation of a solution, it is more practical to use moles of solute per (liter of solution, kilogram of solvent) _____

Answer: kilogram of solvent

60   Molarity is abbreviated by _____ and indicates moles of solute per _____ of _____. Molality is abbreviated by _____ and indicates moles of solute per _____ of _____

Answer: M; liter; solution; m; kilogram; solvent

61   Here is the equation for determining molality.

$$m = \frac{\text{mol of solute}}{\text{kilogram solvent}}$$

Three moles of sugar are added to 1000 grams of H_2O solvent. The molality of the resulting solution is _____ m. (**Hint:** Change weight of solvent to kilograms, abbreviated kg.)

Answer: $m = \dfrac{3 \text{ mol of solute}}{1 \text{ kg solvent}} = 3\,m$

62   Two moles of sugar are added to 500 grams of H_2O solvent. The molality of the resulting solution is _____ m.

Answer: $m = \dfrac{2 \text{ mol of solute}}{0.5 \text{ kg solvent}} = 4.0\,m$

63   Molality can also be determined from the number of grams of solute by substituting the formula for moles in the formula for molality.

$$\text{mol} = \frac{\text{weight (grams)}}{\text{formula weight}}$$

$$m = \frac{\text{mol of solute}}{\text{kilogram solvent}}$$

The new formula becomes:

$$m = \frac{\text{weight of solute in grams} \times \left(\dfrac{1 \text{ mol solute}}{\text{formula weight of solute}} \right)}{\text{kilograms of solvent}}$$

Let's use this formula to determine the molality of a glucose solution. Fifty grams of glucose ($C_6H_{12}O_6$) are added to 0.250 kg of water. Assume the formula weight of glucose ($C_6H_{12}O_6$) is 180 grams. The molality of this glucose solution is _____ m.

$$\text{Answer: } m = \frac{50 \text{ g of glucose} \times \left(\dfrac{1 \text{ mol}}{180 \text{ g}} \right)}{0.250 \text{ kg of solvent}} = 1.1 \, m$$

64 Determine the molality of a solution involving 22.5 grams of glucose dissolved in 500 grams of water. The formula weight of glucose is 180 grams. The molality of this glucose solution (to the nearest hundredth) is_____ m. (Note: Be sure to change grams of solvent to kilograms.)

$$\text{Answer: } m = \frac{22.5 \text{ g glucose} \times \left(\dfrac{1 \text{ mol glucose}}{180 \text{ g glucose}} \right)}{0.500 \text{ kg solvent}} = 0.25 \, m$$

65 Determine the molality of a solution involving 116.8 grams of NaCl dissolved in 500 grams of water. The molality of this NaCl solution is m.

$$\text{Answer: } m = \frac{116.8 \text{ g NaCl} \times \left(\dfrac{1 \text{ mol NaCl}}{58.4 \text{ g NaCl}} \right)}{0.500 \text{ kg solvent}} = 4.0 \, m$$

66 How many grams of solute are necessary to produce a 2.50 m solution of $CaCl_2$ dissolved in 500 grams of water? The formula weight of $CaCl_2$ is 111 grams.

$$\text{Answer: } \text{g } CaCl_2 = 0.500 \text{ kg solvent} \times \frac{2.50 \text{ mol } CaCl_2}{1 \text{ kg solvent}} \times \frac{111 \text{ g } CaCl_2}{1 \text{ mol } CaCl_2} = 138.8 \text{ g } CaCl_2$$

67 In dealing with liquid solutes in liquid solvents, the amount of solute and sometimes the solvent is given in terms of volumes such as liters or milliliters. Volumes must be converted to weights in order to determine the molality of the resulting solution. If we are given the volume of a liquid, we also need the density of the liquid in order to determine its weight. Density is usually expressed as grams per cubic centimeter or grams per milliliter. To find the weight in grams of a liquid, multiply the density times the volume (density × volume = weight). The density of ethyl alcohol is found to be $0.80\,g/mL$. Determine the weight in grams of $23.75\,mL$ of ethyl alcohol. _____

Answer: weight of alcohol = $0.80\,g/mL \times 23.75\,mL = 19.0\,grams$ of alcohol

68 Determine the weight in grams of $300\,mL$ of benzene (C_6H_6) if the density of benzene is $0.90\,g/mL$. _____

Answer: weight of benzene = $0.90\,g/mL \times 300\,mL = 270\,grams$ of benzene

69 A solution is made up of $23.75\,mL$ of ethyl alcohol solute added to $300\,mL$ of benzene solvent. In frames 67 and 68 you have just determined the weight of the ethyl alcohol to be $19.0\,grams$ and the weight of benzene to be $270\,grams$. The formula weight of C_2H_5OH (ethyl alcohol) is $46\,grams$. Determine the molality of this solution.

$$\text{Answer: } m = \frac{19\,\text{g alcohol} \times \left(\dfrac{1\,\text{mol alcohol}}{46\,\text{g alcohol}} \right)}{0.270\,\text{kg benzene}} = 1.5\,m$$

MOLE FRACTION AND VAPOR PRESSURE

70 Placing a solute in a solvent lowers the vapor pressure of the solvent. If both the solvent and the solute are liquids, the **vapor pressure** of the solution is due to both solute and solvent. The amount contributed to the vapor pressure by either solute or solvent depends upon the mole fraction of each within the solution.

$$\text{total mol in a solution} = \text{mol of solute} + \text{mol of solvent}$$

The **mole fraction** is simply the number of moles of one of the components of the solution (either solute or solvent) divided by the total number of moles in the solution.

$$\text{mol fraction of solute} = \frac{\text{mol of solute}}{\text{mol of solute} + \text{mol of solvent}}$$

$$\text{mol fraction of solvent} = \frac{\text{mol of solvent}}{\text{mol of solute} + \text{mol of solvent}}$$

A solution contains 3 mol of solute and 12 mol of solvent. What is the mole fraction of solute? _____?

Answer: $\dfrac{3}{3+12} = \dfrac{1}{5}$ or 0.2

A solution contains 3 mol of solute and 12 mol of solvent. What is the mole fraction of solute? _____

Answer: $\dfrac{12}{(3+12)} = \dfrac{4}{5}$ or 0.8

All mole fractions in a solution must add up to 1. A solution is made up of 23 grams of ethyl alcohol (C_2H_5OH) and 156 grams of benzene (C_6H_6). The formula weight of C_2H_5OH is 46 g/mol. The formula weight of C_6H_6 is 78 g/mol.

(a) What is the mole fraction of ethyl alcohol? _____

(b) What is the mole fraction of benzene? _____

(c) What is the sum of the mole fractions of ethyl alcohol and benzene?

_____

Answer:

(a) mol of $C_2H_5OH = \dfrac{23\,\text{grams}}{46\,\text{grams/mol}} = 0.5\,\text{mol}$

    mol of $C_6H_6 = \dfrac{156\,\text{grams}}{78\,\text{grams/mol}} = 2.0\,\text{mol}$

    mole fraction of $C_2H_5OH = \dfrac{0.5}{0.5+2.0} = 0.2$

(b) mole fraction of $C_6H_6 = \dfrac{2.0}{0.5+2.0} = 0.8$

(c) The combined mole fractions are equal to 1 (0.2 + 0.8 = 1).

A solution can be made of more than two components. To determine the mole fraction of one component in a solution made up of several components, just divide the moles of the component by the total moles in the solution.

    In a solution made up of A, B, and C, suppose you wish to know the mole fraction of A.

$$\text{mole fraction A} = \frac{\text{mol A}}{\text{mol A} + \text{mol B} + \text{mol C}}$$

In this same solution, what is the mole fraction of B? _____

Answer: mole fraction B $= \dfrac{\text{mol B}}{\text{mol A} + \text{mol B} + \text{mol C}}$

74 A solution is made up of 92 grams of ethyl alcohol (C_2H_5OH), 54 grams of H_2O, and 29 grams of acetone (CH_3COCH_3). The formula weight of C_2H_5OH is 46 grams. The formula weight of H_2O is 18 grams. The formula weight of CH_3COCH_3 is 58 grams. Determine the mole fraction of acetone in the solution (to the nearest hundredth). _____

Answer:

$$\text{mol of } C_2H_5OH = \frac{92 \text{ grams}}{46 \text{ grams/mol}} = 2 \text{ mol of } C_2H_5OH$$

$$\text{mol of } H_2O = \frac{54 \text{ grams}}{18 \text{ grams/mol}} = 3 \text{ mol of } H_2O$$

$$\text{mol of } CH_3COCH_3 = \frac{29 \text{ grams}}{58 \text{ grams/mol}} = 0.5 \text{ mol of } CH_3COCH_3$$

$$\text{mole fraction of } CH_3COCH_3 = \frac{0.5}{2 + 3 + 0.5} = 0.09$$

75 A solution made of two or more liquids has a vapor pressure that is determined by the mole fractions of the liquids that make up the solution. This general statement leads to **Raoult's Law**. Raoult's Law is expressed mathematically as:

$$P_A = X_A P_A^0$$

P_A is the vapor pressure contributed by liquid A in the solution. X_A is the mole fraction of liquid A in the solution. P_A^0 is the vapor pressure of liquid A in its pure form.

A solution is made up of toluene and benzene. Toluene is present in a mole fraction of 0.6 and has a vapor pressure of 70 torr in the pure form. We want to find the vapor pressure contributed by toluene in the solution.

$$P_{toluene} = X_{toluene} P_{toluene}^0$$

That is, the vapor pressure contributed by toluene equals the mole fraction of toluene multiplied by the vapor pressure of pure toluene. The vapor pressure contributed by toluene = _____ torr.

Answer:

$$P_{toluene} = X_{toluene} P_{toluene}^0$$
$$P_{toluene} = (0.6)(70 \text{ torr}) = 42 \text{ torr}$$

76 In the same solution of benzene and toluene, benzene also contributes to the vapor pressure of the solution.

$$P_{benzene} = X_{benzene} P_{benzene}^0$$

To determine the vapor pressure contributed by benzene, we must know the mole fraction of benzene in the solution and the vapor pressure of pure benzene. There are only two liquids in the solution. All mole fractions in a solution add up to equal 1. If toluene accounts for a mole fraction of 0.6, what is the benzene mole fraction? _____

Answer: 0.4 (1 – 0.6)

77 In the solution of toluene and benzene, the benzene has a mole fraction of 0.4 and the vapor pressure of benzene in pure form is equal to 190 torr. Use Raoult's Law to calculate the vapor pressure contributed by benzene in this solution. _____

Answer:

$$P_{benzene} = X_{benzene} P^0_{benzene}$$
$$P_{benzene} = (0.4)(190 \text{ torr}) = 76 \text{ torr}$$

78 In our solution made up of toluene and benzene, toluene contributes 42 torr to the vapor pressure of the solution and benzene contributes 76 torr to the vapor pressure of the solution. The total vapor pressure above the solution is the sum of all of the partial vapor pressures (from Dalton's Law of partial pressures, discussed in Chapter 8). The total vapor pressure of the solution is _____ torr.

Answer: 118 (42 + 76)

79 Raoult's Law predicts *ideal* vapor pressures. The actual vapor pressure may be slightly more or less than predicted. Another toluene and benzene solution is prepared with a mole fraction of 0.7 for toluene. Assume the vapor pressure of pure toluene liquid to be 70 torr and the vapor pressure of pure benzene liquid to be 190 torr.

(a) What is the ideal vapor pressure contributed by toluene in the solution?

_____

(b) What is the total ideal vapor pressure of this solution? _____

Answer:

(a) $P_{toluene} = X_{toluene} P^0_{toluene}$

$P_{toluene} = (0.7)(70 \text{ torr}) = 49 \text{ torr}$

(b) The mole fractions of all components in the solution add up to 1. Since toluene accounts for 0.7 mol fraction, the remainder must be attributed to benzene.

$$1 - 0.7 = 0.3$$

$$P_{benzene} = X_{benzene} P^0_{benzene}$$

$$P_{benzene} = (0.3)(190 \, torr) = 57 \, torr$$

The total vapor pressure is the sum of all partial vapor pressures in the solution. The total vapor pressure is 49 torr + 57 torr = 106 torr.

80 Raoult's Law also applies to solutions in which the solute is a solid with no vapor pressure. If solid sugar is dissolved in water, the vapor pressure of the water is lowered according to its mole fraction as predicted by Raoult's Law. The solid sugar had almost no measurable vapor pressure, so the entire vapor pressure of the sugar–water solution is contributed by the water. The vapor pressure of pure water is 23.8 torr at 25°C. Sugar is added to the water until the water makes up a mole fraction of 0.75 in the solution.

(a) What is the vapor pressure of the water in the solution? _____

(b) Since the sugar has no measurable vapor pressure, what is the total vapor pressure of the solution? _____

Answer:

(a) $P_{water} = X_{water} P^0_{water}$

   $P_{water} = (0.75)(23.8 \, torr) = 17.9 \, torr$

(b) The total vapor pressure of the solution is the same as the vapor pressure contributed by the water since the sugar has no vapor pressure. Total vapor pressure = 17.9 torr.

BOILING POINT ELEVATION

81 The vapor pressure of a solvent (such as water) is reduced when a solute with no vapor pressure of its own is added to the solvent. Remember that the boiling point of a liquid is the temperature at which the vapor pressure equals the confining pressure. If a nonvolatile solute (one with no vapor pressure of its own) is added to a solvent, the boiling point of the solvent is raised because the vapor pressure is reduced. The more solute added, the more the vapor pressure is reduced, and the more the boiling point is increased.

Adding 1 mol of a nonvolatile solute that is also a nonelectrolyte to 1 kg of water raises the boiling point by 0.51°C. If the boiling point of pure water is 100.00°C, then adding 1 mol of nonvolatile and nonelectrolytic solute to 1 kg of water raises the boiling point of the solution to the new temperature of _____°C.

Answer: 100.51 (100.00°+0.51°C)

82 The presence of 1 mol of any nonvolatile nonelectrolyte such as sugar added to 1 kg of water raises the boiling point of the water by 0.51°C. This 0.51°C is known as the **molal boiling point constant** for water.

A sugar-water solution that is 1 *m* boils at 100.51°C at standard pressure. A sugar-water solution of 1 *m* raises the boiling point of water by 0.51°C. A sugar-water solution of 3 *m* raises the boiling point by three times the molal boiling point constant (3 × 0.51 = 1.53°C) and raises the boiling point of the solution to what new temperature?_____ °C

Answer: 101.53°C [(3 × 0.51) + 100 = 101.53]

83 The molal boiling point constant is usually abbreviated as K_b. The K_b for water is 0.51°*C/m*. (The K_b varies for other solvents.) A useful formula for predicting the boiling point of solutions is:

$$\Delta T_b = K_b \times m$$

The increase in the boiling point (ΔT_b) equals the molal boiling point constant (K_b) multiplied by the molality (*m*) of the solution. For example, if the molality of a sugar-water solution is 4 *m*, then the increase in boiling point (ΔT_b) equals 0.51°C multiplied by 4, or 2.04°C. Calculate the increase in the boiling point if a sugar-water solution has a molality of 5 *m*.

$\Delta T_b =$ _____

Answer: $K_b \times m = 0.51°C \times 5 = 2.55°C$

84 A sugar-water solution of 7.0 *m* would increase the boiling point by _____°C.

Answer: 7.0 × 0.51°C = 3.57°C

85 We can also modify the formula to determine the molality of a solution, by solving for *m*.

$$m = \frac{\Delta Tb}{K_b}$$

What is the molality of a sugar-water solution that increases the boiling point by 0.255°C? _____

Answer: $m = \dfrac{\Delta T_b}{K_b} = \dfrac{0.255°C}{0.51°C/m} = 0.5\,m$

FREEZING POINT DEPRESSION

 86   A nonvolatile solute not only raises the boiling point of the solution, but it also lowers the freezing point of the solution. Each mole of any nonvolatile nonelectrolyte, such as sugar, in 1 kg of water will lower the freezing point by 1.86°C. Since the normal freezing point of water is 0°C, the new freezing point of a solution of 1 mol of sugar added to 1 kg of water will be _____°C.

Answer: −1.86

87   The molal freezing point constant for water is −1.86°C/m. The molal freezing point constant is abbreviated as K_f and varies for different solvents as does K_b. The formula for determining the freezing point decrease is:

$$\Delta T_f = K_f \times m$$

Determine the value of ΔT_f (the amount of decrease in the freezing point) of a 3 m sugar-water solution. $\Delta T_f =$ _____

Answer: −1.86°C × 3 = −5.58°C

88   The freezing point change for a 2.5 m solution of sugar-water would be _____°C.

Answer: −4.65 (−1.86°C × 2.5 = −4.65°C)

89   As with the formula for boiling point increase, the freezing point decrease formula can also be used to determine the molality of a solution.

$$m = \frac{\Delta T_f}{K_f}$$

The molality of a sugar-water solution that freezes at −3.50°C (to the nearest hundredth) is _____

Answer: The change in freezing point is −3.50°C.

$$m = \frac{\Delta T_f}{K_f} = \frac{-3.50°C}{-1.86°C/m} = 1.88\,m$$

90 The sugar solution has been used in the previous examples because it is a nonelectrolyte and the formulas are valid only for such substances. The formulas involving the molal freezing and boiling point constants are valid for which of the following?

_____ (a) NaCl, a strong electrolyte
_____ (b) sugar, a nonelectrolyte
_____ (c) acetic acid, a weak electrolyte

Answer: only (b)

91 Every mole of a nonelectrolyte will lower the freezing point and raise the boiling point by an amount equal to K_f and K_b. Each mole has a specific number of particles, so we can say that the greater the number of solute particles in a solvent the greater will be the freezing point decrease and boiling point increase. A strong electrolyte such as 1 mol of NaCl in a kilogram of water is actually a solution of 1 mol of Na^+ ions and 1 mol of Cl^- ions, or 2 mol of particles. So a solution of 1 m NaCl would have the same effect as a 2 m solution of sugar. What would be the change in freezing point of a 1 m NaCl solution? _____

Answer: $-1.86°C \times 2 = -3.72°C$

92 K_2SO_4 is a strong electrolyte.

$$K_2SO_4(s) \rightarrow 2\,K^+(aq) + SO_4{}^{2-}(aq)$$

Calculate the change in the boiling point of a 1 m K_2SO_4 solution. _____

Answer: $0.51°C \times 3 = 1.53°C$

93 A 1 m aqueous solution of carbonic acid, H_2CO_3, has lowered the freezing point $-3.00°C$. Is H_2CO_3 a strong electrolyte, weak electrolyte, or nonelectrolyte?

_____

Answer: weak electrolyte (If it were a nonelectrolyte the freezing point would be lowered $-1.86 \times 1 = -1.86°C$. If it were a strong electrolyte and dissociated as $H_2CO_3 \rightarrow H^+ + HCO_3{}^-$, the lowering would be $-1.86 \times 2 = -3.72°C$. It might also dissociate as $H_2CO_3 \rightarrow 2H^+ + CO_3{}^{2-}$, giving a lowering of $-1.86 \times 3 = -5.58°C$. The ΔT_f falls between the expected values for a nonelectrolyte and strong electrolyte and is, therefore, presumed to be a weak electrolyte.)

In reality, 1 m NaCl and 1 m K_2SO_4 do not affect the freezing point and boiling point changes as expected. They do not have two or three times the effect predicted by their 100% dissociation. They have somewhat less effect because the "activity" of the ions is not ideal, but that activity is beyond the scope of this book. In general, nonelectrolytes conform to the expected molal freezing and boiling point constants. Weak electrolytes decrease the freezing point and increase the boiling point only slightly beyond that of nonelectrolytes. Strong electrolytes (which dissociate completely) come close to having each ion act independently in increasing the boiling point and decreasing the freezing points.

You have just learned several things about solutions that chemists and industries use daily in the laboratory or in the factory to produce desired products or results. We could not operate without understanding the solute-solvent properties and relationships we have discussed. You will encounter the concentration terms in the remaining chapters of this book and most certainly in any other chemistry course you might take.

SELF-TEST

This self-test is designed to show how well you have mastered this chapter's objectives. Correct answers and review instructions follow the test. Round answers to the nearest hundredth unless otherwise indicated.

1. Complete and balance the following equations. Indicate if the product(s) of each equation is called an acid or a base.

   (a) $CO_2 + H_2O \rightarrow$ _____

   (b) $Na_2O + H_2O \rightarrow$ _____

2. Complete and balance the following equations. Indicate if the product(s) of each equation is called an acid or a base.

   (a) $Li_2O + H_2O \rightarrow$ _____

   (b) $P_2O_5 + 3H_2O \rightarrow$ _____

3. How many grams of H_2O could you obtain from 1 kg of $CuSO_4 \cdot 5H_2O$ (to the nearest gram)? _____

4. The blue hydrate $CuSO_4 \cdot ?H_2O$ is heated to give:

$$CuSO_4 \cdot ?H_2O \xrightarrow{\text{heat}} CuSO_4 + ?H_2O$$

   A sample of the hydrate weighed 499.36 g before heating. After heating, the anhydrous $CuSO_4$ weighed 319.22 g. Determine:

   (a) the weight of the water formed. _____

   (b) the moles of water formed. _____

(c) the moles of $CuSO_4$ formed. _____

(d) the true hydrate formula. _____

5. Ammonia gas is passed into water yielding a solution whose density is 0.93 g/mL and is 18.6% by weight of NH_3. What is the weight of NH_3 per milliliter of solution?

6. The grocery store sells a solution of vinegar that is 5.0% acetic acid ($HC_2H_3O_2$) by volume. In a 250-milliliter quantity of this solution, there are how many milliliters of $HC_2H_3O_2$? (Hint: The volume of the solute is the unknown.) _____

7. What is the molarity of a solution that contains 16.5 grams $(NH_4)_2SO_4$ in 100 mL of solution? _____

8. The formula weight of NaCl is 58.4 g/mol. How many grams of NaCl are needed to make 4 liters of 0.3 M NaCl solution?

9. Determine the molality of a solution involving 18.6 grams of glucose dissolved in 750 grams of water. The formula weight of glucose is 180 grams. The molality of this glucose solution (to the nearest hundredth) is _____ m.

10. Determine the molality of a solution involving 46.3 grams of NaCl dissolved in 845 grams of water. The formula weight of NaCl is 58.4 grams. The molality of this NaCl solution (to the nearest hundredth) is _____ m.

11. A solution contains 58 grams of acetone (CH_3COCH_3), 69 grams of ethyl alcohol (C_2H_5OH), and 63 grams of H_2O. What is the mole fraction of water in this mixture? _____

12. A solution contains 238.5 g of glucose ($C_6H_{12}O_6$) and 485.5 g of water. What is the mole fraction of glucose in this mixture?

13. What is the ideal vapor pressure contributed by benzene in a toluene-benzene solution prepared with a mole fraction of 0.8 for benzene? Assume pure toluene to be 0.092 atm and the vapor pressure of pure benzene to be 0.250 atm.

14. A solution containing 15 grams of an unknown substance dissolved in 100 grams of water has a boiling point of 100.255°C. What is the molecular weight of the unknown (to the nearest gram)? _____

15. A 1 m solution of HCN has a boiling point of 100.72°C. Is it a strong electrolyte, a weak electrolyte, or nonelectrolyte? Defend your answer. _____

ANSWERS

Compare your answers to the self-test with those given below. If you answer all questions correctly, you are ready to proceed to the next chapter. If you miss any, review the frames indicated in parentheses following the answers. If you miss several questions, you should probably reread the chapter carefully.

1. H_2CO_3 (acid); (b) 2NaOH (base) (frames 5–11)

2. (a) 2LiOH (base) (b) $2H_3PO_4$ (acid) [frames 5-11]

3. 1 kg $CuSO_4 \cdot 5H_2O$ = (1,000 g ÷ 249.6 g/mol) = 4.01 mol $CuSO_4 \cdot 5H_2O$

   1 mol of hydrate would yield 5 mol of H_2O.

   ∴ 4.01 mol of hydrate would yield (4.01) × (5) = 20.0 mol H_2O

   ∴ grams of H_2O obtained = (20.0 mol)(18 g/mol) = 360 grams (frames 16–23)

4. (a) wt. of H_2O = 499.36 − 319.22 g = 180.18 g H_2O

   (b) mol. H_2O = 180.18 g x (1 mol. H_2O/18.0 g H_2O) = 10 mol H_2O

   (c) mol. $CuSO_4$ = 319.22 g x (1 mol. $CuSO_4$ / 169.61 g = 2 mol $CuSO_4$

   (d) mol H_2O / mol $CuSO_4$ = 10/2 = 5

   The formula is $CuSO_4 \cdot 5H_2O$. [frames 24-28]

5. weight NH_3 = density × % by weight × volume = 0.93 g/mL × 0.186 × 1 mL = 0.17 grams (frames 34–41)

6. volume $HC_2H_3O_2$ = 250 mL solution x (5.0 mL $HC_2H_3O_2$ / 100 mL solution) = 12.5 mL $HC_2H_3O_2$ [frames 34-44]

7. $M = \dfrac{\text{mol}}{\text{liter}} = \dfrac{16.5\,g \times \left(\dfrac{1\,\text{mol}}{132\,g}\right)}{0.100\,\text{liter}} = 1.25\,M$

   (frames 45–51)

8. g NaCl = 4 liters of solution x (0.3 mol NaCl / 1 L solution) x (58.4 g NaCl / 1 mol NaCl) = 70.1 g NaCl [frames 45-52]

9. First obtain moles of glucose.

$$18.6 \text{ glucose} \times \frac{1 \text{mol glucose}}{180 \text{ g glucose}} = 0.103 \text{ moles glucose}$$

Next, solve for molality.

molality (m) = moles of solute / kg of solvent = 0.103 mol glucose / 0.750 kg = 0.137 m [frames 58-69]

10. First obtain moles of NaCl.

$$46.3 \text{ g NaCl} \times \frac{1 \text{mol NaCl}}{58.4 \text{ g NaCl}} = 0.793 \text{ moles NaCl}$$

Next, solve for molality.

molality (m) = moles of solute / kg of solvent = 0.793 mol NaCl / 0.845 kg = 0.938 m [frames 58-69]

11. mole fraction water $= \dfrac{n_1}{n_1 + n_2 + n_3}$

$$n_1 = \text{mol of water} = \frac{63 \text{ g}}{18 \text{ g/mol}} = 3.5 \text{ mol } H_2O$$

$$n_2 = \text{mol of acetone} = \frac{58 \text{ g}}{58 \text{ g/mol}} = 1.0 \text{ mol of acetone}$$

$$n_3 = \text{mol of ethyl alcohol} = \frac{69 \text{ g}}{46 \text{ g/mol}} = 1.5 \text{ mol of ethyl alcohol}$$

$$\therefore \text{mol fraction } H_2O = \frac{3.5}{1.0 + 1.5 + 3.5} = \frac{3.5}{6.0} = 0.58 \text{ mole fraction } H_2O$$

(frames 70–74)

12. mole fraction of glucose $= \dfrac{n1}{n_1 + n_2}$

$$n_1 = \text{mol of glucose} = \frac{238.5 \text{ g}}{180 \text{ g/mol}} = 1.3 \text{ mol } C_6H_{12}O_6$$

$$n_2 = \text{mol of } H_2O = \frac{485.5 \text{ g}}{18.0 \text{ g/ mol}} = 27 \text{ mol } H_2O$$

mole fraction glucose $= \dfrac{1.3}{1.3 + 27} = 4.6 \times 10^{-2}$ mole fraction glucose [frames 70 – 74]

13. $P_{\text{benzene}} = X_{\text{benzene}} P^{\circ}{}_{\text{benzene}} = (0.8)(0.250 \text{ atm}) = 0.200 \text{ atm}$ [frames75-79]

14. molecular weight solute $= \dfrac{(\text{weight of solute})(K_b)}{(\Delta T_b)(\text{kg solvent})}$

(Note : Assume substance is a nonelectrolyte.)

$$\therefore \text{molecular weight} = \frac{(15)(0.51)}{(0.255)(0.100)} = 300 \text{ g}$$

(frames 81 – 93)

15. If it is nonelectrolyte, 1 m HCN should boil at 100.51° C (100° C $+ \Delta T_b$).

$\Delta T_b = (1)(0.51) = 0.51°C$

∴boiling point $= 100°C + 0.51°C = 100.51°C$

If it is a strong electrolyte (HCN $\rightarrow$ H$^+$ + CN$^-$), then a 1 m solution should boil at 101.02°C.

$\Delta T_b = (2)(0.51) = 1.02°C$

∴boiling point $= 100°C + 1.02°C = 101.02°C$

Since the observed boiling point is between these two extremes, we would conclude that it is a weak electrolyte, only partially dissociated. (frames 81–93)

12 Chemical Equilibrium

You have just learned several properties of solutions (mixtures of solids, liquids, and gases). We have discussed reactions that go to completion (reactants totally consumed, leaving only new products) in Chapter 5 and electrolytes that dissociate completely in water in Chapter 11. Both of these concepts imply a one-way reaction, continuous movement toward the product side. However, in Chapter 10 we discussed a dynamic equilibrium where the rate of evaporation equals the rate of condensation, that is, the reactions are "reversible."

Many chemical reactions are reversible. The products formed react to give back the original reactants, even as the reactants are forming more products. After some time, both the forward and reverse reactions will be going on at the same rate. When this occurs, the reaction is said to have reached equilibrium. There is no further change in the *amount* of any reactant or product, though both reactions still go on (forever). Since there are many such reactions that appear to go only partway to completion, their study is of major importance to the chemist.

We will discuss several types of equilibrium in this chapter, along with their associated problems and concepts. You will use the concept of molarity you just learned in Chapter 11 to solve equilibrium problems.

OBJECTIVES

After completing this chapter, you will be able to

- recognize and apply or illustrate: equilibrium, equilibrium constant, equilibrant, equilibrium mixture, homogeneous equilibrium, heterogeneous equilibrium, Le Chatelier's Principle, common ion effect, ion product constant;
- define "reversible reaction" and write an example, and recognize at sight whether a chemical equation is written so as to indicate that the reaction is reversible;
- given a chemical equation, write the equilibrium constant expression;
- given the equilibrium constant expression, write the equation for the corresponding chemical reaction;
- given the equilibrium constant expression, state the units of K_{eq};

- predict whether changing the temperature of a system in equilibrium will shift the equilibrium to the right or to the left;

- state the effect on a system in chemical equilibrium of removing (or adding) a portion of one reactant (or product) and increasing (or decreasing) the total pressure on a system in equilibrium;

- given the balanced chemical equation for the ionization of a weak acid (or weak base), write the expression for the equilibrium constant;

- given the equilibrium constant expression and all the terms in the equation but one, solve for the unknown;

- given the data needed to calculate concentration terms for the equilibrium calculation, change the data into concentrations;

- given the degree of ionization (percentage ionization or fraction ionized) of a weak acid or weak base, calculate K_{eq}, or given K_{eq} and the necessary concentrations, calculate the degree of ionization;

- write the chemical equation and the ion product constant expression for the ionization of water.

1   In Chapter 10, you learned about Le Chatelier's Principle in connection with the evaporation and condensation of liquids. Le Chatelier's Principle involved stresses placed upon a **dynamic equilibrium**. In the case of evaporation and condensation, which are two opposite processes, a dynamic equilibrium occurs when the rate of evaporation and the rate of condensation are equal. Both processes continue, but the rates are equal at equilibrium.

$$\text{liquid evaporates} \rightarrow \text{vapor}$$

$$\text{liquid} \leftarrow \text{vapor condenses}$$

In a dynamic equilibrium, two opposite processes continue to occur, but the rates of the opposite processes must be equal. According to **Le Chatelier's Principle**, when a system in equilibrium is subjected to a stress, the system will change its conditions, relieving the stress and forming a new equilibrium.

In a covered jar containing a little water, the rates of evaporation and condensation will be equal after a time and equilibrium is achieved. If the covered jar is suddenly pressurized, the rate of condensation will suddenly be increased.

(a) What is the stress in this example? _____

(b) What will happen to the rate of evaporation to restore dynamic equilibrium?

Answer: (a) pressure (the sudden pressurization of the jar); (b) The rate of evaporation will decrease suddenly. After a time it will again equal the rate of condensation, thus reestablishing equilibrium.

2   The following two reactions are completely opposite processes.

$$CO_2 + H_2 \rightarrow H_2O + CO$$

$$CO + H_2O \rightarrow CO_2 + H_2$$

The reactants of the first reaction are the products in the second reaction and vice versa. Both reactions can be written as one reversible reaction.

$$CO_2 + H_2 \rightleftharpoons H_2O + CO$$

In this reversible reaction, what symbol(s) are used to indicate that the reaction is reversible? _____

Answer: $\rightleftharpoons$ (The two arrows pointing in opposite directions indicate that the reaction is reversible.)

3   Up to this point we have treated all reactions as if they went to completion and continued until one or more of the reactants was consumed. At that point the reaction stopped. However, the reversible reaction in frame 2 proceeds in one direction or the other until an equilibrium is achieved. At equilibrium, all reactants and products are still available in some concentration. Both of the opposite reactions continue at the same rate at equilibrium, but the proportion of reactants and products remains unchanged as long as the equilibrium is maintained.

When a reversible reaction goes to equilibrium, do the opposite reactions stop? _____

Answer: no (Both of the opposite reactions continue at equilibrium but the rates of the two reactions are equal.)

4   The following reversible reaction:

$$CO_2 + H_2 \rightleftharpoons H_2O + CO$$

can also be written as:

$$H_2O + CO \rightleftharpoons CO_2 + H_2$$

By convention, the starting materials of the reaction are written on the left side of the equation. The substances on the left side are also arbitrarily called the reactants, and those on the right side are called the products. The reaction with the arrow pointing to the right ($\rightarrow$) is called the forward reaction. The reaction with the arrow pointing to the left ($\leftarrow$) is called the reverse reaction.

Write a reversible reaction with the starting material of H_2 and I_2 (product would be HI). _____

Answer: $H_2 + I_2 \rightleftharpoons 2HI$

5   Indicate whether each of the following is a forward or reverse reaction.

(a) $PCl_5 \rightarrow PCl_3 + Cl_2$ _____
(b) $PCl_3 + Cl_2 \leftarrow PCl_5$ _____

Answer: (a) forward; (b) reverse

6   Reversible reactions that go to equilibrium (instead of to completion) end up with an **equilibrium mixture** of reactants and products after equilibrium is reached. The equilibrium mixture contains various concentrations of each reactant and product. List all the substances found in the equilibrium mixture of the reversible reaction below. _____

$$CO_2 + H_2 \rightleftharpoons H_2O + CO$$

Answer: CO_2, H_2, CO, and H_2O (all reactants and products)

7   When a reversible reaction reaches chemical equilibrium, the concentrations of the components in the equilibrium mixture (the products and reactants) remain constant as long as experimental conditions are not changed. At chemical equilibrium, how does the rate of the forward reaction compare with the rate of the reverse reaction? _____

Answer: The rates are equal.

EQUILIBRIUM CONSTANT

8   For most reversible reactions going to equilibrium, equilibrium constants have been determined experimentally. An **equilibrium constant** is the ratio of the concentration of the products divided by the concentration of the reactants at equilibrium and at a specified temperature. The equilibrium constant (product concentrations divided by reactant concentrations) is valid only at a specified temperature after the reaction has gone to (completion, equilibrium) _____

Answer: equilibrium

9 Below is a reversible reaction and the expression for the equilibrium constant for this reversible reaction.

$$H_2O + CO \rightleftharpoons CO_2 + H_2$$

$$K_{eq} = \frac{[CO_2][H_2]}{[H_2O][CO]}$$

The symbol K_{eq} represents the equilibrium constant and the brackets [ ] represent the concentration (usually in moles per liter) of each product and reactant. Look at the placement of each reactant and product in the equilibrium constant expression. In the equilibrium constant expression for a reversible reaction, the (products, reactants) _____ are located in the numerator or upper part of the fraction and the (products, reactants) _____ are located in the denominator or lower part of the fraction.

Answer: products; reactants

10 The standard equation, then, for K_{eq} is as follows.

$$K_{eq} = \frac{[product][product]}{[reactant][reactant]}$$

Write the equilibrium constant expression for the following reversible reaction.

$$PCl_5 \rightleftharpoons PCl_3 + Cl_2$$

$K_{eq} =$ _____

Answer: $\dfrac{[PCl_3][Cl_2]}{[PCl_5]}$

(Since there are two products, they should be placed in the upper part of the fraction. The one reactant belongs in the lower part of the fraction.)

11 A reversible reaction such as $H_2 + I_2 \rightleftharpoons 2HI$ has the following equilibrium constant expression.

$$K_{eq} = \frac{[HI]^2}{[H_2][I_2]}$$

Note that the coefficient in front of HI in the reaction becomes an exponent in the equilibrium constant expression.

Write the equilibrium constant expression for the following reversible reaction.

$$N_2 + 3H_2 \rightleftharpoons 2NH_3$$

$K_{eq} =$ _____

Answer: $\dfrac{[NH_3]^2}{[N_2][H_2]^3}$

12 It is easy to prove to yourself that the coefficients in the reaction should be exponents in the expression for an equilibrium constant.

$$N_2 + 3H_2 \rightleftharpoons 2NH_3$$

This reaction can also be written as:

$$N_2 + H_2 + H_2 + H_2 \rightleftharpoons NH_3 + NH_3$$

Write the equilibrium constant expression for this last reaction.

$K_{eq} =$ _____

Answer: $\dfrac{[NH_3][NH_3]}{[N_2][H_2][H_2][H_2]}$

The concentrations of the reactants are multiplied together: $[H_2][H_2][H_2] = [H_2]^3$. The concentrations of the products are multiplied together: $[NH_3][NH_3] = [NH_3]^2$. Therefore,

$$K_{eq} = \frac{[NH_3]^2}{[N_2][H_2]^3}$$

13 Write the correct equilibrium constant expression for the following reversible reaction.

$$2SO_2 + O_2 \rightleftharpoons 2SO_3$$

$$K_{eq} = \underline{\hspace{4cm}}$$

Answer: $\dfrac{[SO_3]^2}{[SO_2]^2[O_2]}$

CALCULATING EQUILIBRIUM CONSTANTS

14 Now let's use this general equation to find some actual equilibrium constants. The equilibrium constant expression for $H_2 + I_2 \rightleftharpoons 2HI$ is:

$$K_{eq} = \frac{[HI]^2}{[H_2][I_2]}$$

At 458°C, the equilibrium mixture consists of:

$$[HI] = 1.5 \times 10^{-2} \text{ mol/liter}$$

$$[H_2] = 4.6 \times 10^{-3} \text{ mol/liter}$$

$$[I_2] = 1.0 \times 10^{-3} \text{ mol/liter}$$

Calculate the value of K_{eq} at 458°C (to the nearest tenth).

$$K_{eq} = \underline{\hspace{4cm}}$$

Answer: $\dfrac{[HI]^2}{[H_2][I_2]} = \dfrac{(1.5 \times 10^{-2} \text{ mol/liter})^2}{(4.6 \times 10^{-3} \text{ mol/liter})(1.0 \times 10^{-3} \text{ mol/liter})} = 48.9$

15 At 458°C, the equilibrium constant (K_{eq}) for the reaction $H_2 + I_2 \rightleftharpoons 2HI$ is 48.9. If you were to repeat the reaction $H_2 + I_2 \rightleftharpoons 2HI$ using three times as much starting material (H_2 and I_2) but kept the temperature at 458°C, what would be the value of K_{eq}? _____

(Hint: This question is tricky. Remember the meaning of the word constant.)

Answer: 48.9 (Any amount of starting material for this reaction under normal conditions will produce a chemical equilibrium with an equilibrium constant of 48.9. The equilibrium constant for this reaction is a constant at the specified temperature. It does not change.)

16 After the equilibrium for a reaction is determined experimentally, it is useful for predicting the amount of products at equilibrium or the amount of reactants needed for a different situation involving the same reaction at the specified temperature.

$$K_{eq} = \frac{[HI]^2}{[H_2][I_2]}$$

Determine [HI] (the concentration of hydrogen iodide) at equilibrium if:

$$[H_2] = 1.24 \times 10^{-2} \text{ mol/liter}$$

$$[I_2] = 2.46 \times 10^{-2} \text{ mol/liter}$$

$$K_{eq} = 48.9$$

$$[HI] = \underline{\hspace{4cm}}$$

Answer:

$$K_{eq} = \frac{[HI]^2}{[H_2][I_2]}$$

$$[HI]^2 = K_{eq}[H_2][I_2]$$

$$[HI] = \sqrt{K_{eq}[H_2][I_2]} = \sqrt{(48.9)(1.24 \times 10^{-2} \text{ mol/liter})(2.46 \times 10^{-2} \text{ mol/liter})}$$

$$= 1.22 \times 10^{-1} \text{ mol/liter of hydrogen iodine}$$

17   An equilibrium constant expression is valid only if the reaction is balanced. Write the equilibrium constant expression for the following unbalanced reversible reaction. Make sure to balance the reaction equation first. (Refer to Chapter 6, if necessary, for review.)

$$HCl + O_2 \rightleftharpoons Cl_2 + H_2O$$

$$K_{eq} = \underline{\hspace{4cm}}$$

Answer: Balance the reaction equation first (equal numbers of atoms on both sides).

$$4HCl + O_2 \rightleftharpoons 2Cl_2 + 2H_2O$$

$$K_{eq} = \frac{[Cl_2]^2[H_2O]^2}{[HCl]^4[O_2]}$$

18   So far, the values of the equilibrium constant (K_{eq}) in all examples have been just numbers because all of the mol/liter concentration terms have canceled out. In determining the value of K_{eq} for the following expression, all mol/liter terms will cancel and K_{eq} will be a unitless number.

$$K_{eq} = \frac{[\ ]^2}{[\ ][\ ]} = \frac{(mol/liter)^2}{(mol/liter)(mol/liter)}$$

Suppose that for a different reaction, the equilibrium constant expression is as follows.

$$K_{eq} = \frac{[\ ]^2[\ ]}{[\ ]^2} = \frac{(mol/liter)^2(mol/liter)}{(mol/liter)^2}$$

All mol/liter *do not* cancel in this case. Instead of being just a number, K_{eq} will have a value in terms of _____ .

Answer: mol/liter

$$K_{eq} = \frac{(mol/liter)^2(mol/liter)}{(mol/liter)^2} = \frac{(mol/liter)^3}{(mol/liter)^2} = mol/liter$$

 19 For the following equilibrium constant expression, the value of K_{eq} will not be just a number.

$$K_{eq} = \frac{[\ ]^2[\ ]^2}{[\ ]^3} = \frac{(mol/liter)^2(mol/liter)^2}{(mol/liter)^3}$$

K_{eq} will be expressed in terms of _____

Answer: mol/liter

$$K_{eq} = \frac{(mol/liter)^2(mol/liter)^2}{(mol/liter)^3} = \frac{(mol/liter)^4}{(mol/liter)^3} = mol/liter$$

20 For the following equilibrium constant expression, the value of K_{eq} is not just a number. This example is a bit more difficult than the previous examples since the denominator has the larger exponent.

$$K_{eq} = \frac{[\ ]^2}{[\ ]^3} = \frac{(mol/liter)^2}{(mol/liter)^3}$$

The value of K_{eq} will be in terms of _____

Answer: liters/mol

$$K_{eq} = \frac{(mol/liter)^2}{(mol/liter)^3} = \frac{1}{(mol/liter)} = liters/mol$$

To divide by a fraction, invert the fraction and multiply:

$$1 \div \frac{mol}{liter} = 1 \times \frac{liter}{mol} = liters/mol$$

 21 For the following equilibrium constant expression, the value of K_{eq} will not be a pure number.

$$K_{eq} = \frac{[\ ]^2}{[\ ]^{22}}$$

The value of K_{eq} will be in terms of _____.

Answer: $K_{eq} = \frac{(mol/liter)^2}{(mol/liter)^2(mol/liter)^2} = \frac{1}{(mol/liter)^2} = liters^2/mol^2$

To divide by a fraction, invert the fraction and multiply:

$$1 \div \left(\frac{mol}{liter}\right)^2 = 1 \times \left(\frac{liters}{mol}\right)^2 = (liters/mol)^2 = liters^2/mol^2$$

22 For the following equilibrium constant expression, the value of K_{eq} will not be a pure number.

$$K_{eq} = \frac{[Cl]^2[H_2O]^2}{[HCl]^4[O_2]}$$

The value of K_{eq} will be in terms of _____.

Answer: $K_{eq} = \frac{[Cl]^2[H_2O]^2}{[HCl]^4[O_2]} = \frac{(mol/liter)^2(mol/liter)^2}{(mol/liter)^4(mol/liter)} = \frac{(mol/liter)^4}{(mol/liter)^5} = \frac{1}{mol/liter} = liters/mol$

23 A value of K_{eq} should include proper units such as mol/liter or liters/mol so that further calculations using K_{eq} will have the correct units. Calculate the value of the equilibrium constant for the following reversible reaction.

$$N_2 + 3H_2 \rightleftharpoons 2NH_3$$

The concentrations of the equilibrants (reactants and products) at equilibrium at 400°C are as follows.

$$N_2 = 1.20 \text{ mol/liter}$$

$$H_2 = 0.80 \text{ mol/liter}$$

$$NH_3 = 0.28 \text{ mol/liter}$$

In addition to the numerical value of K_{eq} we must determine the proper units for K_{eq}. After checking the equation for proper balancing, write the expression for the equilibrium constant. After determining the proper expression, you may find it easiest to calculate the numerical value of K_{eq} first and determine the proper units in a separate step.

The value of K_{eq} at 400°C is _____.

Answer: $K_{eq} = \dfrac{[NH_3]^2}{[N_2][H_2]^3} = \dfrac{(0.28)^2}{(1.20)(0.80)^3} = \dfrac{(0.0784)}{(1.20)(0.512)} = 0.128$

The proper unit can be determined as follows.

$$K_{eq} = \frac{[NH_3]^2}{[N_2][H_2]^3} = \frac{(\text{mol/liter})^2}{(\text{mol/liter})(\text{mol/liter})^3} = \frac{(\text{mol/liter})^2}{(\text{mol/liter})^4}$$

$$= \frac{1}{(\text{mol/liter})^2} = (\text{liters/mol})^2 = \text{liters}^2/\text{mol}^2$$

The actual value of K_{eq} is 0.128 liters²/mol² at 400°C.

TYPES OF EQUILIBRIUM

24 All of the reversible reactions used as examples have so far included only reactions in which all **equilibrants** (reactants and products) are of the same state (all gases or all soluble solids). A chemical equilibrium in which all equilibrants are of the same state is called a **homogeneous equilibrium**. A chemical equilibrium in which the equilibrants are of different states is called a **heterogeneous equilibrium**.

A chemical equilibrium of a reaction in which a solid and a gas are reactants and a gas is the product is what kind of equilibrium? _____

Answer: heterogeneous

25   For the following reaction, an abbreviation has been added to indicate the state of each reactant or product.

$$N_2(g) + 3H_2(g) \rightleftharpoons 2NH_3(g)$$

- The letter "g" following a substance indicates that the substance is a gas.
- The letter "*l*" indicates that a substance is a liquid.
- The letter "s" indicates that a substance is a solid.
- Finally, "aq" indicates that a substance is an ion in aqueous (water) solution.

(a) The chemical equilibrium involving the reaction above is what kind of equilibrium? _____

(b) An equilibrium involving the following reaction would be what kind of equilibrium? _____

$$NH_3(g) + H_2O \rightleftharpoons NH_4^+(aq) + OH^-(aq)$$

Answer: (a) homogeneous; (b) heterogeneous

26   Some heterogeneous equilibrium reactions are special cases when the equilibrium constant expression is written for them.

$$NH_3(g) + H_2O \rightleftharpoons NH_4^+(aq) + OH^-(aq)$$

The correct equilibrium constant expression is:

$$K_{eq} = \frac{[NH_4^+][OH^-]}{[NH_3]}$$

What reactant or product has not been included in the above expression for the equilibrium constant? _____

Answer: H_2O

27 H_2O was not included in the equilibrium expression in frame 26 because it is a pure liquid whose concentration does not change significantly. Pure liquids as well as pure solids should *not* be included in the expression for an equilibrium constant because the concentration of a pure liquid or pure solid does not change if the temperature does not change.

Because the concentrations of pure solids and liquids stay constant, it would be redundant to include them in the equilibrium constant expression. ("Pure" indicates that the compound or element is not mixed with another compound or element. Water is a pure liquid if some other substance is not dissolved or mixed with it.) Gases, however, change concentration with pressure even if the temperature is constant. Only those equilibrants whose concentration is variable should be included in the equilibrium constant expression. Unless otherwise stated, any reactant or product that is a liquid (l) or solid (s) should not be included in the equilibrium constant expression.

Write the equilibrium constant expression for the following reversible reaction.

$$CuO(s) + H_2(g) \rightleftharpoons Cu(s) + H_2O(g)$$

(Note that H_2O is a vapor that behaves like a gas and not a liquid in this reaction.)

$K_{eq} = $ _____

Answer: $\dfrac{[H_2O]}{[H_2]}$

(The CuO and Cu are solids, as indicated by (s) that follows these solids in the equation, and should not be included in the expression.)

28 Write the equilibrium constant expression for the following reaction.

$$C(s) + CO_2(g) \rightleftharpoons 2CO(g)$$

$K_{eq} = $ _____

Answer: $\dfrac{[CO]^2}{[CO_2]}$

(The solid does not appear in the equilibrium constant expression, gases do.)

29 Write the equilibrium constant expression for the following reaction.

$$Fe_3O_4(s) + 4H_2(g) \rightleftharpoons 3Fe(s) + 4H_2O(g)$$

(Note that H_2O is not a liquid in this reaction.)

$K_{eq} = $ _____

Answer: $\dfrac{[H_2O]^4}{[H_2]^4}$

30 Write the equilibrium constant expression for the following reaction.

$$NH_4HS(s) \rightleftharpoons NH_3(g) + H_2S(g)$$

$K_{eq} = \underline{\hspace{3cm}}$

Answer: [NH$_3$][H$_2$S] (The only reactant is a solid, therefore the denominator is eliminated.)

STRESS IN EQUILIBRIUM

31 Le Chatelier's Principle is applicable to all chemical equilibria. Such stresses as adding more of a reactant or product to a reaction, or changing the pressure of a reaction involving gases, will have effects that can be predicted through Le Chatelier's Principle. Le Chatelier's Principle states that when a stress is placed on an equilibrium system, the equilibrium will shift so that the stress is (increased, relieved) _____.

Answer: relieved

32 The following reaction can serve as an example of what happens when a stress is placed upon a chemical equilibrium.

$$PCl_5(g) \rightleftharpoons PCl_3(g) + Cl_2(g)$$

First, write the equilibrium constant expression for this reaction.

$K_{eq} = \underline{\hspace{3cm}}$

Answer: $\dfrac{[PCl_3][Cl_2]}{[PCl_5]}$

33 Assume that at a certain temperature, the concentration in mol/liter of each equilibrant in the reaction from frame 32 is as follows.

$$PCl_3 = 6 \, mol/liter$$
$$Cl_2 = 6 \, mol/liter$$
$$PCl_5 = 9 \, mol/liter$$

Calculate the value of K_{eq} _____

Answer: $K_{eq} = \dfrac{(6 \, mol/liter)(6 \, mol/liter)}{(9 \, mol/liter)} = 4 \, mol/liter$

34 To observe the effects of Le Chatelier's Principle, we now add an additional 6 mol/liter of Cl$_2$ gas to the equilibrium. We suddenly have a total of 12 mol/liter of Cl$_2$ as well as the original amount of PCl$_3$ and PCl$_5$. The system is no longer at equilibrium. Immediately, one of the reaction rates (forward or reverse) increases so as to relieve the stress. Look carefully at the reversible reaction.

$$PCl_5 \rightleftharpoons PCl_3 + Cl_2$$

Since the concentration of Cl_2 has just doubled, the reaction will try to relieve that stress. As a response to all that extra Cl_2, which of the following would you expect to happen?

(a) The forward reaction rate increases (producing even more Cl_2 and PCl_3 while reducing the PCl_5).

(b) The reverse reaction increases (reducing the supply of Cl_2 and PCl_3 while increasing the PCl_5).

Answer: (b) (This is logical since there is a surplus of Cl_2. The reverse reaction rate increases until equilibrium is again restored.)

35 After relieving the stress, the reaction from frames 32–34 will again be at equilibrium, although the concentration of each equilibrant has changed.

$$PCl_5 \rightleftharpoons PCl_3 + Cl_2$$

(1) Original K_{eq}:

$$K_{eq} = \frac{[PCl_3][Cl_2]}{[PCl_5]} = \frac{(6\ mol/liter)(6\ mol/liter)}{(9\ mol/liter)} = 4\ mol/liter$$

(2) Doubling the Cl_2 concentration, the system is no longer at equilibrium:

$$\frac{(6\ mol/liter)(12\ mol/liter)}{(9\ mol/liter)}$$

After doubling the Cl_2 concentration, the system returns to equilibrium.

The concentration of each equilibrant at the new equilibrium has been calculated for you using a mathematical procedure called the quadratic equation. You will not be required to use quadratic equations in this text.

(3) New K_{eq}:

$$K_{eq} = \frac{[PCl_3][Cl_2]}{[PCl_5]} = \frac{(4.22\ mol/liter)(10.22\ mol/liter)}{(10.78\ mol/liter)} = 4.0\ mol/liter$$

Note that after equilibrium is restored, the value for K_{eq} remains the same. Since the temperature has remained the same, should you have expected the value for K_{eq} to change? _____

Answer: no (If the temperature does not change, K_{eq} will not change because it is a *constant* for the reaction at equilibrium.)

36 Compare the new equilibrium concentrations to those of the original equilibrium in frame 35. Besides the doubled Cl_2 concentration (which was the stress applied), which equilibrant

(a) increased in concentration? _____

(b) decreased in concentration? _____

Answer: (a) The concentration of PCl_5 increased from 9 mol/liter to 10.78 mol/liter; (b) The concentration of PCl_3 decreased from 6 mol/liter to 4.22 mol/liter.

37 Adding more of one of the products (Cl_2) in the above reaction resulted in a shift of the equilibrium that favored the reverse reaction, as predicted by Le Chatelier's Principle.

After equilibrium is again restored, the concentration of the other product (PCl_3) is (decreased, increased) _____. The concentration of the reactant (PCl_5) is (decreased, increased) _____.

Answer: decreased; increased

38 As a general rule (because of Le Chatelier's Principle), increasing the concentration of one of the products of a reaction favors the reverse reaction and results in a decrease in the concentration of any other product and an increase in the concentration of the reactants. This reaction represents an ordinary reversible reaction at equilibrium.

$$A + B \rightleftharpoons C + D$$

If the concentration of product D is suddenly increased, label each of the following as increased or decreased.

(a) the concentration of A _____

(b) the concentration of B _____

(c) the concentration of C _____

Answer: (a) increased; (b) increased; (c) decreased (The equilibrium is shifted to favor the reactants.)

39 The reversible reaction $A + B \rightleftharpoons C + D$ can also be written $C + D \rightleftharpoons A + B$. Suppose the concentration of A were increased suddenly. After equilibrium is again restored, which of the following increases and which decreases?

(a) the concentration of B _____

(b) the concentration of C _____

(c) the concentration of D _____

Answer: (a) decreases; (b) increases; (c) increases

40 According to Le Chatelier's Principle:

- If the concentration of one reactant is suddenly increased, the equilibrium will shift so that all products are increased and the other reactant will be decreased.

- If the concentration of one product is suddenly increased, the equilibrium will shift so that all reactants are increased and the other product is decreased. The brackets [ ] indicate concentration (in mol/liter).

$$A + B \rightleftharpoons C + D$$

If $[C]$ is increased, then $[A]$ is _____, $[B]$ is _____, and $[D]$ is _____.

Answer: increased; increased; decreased

41 Decreasing a reactant or product can also be a stress.

$$A + B \rightleftharpoons C + D$$

If C were to be suddenly decreased by removing some from the reaction system, the forward reaction would be favored and more product would be produced until equilibrium was again established. To produce more product would require a decreased concentration of reactants. A decrease in $[C]$ causes the exact opposite of an increase in $[C]$.

(a) If $[A]$ is increased (by adding some A to the reaction mixture), then $[B]$ is _____, $[C]$ is _____, and $[D]$ is _____.

(b) If $[A]$ is decreased (by removing some A from the reaction mixture), then $[B]$ is _____, $[C]$ is _____, and $[D]$ is _____.

Answer: (a) decreased; increased; increased (The forward reaction is favored.); (b) increased; decreased; decreased (The reverse reaction is favored.)

42   Whenever a quantity of a reactant or product is added to or removed from a chemical equilibrium system, the equilibrium responds according to Le Chatelier's Principle. The stress in this case is the addition or removal of some quantity of reactant or product. A chemical equilibrium will also respond to a change in temperature.

A chemical reaction may either require heat to continue the reaction or give off heat during the reaction. A chemical reaction requiring heat (energy) to proceed is called **endothermic**. A chemical reaction giving off heat (energy) during a reaction is **exothermic**.

The forward reaction $CaCO_3(s) \rightleftharpoons CaO(s) + CO_2(g)$ requires heat energy. It is an (exothermic, endothermic) _____ reaction.

Answer: endothermic

43   The forward reaction of $N_2(g) + 3H_2(g) \rightleftharpoons 2NH_3(g)$ is exothermic. It (gives off, requires) _____ heat energy.

Answer: gives off

44   A forward exothermic reaction can be written to show heat as a product.

$$N_2(g) + 3H_2(g) \rightleftharpoons 2NH_3(g) + heat$$

A forward endothermic reaction can be written to show heat as a requirement on the reactant side. The following forward reaction is endothermic. Which equation is correctly written?

_____ (a) $heat + CaCO_3(s) \rightleftharpoons CaO(s) + CO_2(g)$
_____ (b) $CaCO_3(s) \rightleftharpoons Cao + CO_2(g) + heat$

Answer: (a)

45   The equilibrium constant for an endothermic reaction increases with rising temperature. The equilibrium constant for an exothermic reaction decreases with rising temperature. This means that for an endothermic reaction, increasing the temperature will increase the concentration of products (at the expense of reactants). For an exothermic reaction, the opposite holds true.

A simple way to remember these effects is to treat the heat energy as a reactant or product and apply what you have learned from Le Chatelier's Principle.

$$\text{heat} + CaCO_3(s) \rightleftharpoons CaO(s) + CO_2(g)$$

Treating heat as a required item on the reactant side, apply Le Chatelier's Principle. If the temperature is raised (heat is increased), indicate whether the concentrations of the following reactant and products increase or decrease.

(a) $[CaCO_3]$ _____

(b) $[CaO]$ _____

(c) $[CO_2]$ _____

(d) Is the reaction endothermic or exothermic? _____

Answer: (a) decreases; (b) increases; (c) increases; (d) endothermic (Increasing the required heat energy favors the forward reaction.)

46 Indicate whether increasing the temperature for the reaction expressed below would increase or decrease the concentrations of the reactants and product.

$$N_2(g) + 3H_2(g) \rightleftharpoons 2NH_3(g) + \text{heat}$$

(a) $[NH_3]$ _____

(b) $[N_2]$ _____

(c) $[H_2]$ _____

(d) Is this reaction endothermic or exothermic? _____

Answer: (a) decreases; (b) increases; (c) increases; (d) exothermic (Increasing one of the products would favor the reverse reaction. Since heat energy is being treated as a product, increasing it would have the same effect.)

47 A decrease in temperature (some of the heat is removed) has an effect opposite to that of an increase.

$$N_2(g) + 3H_2(g) \rightleftharpoons 2NH_3(g) + \text{heat}$$

If the temperature is decreased, indicate whether the concentrations of the reactants and product would increase or decrease.

(a) $[NH_3]$ _____

(b) $[N_2]$ _____

(c) $[H_2]$ _____

Answer: (a) increases; (b) decreases; (c) decreases (Decreasing one of the products favors the forward reaction. Since heat is being treated as a product, decreasing it has the same effect.)

48 The effects of increasing or decreasing the pressure surrounding a chemical equilibrium involving gases can also be predicted on the basis of Le Chatelier's Principle. Increasing the pressure favors the side of the reaction (reactant side or product side) containing the fewest moles of gas.

For the following reaction, $N_2(g) + 3H_2(g) \rightleftharpoons 2NH_3(g)$, the reactant side has 4 mol of gas and the product side has 2 mol of gas. If the pressure is increased, which side will be favored and end up with an increase in concentration? _____

Answer: the product side (Increased pressure favors the reaction side with the fewest moles of gas.)

49 In this reaction, only the CO_2 product is a gas. The other equilibrants are solids that cannot be compressed.

$$CaCO_3(s) \rightleftharpoons CaO(s) + CO_2(g)$$

An increase in pressure causes the reaction equilibrium to shift in favor of the side of the reaction with the least moles of gas. In this reaction, the reactant side has no moles of gas while the product side has 1 mol of gas. An increase in pressure would favor which side of the reaction? _____

Answer: the reactant side (Zero moles of gas is less gas than 1 mol of gas.)

50 Now look at this reaction.

$$PCl_5(g) \rightleftharpoons PCl_3(g) + Cl_2(g)$$

(a) The reactant side of this reaction has how many moles of gas? _____

(b) The product side has how many moles of gas? _____

(c) Which side of the reaction would be favored by an increase in pressure?

(d) Which side of the reaction would be favored by a decrease in pressure? (Remember, decreasing the pressure has the opposite effect of increasing the pressure.) _____

Answer: (a) 1 mol; (b) 2 mol (PCl$_3$ and Cl$_2$ each represent 1 mol of gas, which adds up to 2 mol on the product side.); (c) reactant side; (d) product side

51 Here's a tricky question. Remember, increasing the pressure favors the side of the reaction with the least moles of gas, provided there *is* a side with the least moles of gas. For the following reversible reaction, increasing the pressure will favor which side of the reaction? _____

$$H_2(g) + I_2(g) \rightleftharpoons 2HI(g)$$

Answer: Neither side is favored because there are 2 mol of gas on each side of the reaction. (Increasing the pressure has no effect if both sides of a reaction have an equal number of moles of gas.)

52 The reversible reaction below is presently at equilibrium. You desire to shift the equilibrium to favor the product side in order to increase the concentration of CO_2 as much as possible.

$$2CO(g) + O_2(g) \rightleftharpoons 2CO_2(g) + heat$$

To produce an equilibrium with a large concentration of CO_2, indicate whether you would increase or decrease the following conditions.

(a) pressure _____

(b) temperature _____

(c) concentration of $O_2(g)$ _____

Answer: (a) increase (This favors the product side since it has fewer moles of gas.); (b) decrease (Treating heat as a product, this removes some heat favoring the production of more product.); (c) increase (Increasing the concentration of a reactant results in an increase in products as well as a decrease in other reactants.)

We have discussed the concept of equilibrium and how it applies to those chemical reactions that do not go to completion. You have learned what happens to a reaction at equilibrium when a "stress" is applied to the reaction. Most of the equilibria discussed involved reactants and products that were all in the gaseous state or equilibria between gases and pure solids and liquids.

Chemists have learned that many reactions that occur between ions in aqueous solutions also do not go to completion and reach a state of equilibrium. Ionic equilibria are also affected by stresses according to Le Chatelier's Principle.

We now discuss several special cases of ionic equilibria and how they are used to determine the degree of dissociation of weak electrolytes and the concentration of ions in aqueous solutions. We also discuss how slightly soluble compounds behave and how their solubility is determined in aqueous solutions and in the presence of other ions.

IONIC EQUILIBRIA

 53

You have used Le Chatelier's Principle and the equilibrium constant for reversible reactions. Equilibria also exist for such things as salts and their ions in solution, and acids and bases and their dissociated ions. Other substances, even though they may dissociate only very slightly into ions, are at equilibrium with those ions. An ionic equilibrium constant expression is written just like those of the reversible reactions previously encountered. Write the ionic equilibrium constant for the following reaction.

$$HC_2H_3O_2(aq) \rightleftharpoons H^+(aq) + C_2H_3O_2^-(aq)$$

$K_{eq} =$ _____

Answer: $\dfrac{[H^+][C_2H_3O_2^-]}{[HC_2H_3O_2]}$

54

Instead of K_{eq}, the equilibrium constant may be K_a if it represents the dissociation of an acid, K_b if it represents the dissociation of a base, K_i for a general ionization constant, or K_{sp} for a solubility product constant. We will discuss K_{sp} now and these other constants later. The equilibrium constant expression is the same for any of these as for the reactions dealt with previously. Write the equilibrium constant expression for the solubility product of the very slightly soluble salt AgBr.

$$AgBr(s) \rightleftharpoons Ag^+(aq) + Br^-(aq)$$

$K_{sp} =$ _____

Answer: $[Ag^+][Br^-]$ (The solid is not included since its concentration is constant at constant temperature.)

55 The abbreviation K_{sp} stands for the **solubility product constant**. It represents the solubility of a salt in a saturated solution and is called a product because the negative and positive ion concentrations are multiplied together to determine the value of the constant. Write the solubility product expression for the salt CaF_2.

$$CaF_2(s) \rightleftharpoons Ca^{2+}(aq) + 2F^-(aq)$$

$K_{sp} = $ _____

Answer: $[Ca^{2+}][F^-]^2$ **(Don't forget the exponent.)**

56 The solubility product equation represents the dynamic equilibrium between a solid and its dissociated ions in a saturated solution.

$$Bi_2S_3(s) \rightleftharpoons 2Bi^{3+}(aq) + 3S^{2-}(aq)$$

Write the solubility product expression for the saturated aqueous solution of Bi_2S_3.

$K_{sp} = $ _____

Answer: $[Bi^{3+}]^2[S^{2-}]^3$

57 Write the K_{sp} expression for the saturated aqueous solution of AgCl. (AgCl is only slightly soluble.)

$$AgCl(s) \rightleftharpoons Ag^+(aq) + Cl^-(aq)$$

$K_{sp} = $ _____

Answer: $[Ag^+][Cl^-]$

58 In a saturated solution of slightly soluble AgCl in water, an equilibrium exists between the AgCl solid and its dissociated aqueous ions.

$$AgCl(s) \rightleftharpoons Ag^+(aq) + Cl^-(aq)$$

Suppose that extra Cl^- ion is added to this equilibrium. From Le Chatelier's Principle, you know that the equilibrium would favor the left side of the equation. This situation is like the general reversible reaction $A + B \rightleftharpoons C + D$. Adding more Cl^- ion to the equilibrium is like adding more of product D to the reaction. Adding more of product D causes an increase in the concentration of the reactants and a decrease in the concentration of product C. Indicate whether adding more Cl^- ion to the equilibrium causes an increase or decrease in the following:

(a) the formation of $AgCl(s)$ _____

(b) the concentration of Ag^+ _____

Answer: (a) increase; (b) decrease

59 The effect of adding more Cl^- to the saturated solution of AgCl can also be shown by using the solubility product constant. The solubility of AgCl is 1.2×10^{-5} mol/liter.

$$AgCl \rightleftharpoons Ag^+(aq) + Cl^-(aq)$$

$$K_{sp} = [Ag^+][Cl^-] = (1.2 \times 10^{-5} \text{ mol/liter}) \times (1.2 \times 10^{-5} \text{ mol/liter})$$

Assume the K_{sp} to equal 1.44×10^{-10} mol^2/liter^2. If the Cl^- concentration were to be increased, what must happen to the Ag^+ concentration? (Remember that K_{sp} is a constant and does not change.) _____

Answer: The Ag^+ concentration must decrease. (When the Ag^+ concentration and Cl^- concentration are multiplied together, they must equal the K_{sp} constant. When the concentration of one of the ions is increased, the other must decrease to keep the K_{sp} constant.)

60 The extra Cl^- ion that is added to the saturated solution of AgCl of frames 58 and 59 might have come from NaCl or some other strong electrolyte having a Cl^- ion. This is an example of the **common ion effect**. The slightly soluble AgCl and the strong electrolyte NaCl both have the Cl^- ion in common. Adding some NaCl to a saturated solution of AgCl forces the equilibrium to shift so that some of the $Ag^+(aq)$ and the extra $Cl^-(aq)$ come out of the solution and precipitate. In this example of the common ion effect, which is the "common" ion? _____

Answer: Cl^- (from both NaCl and AgCl)

61 If some of the strong electrolyte $AgNO_3$ is added to a saturated solution of AgCl, the common ion effect can also be observed.

$$AgCl(s) \rightleftharpoons Ag^+(aq) + Cl^-(aq)$$

In this example of the common ion effect, which would be the "common" ion?

_____

Answer: Ag^+ (from both AgCl and $AgNO_3$)

62 If you added some strong electrolyte such as $NaNO_3$ to a saturated solution of AgCl, would you expect to observe the common ion effect? _____

Answer: no (There are not two ions alike in AgCl and $NaNO_3$.)

63 The common ion effect is observed when a strong electrolyte is added to a saturated solution of a very slightly soluble salt, weak acid, or weak base, and one ion of the strong electrolyte is the same as one of the ions in the original solution.

You may have wondered why AgCl is used as an example since we have previously assumed that all of AgCl precipitated out of solution. In actuality, all salts are soluble to some degree. The solubility of some salts such as AgCl is so minimal that very little error is introduced by treating it elsewhere as not soluble at all.

Is there a salt that is absolutely not soluble even to a small degree? _____

Answer: no (All salts are soluble to some slight degree. For practical purposes in some experiments, we can often ignore the dissolved portion of a very slightly soluble salt. For K_{sp} calculations and the common ion effect, we cannot ignore it.)

64 The value of the solubility product constant (K_{sp}) can easily be determined if the solubility of a salt is known.

$$PbS(s) \rightleftharpoons Pb^{2+}(aq) + S^{2-}(aq)$$

Lead sulfide (PbS) is a slightly soluble salt. Only 1×10^{-14} mol of the salt will dissolve in a liter of water at $20\,°C$. In pure water, at $20\,°C$, this means that the concentration of Pb^{2+} ion will be 1×10^{-14} mol/liter and the concentration of S^{2-} ion will also be 1×10^{-14} mol/liter. Calculate K_{sp}

$K_{sp} = [Pb^{2+}][S^{2-}] = $ _____

Answer: $(1 \times 10^{-14}$ mol/liter$)(1 \times 10^{-14}$ mol/liter$) = 1 \times 10^{-28}$ mol^2/liter^2

65 Up to 7×10^{-5} mol of calcium carbonate will dissolve in a liter of pure water at $20°C$.

$$CaCO_3(s) \rightleftharpoons Ca^{2+}(aq) + CO_3^{2-}(aq)$$

Calculate the K_{sp} of $CaCO_3$ at $20°C$.

$K_{sp} =$ _____

Answer: $[Ca^{2+}][CO_3{}^{2-}] = (7 \times 10^{-5}$ mol/liter$)(7 \times 10^{-5}$ mol/liter$) = 4.9 \times 10^{-9}$ mol^2/liter^2

66 At a certain temperature, up to 2×10^{-4} mol of CaF_2 will dissolve in a liter of pure water. Calculate the K_{sp} of CaF_2 at that temperature.

$$CaF_2(s) \rightleftharpoons Ca^{2+}(aq) + 2F^-(aq)$$

(Note: The fluoride ion has a coefficient of 2 in front of it. This means that two fluoride ions are produced for every calcium ion when CaF_2 dissociates. It also means that the fluoride ion concentration must be squared in the K_{sp} expression.)

$K_{sp} =$ _____

Answer: $[Ca^{2+}] = 2 \times 10^{-4}$ mol/liter

Two fluoride ions are formed for every calcium ion.

$$[F^-] = (2)(2 \times 10^{-4} \text{ mol/liter}) = 4 \times 10^{-4} \text{ mol/liter.}$$

$$K_{sp} = [Ca^{2+}][F^-]^2 = (2 \times 10^{-4}\text{mol/liter})(4 \times 10^{-4}\text{mol/liter})^2$$

$$= (2 \times 10^{-4}\text{mol/liter})(16 \times 10^{-8}\text{mol}^2/\text{liter}^2) = 32 \times 10^{-12} \text{ mol}^3/\text{liter}^3$$

$$= 3.2 \times 10^{-11}\text{mol}^3/\text{liter}^3$$

67 All of the K_{sp} solubilities calculated in the examples assume that the salts are dissolved in pure water (with no common ions present). Once the K_{sp} is determined, however, we can use the constant to determine the concentration of one ion of the salt if the concentration of the other ion is known.

From a table of K_{sp} values, the value of the K_{sp} for iron sulfide (FeS) is 1×10^{-22} mol^2/liter^2 at a certain temperature.

$$FeS(s) \rightleftharpoons Fe^{2+}(aq) + S^{2-}(aq)$$

First, write the expression for the K_{sp} of FeS.

$K_{sp} =$ _____

Answer: $[Fe^{2+}][S^{2-}] = 1 \times 10^{-22}$ mol^2/liter^2

68 A liter of water solution already contains some small quantity of Fe^{2+} in it. The water is at the proper temperature. By experiment, it is determined that a maximum of 1×10^{-3} mol of S^{2-} can be mixed in this liter of water solution before the FeS precipitate is formed. Since the value for K_{sp} is known and the concentration of S^{2-} is 1×10^{-3} mol/liter, we can determine the concentration of Fe^{2+} in the water sample by modifying the K_{sp} expression.

$$[Fe^{2+}] = \frac{K_{sp}}{[S^{2-}]}$$

Calculate the concentration of Fe^{2+}. _____

Answer: $[Fe^{2+}] = \dfrac{K_{sp}}{[S^{2-}]} = \dfrac{1 \times 10^{-22} \text{ mol}^2/\text{liter}^2}{1 \times 10^{-3} \text{ mol/liter}} = 1 \times 10^{-19}$ mol/liter

69 The K_{sp} for $BaCO_3$ is 5×10^{-9} mol^2/liter^2 at 20°C.

$$K_{sp} = [Ba^{2+}][CO_3{}^{2-}] = 5 \times 10^{-9} \text{ mol}^2/\text{liter}^2$$

An aqueous solution contains a Ba^{2+} concentration of 2×10^{-4} mol/liter. What is the maximum possible $CO_3{}^{2-}$ concentration in that solution just before precipitation of $BaCO_3$ occurs? The solution temperature is kept at 20°C.

$[CO_3{}^{2-}] =$ _____

Answer: $\dfrac{K_{sp}}{[Ba^{2+}]} = \dfrac{5 \times 10^{-9} \text{ mol}^2/\text{liter}^2}{2 \times 10^{-4} \text{ mol/liter}} = 2.5 \times 10^{-5}$ mol/liter

70 Determine the maximum possible F^- ion concentration in an aqueous solution containing 1×10^{-3} mol/liter of Ca^{2+} ion. The K_{sp} for CaF_2 is equal to 4×10^{-11} mol^3/liter^3 at the same temperature as the solution. (This one is more difficult. Remember to square the F^- in the K_{sp} expression.)

$$CaF_2(s) \rightleftharpoons Ca^{2+}(aq) + 2F^-(aq)$$

$[F^-] =$ _____

Answer:
$$K_{sp} = [Ca^{2+}][F^-]^2$$

$$[F^-]^2 = \frac{K_{sp}}{[Ca^{2+}]}$$

$$[F^-] = \sqrt{\frac{K_{sp}}{[Ca^{2+}]}} = \sqrt{\frac{4 \times 10^{-11} \text{ mol}^3/\text{liter}^3}{1 \times 10^{-3} \text{ mol/liter}}} = \sqrt{4 \times 10^{-8} \text{ mol}^2/\text{liter}^2} = 2 \times 10^{-4} \text{ mol/liter}$$

71 Water is usually considered to be completely a nonelectrolyte. Even water, however, dissociates to a very small degree. This is called self-ionization and will be covered in more detail in the next chapter. The dissociation of water is represented by the following equation.

$$H_2O(l) \rightleftharpoons H^+(aq) + OH^-(aq)$$

Write the ion product constant expression (K_w) for the dissociation of water.

$K_w =$ _____

Answer: [H⁺][OH⁻] (Liquid water is left out since its concentration essentially does not change.)

72 The actual value of the ion product constant for water (K_w) is 1×10^{-14} mol²/liter² (at 25°C). You should memorize the value of K_w since it is very commonly used in aqueous acid and base calculations. In water with no impurities (at 25°C) the concentration of H^+ is 1×10^{-7} mol/liter and the concentration of OH^- is also 1×10^{-7} mol/liter. Calculate the value of K_w with these concentrations.

$K_w = (1 \times 10^{-7}$ mol/liter$)(1 \times 10^{-7}$ mol/liter$) =$ _____

Answer: 1×10^{-14} mol²/liter²

73 The most common definition of an acid or base depends upon the concentration of the hydrogen ion in an aqueous solution. In pure water, the concentrations of H^+ and OH^- are equal. It is neutral (neither acidic nor basic). If the H^+ ion concentration is greater than 1×10^{-7} mol/liter, the solution is acidic. If the H^+ ion concentration is less than 1×10^{-7} mol/liter, the solution is basic.

A water solution is found with [H⁺] = 1×10^{-5} mol/liter. The solution is (acidic, basic) _____.

Answer: acidic (A concentration of 1×10^{-5} is greater than 1×10^{-7}.)

74 If the [H⁺] is 1×10^{-5} mol/liter for a particular solution, the [OH⁻] can be calculated.

$$K_w = [H^+][OH^-] = 1 \times 10^{-14} \text{ mol}^2/\text{liter}^2$$

$$[OH^-] = \frac{(K_w)}{[H^+]}$$

Calculate the concentration of OH^- for this solution.

$[OH^-] = $ _____

Answer: $\dfrac{K_w}{[H^+]} = \dfrac{1 \times 10^{-14} \text{ mol}^2/\text{liter}^2}{1 \times 10^{-5} \text{ mol/liter}} = 1 \times 10^{-9} \text{ mol/liter}$

75 In Chapter 13, acids and bases will be covered in greater detail. For now, an acid is considered to be a substance that increases the H^+ concentration of neutral water. A base is considered to be a substance that decreases the H^+ concentration of neutral water. The strength of an acid or base depends only upon the extent of dissociation when in water (aqueous) solution. The same definition is already familiar to you as a strong electrolyte or a weak electrolyte. A **strong acid (or base)** is a strong electrolyte that dissociates completely into its ions in aqueous solution. A **weak acid (or base)** is a weak electrolyte that dissociates only partially in aqueous solution.

HCl is an acid that is a strong electrolyte.

$$HCl \rightarrow H^+ + Cl^-$$

Remember that M indicates molarity (moles/liter). If a 0.1 M solution of HCl dissociates, it produces a _____M solution of H^+ and a _____M solution of Cl^-.

Answer: 0.1; 0.1

76 HCl is such a strong electrolyte that in an aqueous solution, there are only H^+ and Cl^- ions. The equation for the dissociation of HCl is:

$$HCl \rightarrow H^+(aq) + Cl^-(aq)$$

In aqueous solution, there is no more HCl; only H^+ and Cl^- ions can be found. The equation for the dissociation of HCl goes to completion. Only the products are left. The reactant (HCl) is completely used up to form products. An equation that goes to completion must be distinguished from an equilibrium equation. In an equilibrium equation, both the products and the reactants are continually present and the forward and reverse reactions continue at equal rates.

An aqueous solution of 0.001 M HCl is actually dissociated into a solution of 0.001 M H^+ and 0.001 M Cl^-.

Ignoring the chloride ion, a solution of 0.001 M H^+ is the same as $[H^+] = 1 \times 10^{-3}$ mol/liter by definition.

(a) A solution of 0.01 M HCl results in $[H^+] = $ _____

(b) After being dissociated into ions in an aqueous solution, would you expect to find any undissociated HCl in the 0.01 M solution? _____

Answer: (a) 0.01 M, or 1×10^{-2} M, or 1×10^{-2} mol/liter (all are correct answers); (b) No, the HCl dissociates completely into ions.

77 Not all acids are strong electrolytes. Some are weak electrolytes that only dissociate partly into ions. Part of the weak electrolyte remains undissociated. This part establishes equilibrium with the dissociated ions. The equilibrium constant for a weak acid is abbreviated K_a.

Acetic acid is an example of a weak acid. What is the acetic acid equilibrium constant expression?

$$HC_2H_3O_2(aq) \rightleftharpoons H^+(aq) + C_2H_3O_2^-(aq)$$

$K_a = $ _____

Answer: $\dfrac{[H^+][C_2H_3O_2^-]}{[HC_2H_3O_2]}$

78 At 0.02 mol/liter concentration, acetic acid is 3% dissociated. Therefore, the hydrogen ion concentration of the solution equals 3% multiplied by 0.02 mol/liter. (The 3% becomes 0.03 in decimal form.)

$$HC_2H_3O_2 \rightleftharpoons H^+ + C_2H_3O_2^-$$

The acetate ion concentration $[C_2H_3O_2^-]$ equals the hydrogen ion concentration $[H^+]$ since one acetate ion is formed for every hydrogen ion. What is the concentration of the hydrogen ion?

$$[H^+] = (0.03)(0.02\,\text{mol/liter}) = \underline{\qquad}$$

Answer: (The solution is made up of 0.02 mol/liter $HC_2H_3O_2$ acid. If the acid were to dissociate completely (100%), the value of $[H^+]$ would also be 0.02 mol/liter. However, the acid only dissociates 3%. Therefore, the hydrogen ion concentration will be 3% of 0.02 mol/liter.)

$$[H^+] = (0.03)(0.02\,\text{mol/liter}) = 0.0006\,\text{mol/liter or } 6 \times 10^{-4}\,\text{mol/liter}$$

79 Acetic acid is 3% dissociated in a 0.02 mol/liter solution. Therefore,

$$[H^+] = (0.03)(0.02\,\text{mol/liter}) = 6 \times 10^{-4}\,\text{mol/liter}$$

$$[C_2H_3O_2^-] = (0.03)(0.02\,\text{mol/liter}) = 6 \times 10^{-4}\,\text{mol/liter}$$

The remaining 97% is molecular acetic acid. The 97% becomes 0.97 in decimal form. The concentration of the remaining molecular acetic acid can be found by multiplying 0.97 times 0.02 mol/liter.

$$[HC_2H_3O_2] = (0.97)(0.02 \text{ mol/liter}) = \underline{\hspace{2cm}}$$

Answer: 0.0194 mol/liter or 1.94×10^{-2} mol/liter

 Calculate the ionization constant (K_a) for acetic acid. (Round to the nearest hundredth.)

$$[H^+] = (0.03)(0.02 \text{ mol/liter}) = 6 \times 10^{-4} \text{ mol/liter}$$

$$[C_2H_3O_2{}^-] = (0.03)(0.02 \text{ mol/liter}) = 6 \times 10^{-4} \text{ mol/liter}$$

$$[HC_2H_3O_2] = (0.97)(0.02 \text{ mol/liter}) = 1.94 \times 10^{-2} \text{ mol/liter}$$

$$K_a = \frac{[H^+][C_2H_3O_2{}^-]}{[HC_2H_3O_2]} = \underline{\hspace{5cm}}$$

Answer: $\dfrac{(6 \times 10^{-4} \text{ mol/liter})(6 \times 10^{-4} \text{ mol/liter})}{1.94 \times 10^{-2} \text{ mol/liter}} = 1.86 \times 10^{-5} \text{ mol/liter}$

81 A solution of $0.050\,M$ hydrofluoric acid (HF) is found to be 10% dissociated (or ionized). If the acid were to dissociate completely (100%), the value of $[H^+]$ would also be $0.050\,M$. However, the acid only dissociates 10%. Therefore, the hydrogen ion concentration will be 10% of $0.050\,M$. The 10% converts to a decimal of 0.10 in the calculation. The portion of HF that dissociates produces one H^+ ion for each F^- ion.

$$HF(aq) \rightleftharpoons H^+(aq) + F^-(aq)$$

Determine the concentrations (expressed as M, molarity) of H^+ and F^- in the solution.

$$[H^+] = \underline{\hspace{2.5cm}}$$

$$[F^-] = \underline{\hspace{2.5cm}}$$

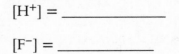

Answer: (0.10)(0.050 M) = 0.005 M or 5×10^{-3} M
(0.10)(0.050 M) = 0.005 M or 5×10^{-3} M

82 The HF acid solution is 0.050 M. If 90% of the acid remains in molecular form, calculate the concentration of the acid remaining in molecular form. (The 90% becomes 0.90 in decimal form.)

[HF] = _____

Answer: (0.90)(0.050 M) = 4.5×10^{-2} M

83 Write the ionization constant expression (K_a) for hydrofluoric acid and calculate the value of K_a using the various concentrations just calculated in the previous two frames. (Round to the nearest tenth.)

$$HF(aq) \rightleftharpoons H^+(aq) + F^-(aq)$$

$K_a =$ _____

Answer: 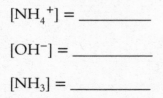 $\dfrac{[H^+][F^-]}{[HF]} = \dfrac{(5 \times 10^{-3}\ M)(5 \times 10^{-3}\ M)}{(4.5 \times 10^{-2}\ M)} = 5.6 \times 10^{-4}\ M$

84 Some bases are also weak electrolytes. The ionization constants of weak bases are calculated in the same way as those of weak acids. An ammonium hydroxide solution is 0.05 M. Assume that the ammonium hydroxide is 2% ionized (dissociated).

$$NH_3(aq) + H_2O(l) \rightleftharpoons NH_4^+(aq) + OH^-(aq)$$

Determine the concentration of NH_4^+ ions and OH^- ions in the solution. Also determine the concentration of the $NH_3(aq)$ that did not dissociate.

[NH$_4^+$] = _____

[OH$^-$] = _____

[NH$_3$] = _____

Answer: The concentrations of NH_4^+ ions and OH^- ions are equal. The solution is 0.05 M but only 2% ionized.

$$[NH_4^+] = (0.02)(0.05\ M) = 0.001\ M \text{ or } 1 \times 10^{-3}\ M$$

$$[OH^-] = (0.02)(0.05\ M) = 0.001\ M \text{ or } 1 \times 10^{-3}\ M$$

If 2% is ionized, that leaves 98% in molecular form.

$$[NH_3] = (0.98)(0.05\ M) = 0.049\ M = 4.9 \times 10^{-2}\ M$$

85 Determine the value of the ionization constant (K_b) for the base ammonium hydroxide. (Round to the nearest tenth.)

$$[NH_4^+] = 1 \times 10^{-3} M$$

$$[OH^-] = 1 \times 10^{-3} M$$

$$[NH_3] = 4.9 \times 10^{-2} M$$

$$K_b = \underline{\hspace{2in}}$$

Answer: $\dfrac{[NH_4^+][OH^-]}{[NH_3]} = \dfrac{(1 \times 10^{-3}\ M)(1 \times 10^{-3}\ M)}{(4.9 \times 10^{-2}\ M)} = 2.0 \times 10^{-5}\ M$

86 Calculate the ionization constant (K_a) for formic acid (HCOOH) that ionizes (dissociates) 4% in a 0.10 M solution. (Round to the nearest tenth.)

$$HCOOH \rightleftharpoons H^+ + COOH^-$$

$$K_a = \underline{\hspace{2in}}$$

Answer: If 4% is ionized, then 96% must be in molecular form.

$$[H^+] = (0.04)(0.10\ M) = 0.004\ M = 4 \times 10^{-3}\ M$$

$$[COOH^-] = (0.04)(0.10\ M) = 0.004\ M = 4 \times 10^{-3}\ M$$

$$[HCOOH] = (0.96)(0.10\ M) = 0.096\ M = 9.6 \times 10^{-2}\ M$$

$$K_a = \frac{[H^+][COOH^-]}{[HCOOH]} = \frac{(4 \times 10^{-3})(4 \times 10^{-3})}{(9.6 \times 10^{-2}\ M)} = 1.7 \times 10^{-4}\ M$$

87 The weak acid H_3PO_4 is different from previous weak acids because three hydrogen ions are formed for every phosphate (PO_4^{3-}) ion. An acid with more than one hydrogen ion for each negative ion is called a **polyprotic acid**. A polyprotic acid loses its hydrogen ions one at a time. The three steps for the dissociation of H_3PO_4 are:

(1) $H_3PO_4 \rightleftharpoons H^+ + H_2PO_4^-$

(2) $H_2PO_4^- \rightleftharpoons H^+ + HPO_4^{2-}$

(3) $HPO_4^{2-} \rightleftharpoons H^+ + PO_4^{3-}$

Write the equilibrium constant expressions for each of these steps.

$$K_{a_1} = \underline{\hspace{3cm}}$$

$$K_{a_2} = \underline{\hspace{3cm}}$$

$$K_{a_3} = \underline{\hspace{3cm}}$$

Answer:

$$K_{a_1} = \frac{[H^+][H_2PO_4^-]}{[H_3PO_4]}$$

$$K_{a_2} = \frac{[H^+][HPO_4^{2-}]}{[H_2PO_4^-]}$$

$$K_{a_3} = \frac{[H^+][PO_4^{3-}]}{[HPO_4^{2-}]}$$

88 Up to this point, all solutions involving equilibrium have been aqueous. Solvents other than water are also possible since the solvent makes no difference in the calculations. Gases are also possible solutes. One example could be N_2O_4 in equilibrium with NO_2 in a chloroform solvent. Write the equilibrium constant expression (K_{eq}) for $N_2O_4 \rightleftharpoons 2NO_2$.

$$K_{eq} = \underline{\hspace{3cm}}$$

Answer: $\frac{[NO_2]^2}{[N_2O_4]}$

89 The equilibrium constant (K_{eq}) for the equilibrium of $N_2O_4 \rightleftharpoons 2NO_2$ is equal to 1.1×10^{-5} mol/liter at a specified temperature. A solution of N_2O_4 in chloroform solvent is determined to be 0.02 mol/liter. Determine the concentration of NO_2 at the specified temperature. (Round to the nearest tenth.)

$$[NO_2] = \underline{\hspace{4cm}}$$

Answer: The equilibrium expression found in frame 88 can be modified to solve for $[NO_2]$.

$$[NO_2]^2 = (K_{eq})[N_2O_4]$$

$$[NO_2] = \sqrt{(K_{eq})[N_2O_4]}$$

$$= \sqrt{(1.1 \times 10^{-5} \text{ mol/liter})(0.02 \text{ mol/liter})}$$

$$= \sqrt{2.2 \times 10^{-7} \text{ mol}^2/\text{liter}^2}$$

$$= 4.7 \times 10^{-4} \text{ mol/liter}$$

The concepts of chemical and ionic equilibria are very important to you and the chemist. Many useful industrial products, such as NH_3 for fertilizers and explosives, require the manipulation of an equilibrium situation to achieve their production. Many physiological processes rely upon equilibria between dissolved salts and weak acids and bases that are present in our bodies. Blood cells must be surrounded by solutions with proper electrolyte balance and H^+ ion concentration similar to that found within the cell, or the cell may be damaged. In the next chapter, we say more about acids and bases and their equilibria.

SELF-TEST

This self-test is designed to show how well you have mastered this chapter's objectives. Correct answers and review instructions follow the test.

1. Indicate whether each reaction is a forward or reverse reaction.

   (a) $N_2 + 3H_2 \rightarrow 2NH_3$

   (b) $3O_2 \leftarrow 2O_3$

   (c) $H_2CO_3 \rightarrow CO_2 + H_2O$

   (d) $H_2SO_4 \leftarrow H_2O + SO_3$

2. Write the equilibrium constant expressions for the following reversible reactions.

   (a) $2HI \rightleftharpoons H_2 + I_2$

   (b) $2SO_2 + O_2 \rightleftharpoons 2SO_3$

3. Write the equilibrium constant expressions for the following reversible reactions.

   (a) $N_2 + O_2 \rightleftharpoons 2NO$

   (b) $2NO + O_2 \rightleftharpoons 2NO_2$

4. Write the equilibrium constant expression for the following reaction.

$$4HCl(g) + O_2\ (g) \rightleftharpoons 2Cl_2\ (g) + 2H_2O(g)$$

   (a) $K_{eq} = $ _____

   (b) What units will K_{eq} have? _____

5. Calculate how many moles of HI would be present at equilibrium when $[H_2] = 1 \times 10^{-2}$, $[I_2] = 2.5 \times 10^{-2}$, and $K_{eq} = 50$ for this equilibrium mixture. Assume system volume is 1 liter. (Round to the nearest hundredth.)
_____

$$H_2(g) + I_2(g) \rightleftharpoons 2HI(g)$$

6. Write the equilibrium constant expression for the following equilibrium reaction.
$$Pb1_2(s) \rightleftharpoons Pb^{2+}(aq) + 2I^-(aq)$$
$K_{eq} = $ _____

7. Write the equilibrium constant expression for the following equilibrium reaction.
$$CH_3CO_2H(aq)+ \rightleftharpoons CH_3CO_2^-(aq) + H^+(aq)$$
$K_{eq} = $ _____

8. Write the equilibrium constant expression for the following equilibrium reaction. $Al_2Cl_6(aq) + 12H_2O(l) \rightleftharpoons 2Al(H_2O)_6^{3+}(aq) + 6Cl^-(aq)$
$K_{eq} = $ _____

9. In what direction will the equilibrium be shifted (left toward the reactant side or right toward the product side) by the following changes?

$$heat + N_2(g) + O_2(g) \rightleftharpoons 2NO(g)$$

(a) an increase in pressure _____

(b) a decrease in temperature _____

(c) a decrease in N_2 concentration _____

10. In what direction will the equilibrium be shifted (left towards the reactant side or right toward the product side) by the following changes?
$$HCO_2H(aq) \rightleftharpoons HCO_2^-(aq) + H^+(aq)$$

(a) An increase in the concentration H+_____

(b) A decrease in the concentration of HCO_2H_____

(c) A decrease in the concentration of HCO_2^-_____

11. In what direction will the equilibrium be shifted (left towards the reactant side or right toward the product side) by the following changes?
$$NH_3(aq) + HCL(aq) \rightleftharpoons NH_4Cl(qa) + heat$$

(a) A decrease in temperature_____

(b) An increase in the concentration of NH_3_____

(c) A decrease in the concentration of HCl_____

12. Calculate (to the nearest hundredth) the molar solubility of $BaSO_4$ if

$$K_{sp} \text{ of } BaSO_4 = 1.1 \times 10^{-10}$$

13. Calculate (to the nearest hundredth) the molar solubility of PbS if K_{sp} of PbS = 9.04 x 10^{-29}.
$$PbS(s) \rightleftharpoons Pb^{2+}(aq) + S^{2-}(aq)$$

14. A 0.10 *M* solution of ammonium hydroxide is found to be 1% ionized. Calculate K_b (to the nearest tenth). _____

15. A 0.15 *M* solution of hydrocyanic acid (HCN) is found to be 1% ionized. Calculate K_a (to the nearest tenth). The K_a for HCN is 4.9 x 10^{-10}.

ANSWERS

Compare your answers to the self-test with those given below. If you answer all questions correctly, you are ready to proceed to the next chapter. If you miss any, review the frames indicated in parentheses following the answers. If you miss several questions, you should probably reread the chapter carefully.

1. (a) Forward

   (b) Reverse

   (c) Forward

   (d) Reverse [frames 1-5]

2. (a) $K_{eq} = ([H_2][I_2]) / [HI]^2$ [frames 8-13]

   (b) $K_{eq} = [SO_3]^2 / ([SO_2]^2[O_2])$ [frames 8-13]

3. (a) $K_{eq} = [NO]^2 / ([N_2][O_2])$ [frames 8-13]

   (b) $K_{eq} = [NO_2]^2 / ([NO]^2[O_2])$ [frames 8-13]

4. (a) $\dfrac{[Cl_2]^2[H_2O]^2}{[HCl]^4[O_2]}$ (frames 9–13)

   (b) liters/mol (frames 18–22)

5. $$K_{eq} = \frac{[HI]^2}{[H_2][I_2]} = 50$$

   $$[HI] = \sqrt{50 \times [H_2] \times [I_2]}$$

   $$[HI] = \sqrt{(50)(1 \times 10^{-2})(2.5 \times 10^{-2})} = \sqrt{1.25 \times 10^{-2}} = 1.12 \times 10^{-1} \text{ mol}$$

   (frames 14–16)

6. $K_{eq} = [Pb^{2+}][I^-]^2$ [frames 9-13, 42–47]

7. $([CH_3CO_2^-][H^+]) / [CH_3CO_2H]$ [frames 9-13, 42–47]

8. $\dfrac{[Al(H_2O)_6^{3+}]^2[Cl^-]^6}{[Al_2Cl_6]}$ (frames 9–13, 25–30)

9. (a) no change (frames 48–51)

   (b) left toward the reactant side (frames 42–47)

   (c) left toward the reactant side (frames 34–41)

10. (a) Left toward the reactant side [frames 53-58]

    (b) Left toward the reactant side [frames 53-58]

    (c) Right toward the product side [frames 53-58]

11. (a) Right toward the product side [frames 42-47]

    (b) Right toward the product side [frames 42-47]

    (c) Left toward the reactant side [frames 42-47]

12.
$$BaSO_4(s) \rightleftharpoons Ba^{2+} + SO_4^{2-}$$

if S = molar solubility

$$\therefore [Ba^{2+}] = S$$

$$[SO_4^{2-}] = S$$

$$K_{sp} = [Ba^{2+}][SO_4^{2-}] = 1.1 \times 10^{-10}$$

$$\therefore S^2 = 1.1 \times 10^{-10}$$

$$\therefore S = 1.05 \times 10^{-5} \text{ mol/liter (frames 55, 64, 65)}$$

13. $PbS(s) \rightleftharpoons Pb^{2+}(aq) + S^{2-}(aq)$
    If S = molar solubility, $[Pb^{2+}] = S$ and $[S^{2-}] = S$
    $K_{sp} = [Pb^{2+}][S^{2-}] = 9.04 \times 10^{-29}$
    $S^2 = 9.04 \times 10^{-29}$
    $S = 9.51 \times 10^{-15}$ mol/liter [frames 55, 64, 65]

14.
$$NH_3(g) + H_2O(l) \rightleftharpoons NH_4^+(aq) + OH^-(aq)$$

$$[NH_4^+] = (0.1)(0.01) = 0.001$$

$$[OH^-] = (0.1)(0.01) = 0.001$$

$$[NH_3] = (0.10 - 0.001) = 0.099$$

$$K_b = \frac{(0.001)^2}{0.099} = \frac{(1 \times 10^{-3})^2}{9.9 \times 10^{-2}} = 1.0 \times 10^{-5}$$

(frames 84, 85)

15. $HCN(aq) \rightleftharpoons CN^-(aq) + H^+(aq)$
$[CN^-] = (0.15)(0.01) = 0.0015$
$[H^+] = (0.15)(0.01) = 0.0015$
$[HCN] = (0.15 - 0.0015) = 0.1485$
$K_a = ([CN^-][H^+])/[HCN] = (0.0015)^2 / 0.1485 = 1.5 \times 10^{-5}$ [frames 84–86]

EVERYDAY CHEMISTRY

Neutralizing battery corrosion

A friend had purchased a measuring device that used an internal "D" cell battery. The cell had been left in the device too long. The electrolyte had leaked out of the battery and corroded the contacts and part of the inside of the measuring device. The friend had asked for a suggestion to neutralize the corrosion chemical. He did not know whether the "D" cell battery, which had been removed before he purchased it, was an alkaline cell or a low drain carbon zinc cell often labeled as a "heavy duty" cell. We suggested starting with a small amount of vinegar or lemon juice followed by a baking soda mixture (sodium hydrogen carbonate $NaHCO_3$, also known as sodium bicarbonate) and, of course, protecting eyes and skin from any possible contact. The electrolyte in an alkaline cell is KOH, a strong base. The electrolyte in a "heavy duty" carbon zinc cell, meant for low-drain devices, is an acidic mixture typically containing ammonium chloride and zinc chloride.

The friend asked why we had suggested baking soda. We told him that baking soda ($NaHCO_3$) is **amphiprotic** in an aqueous solution which means it can neutralize either an acidic substance or a base. If the corrosion had been from an alkaline battery, the vinegar which is dilute acetic acid would help neutralize the corrosion. If the corrosion had been from a carbon zinc cell, the vinegar might have added somewhat to the acidic substance but then be quickly neutralized by the baking soda slurry. The "amphi-" part of the word means "both" or "on both sides". In the next chapter you will learn about the "-protic" part of the word amphiprotic and why it is called that, about something called a "buffer" solution, and about two chemists named Brønsted and Lowry who were independently responsible for a related theory of acids and bases.

13 Acids and Bases

Chapters 11 and 12 gave you an indication that all acids have certain properties in common and all bases have certain properties in common. The major common property is that acids react with bases (and vice versa) to produce salts. For example, if solutions of HCl (an acid) and KOH (a base) are mixed, the following reaction occurs.

$$HCl + KOH \rightarrow KCl + H_2O$$

$$acid + base \; \rightarrow salt \; + water$$

Such a reaction gives a solution that no longer has the acidic or basic properties that were evident before mixing, provided the correct volumes and concentrations were used.

What then is an acid? What is a base? There are three definitions that have been developed through the years. Each has its own particular usefulness, depending upon the nature of the reactants and the conditions of the reaction. In this chapter we will discuss each of the definitions and their particular usefulness.

Our discussion of acids and bases will touch on several other important concepts, including reactions of salts with water, another concentration term specially developed for acid–base solutions, and the importance of acid–base chemistry to physiological and industrial processes.

OBJECTIVES

After completing this chapter, you will be able to

- recognize and apply or illustrate the following: Arrhenius, Brønsted–Lowry, and Lewis acids and bases, neutralization, hydrolysis, pH, buffer solution, titration, conjugate acid or base, amphiprotic, indicator, hydronium ion, and hydrated;

- write a chemical equation for a neutralization reaction between any acid and base;

- predict whether a solution of a given salt will be acidic, basic, or neutral;

- calculate the pH of a solution when given:
  - (a) the degree of ionization of a weak acid or base and vice versa,
  - (b) K_a or K_b of the acid or base and vice versa,
  - (c) the concentration of a solution of a strong acid or base;
- solve titration problems.

ARRHENIUS ACIDS AND BASES

1 There are several chemical theories of acids and bases. The most familiar is that of Arrhenius. According to the Arrhenius definition, an **acid** is any substance that produces or increases the H^+ ion concentration in an aqueous solution (remember, water is the solvent). Both HCl and $HC_2H_3O_2$ can increase the H^+ ion concentration of water. Do HCl and $HC_2H_3O_2$ qualify as acids according to Arrhenius? _____

Answer: yes (because they increase the H^+ concentration of water)

2 An Arrhenius **base** is any substance that produces or increases the concentration of OH^- ion in aqueous solutions. All of the substances that have been named acids and bases so far in this book qualify as Arrhenius acids and bases.

NaOH is a strong electrolyte and NH_4OH is a weak electrolyte. In pure water, they both increase the OH^- concentration. Do NaOH and NH_4OH qualify as Arrhenius bases? _____

Answer: yes (because they both increase the OH^- ion concentration in aqueous solution)

3 A **neutralization** reaction takes place between an acid and a base in aqueous solution if the moles of H^+ ion equal the moles of OH^- ion. The general equation (net ionic) for such a neutralization is as follows.

$$H^+ + OH^- \rightarrow H_2O$$

Complete and balance the following neutralization reaction, which takes place in aqueous solution. (You may complete the reaction in molecular form.)

$$Ba(OH)_2 + HCl \rightarrow \underline{\hspace{2cm}}$$

Answer: $BaCl_2 + 2H_2O$

4 Here is another typical neutralization reaction.

$$H_2SO_4 + 2NaOH \rightarrow Na_2SO_4 + 2H_2O$$

In each of these neutralization reactions, the product that all these equations have in common is _____.

Answer: H$_2$O

5 In any neutralization reaction, regardless of the acid–base definition we are using, one of the products is the same as the solvent. In a neutralization reaction taking place between an Arrhenius acid and an Arrhenius base, one product will always be _____.

Answer: H$_2$O

6 If the concentration of an acid solution is unknown, a basic solution of known concentration can be added slowly in measured amounts until the solution is neutralized. The concentration of the acid can then be determined. It is also possible to do the reverse, adding an acid solution to a base. This process of adding a basic solution to an acid or vice versa for neutralization is called **titration**. An **indicator** is used in the solution being titrated. The indicator shows one color at a specific level of hydrogen ion concentration and another color at another level of hydrogen ion concentration.

The concentration of a basic solution can be determined by a process called

_____.

Answer: titration

7 The hydrogen ions of an acid react with the hydroxide ions of a base during neutralization to produce water.

$$H^+ + OH^- \rightarrow H_2O$$

The ions other than H$^+$ and OH$^-$ that make up an acid or a base are **spectator ions** that do not enter the reaction.

(a) Complete and balance the following neutralization reaction.

$HCl + Ca(OH)_2 \rightarrow$ _____.

(b) The complete ionic equation for this reaction is _____.

(c) The spectator ions are _____ and _____.

Answer:

(a) $CaCl_2 + 2H_2O$

(b) $2H^+ + 2Cl^- + Ca^{2+} + 2OH^- \rightarrow Ca^{2+} + 2Cl^- + 2H_2O$

(c) Ca^{2+}; $2Cl^-$

8 The net ionic equation for a neutralization reaction is simply:

$$H^+ + OH^- \rightarrow \underline{\hspace{2cm}}.$$

Answer: H_2O

9 In a neutralization reaction, one H^+ ion reacts with one OH^- ion to form one molecule of H_2O. (The ratio of H^+ to OH^- is 1 to 1.)

Molarity is a measure of concentration. Molarity equals moles of solute per liter of solution. Remember that M indicates moles/liter (this was introduced in Chapter 11). A 1 liter solution of 1 M H^+ ions could be expected to completely neutralize 1 liter of 1 M OH^- ions. (One mole of H^+ ions completely neutralizes 1 mol of OH^- ions.) It would take how many liter(s) of 0.5 M H^+ ion solution to neutralize 2 liters of 0.5 M OH^- ion solution? _____

Answer: 2 liters (Neutralization of H^+ and OH^- takes an equal number of moles of each. Two liters of 0.5 M H^+ solution represents 1 mol of H^+ ions (Chapter 11). Two liters of 0.5 M OH^- solution represents 1 mol of OH^- ions.)

10 The neutralization of H^+ and OH^- to form H_2O is possible when an equal number of moles of each type of ion are present. A 1 liter solution of 0.1 M H^+ ions represents 0.1 mol of H^+ ions and can be neutralized (through titration) by any of the following:

1 liter of 0.1 M OH^- ion solution, which is 0.1 mol of OH^- ions, or
100 milliliters of 1.0 M OH^- solution (also 0.1 mol of OH^- ions), or
200 milliliters of 0.5 M OH^- solution (also 0.1 mol of OH^- ions), or
500 milliliters of _____ M OH^- ion solution (also 0.1 mol of OH^- ions).

Answer: 0.2 $\left(M = \dfrac{0.1 \text{ mol}}{0.5 \text{ liter}} = 0.2 \text{ mol/liter} \right)$

11 How many liter(s) of 0.4 M H^+ solution are needed to neutralize 3 liters of 0.2 M OH^- solution? _____

Answer:

$$\text{mol } H^+ \text{ soln} = 3 \text{ liters } OH^- \times \frac{0.2 \text{ mol } OH^-}{1 \text{ liter soln}} \times \frac{1 \text{ mol } H^+}{1 \text{ mol } OH^-} \times \frac{1 \text{ liter soln}}{0.4 \text{ mol } H^+} = 1.5 \text{ liters } H^+ \text{ soln}$$

(As you probably guessed, "soln" is the abbreviation for "solution.")

 How many milliliters of $0.6 \ M \ H^+$ solution are needed to titrate (neutralize) 250 milliliters of $1.8 \ M \ OH^-$ solution? _____

Answer:

$$\text{mL } H^+ \text{ soln} = 250 \text{ mL } OH^- \text{ soln} \times \frac{1 \text{ liter soln}}{1000 \text{ mL soln}} \times \frac{1.8 \text{ mol } OH^-}{1 \text{ liter soln}} \times \frac{1 \text{ mol } H^+}{1 \text{ mol } OH^-}$$

$$\times \frac{1 \text{ liter soln}}{0.6 \text{ mol } H^+} \times \frac{1000 \text{ mL soln}}{1 \text{ liter soln}}$$

$$= 750 \text{ mL } H^+ \text{ soln}$$

 Let's use H_2SO_4 to titrate an OH^- solution. H_2SO_4 dissociates to form two H^+ ions.

$$H_2SO_4 \rightarrow 2H^+ + SO_4^{2-}$$

Therefore, a $1 \ M$ solution of H_2SO_4 will dissociate to form a $2 \ M \ H^+$ solution. (Ignore the SO_4^{2-} ion since it is a spectator ion.)

How many liters of $1 \ M \ H_2SO_4$ are needed to neutralize 1 liter of $1 \ M \ OH^-$ solution? _____

Answer:

$$\text{liters } H_2SO_4 \text{ soln} = 1 \text{ liter } OH^- \text{ soln} \times \frac{1 \text{ mol } OH^-}{1 \text{ liter soln}} \times \frac{1 \text{ liter soln}}{1 \text{ mol } H_2SO_4} \times \frac{1 \text{ mol } H_2SO_4}{2 \text{ mol } H^+}$$

$$\times \frac{1 \text{ mol } H^+}{1 \text{ mol } OH^-} = 0.50 \text{ liter } 1 \ M \ H_2SO_4 \text{ soln}$$

 How many milliliters of $0.1 \ M \ KOH$ are needed to titrate (neutralize) 25 milliliters of $0.2 \ M \ H_3PO_4$ solution?

Answer:

$$\text{mL KOH soln} = 25 \text{ mL } H_3PO_4 \times \frac{1 \text{ liter soln}}{1000 \text{ mL soln}} \times \frac{0.2 \text{ mol } H_3PO_4}{1 \text{ liter soln}} \times \frac{3 \text{ mol } H^+}{1 \text{ mol } H_3PO_4}$$

$$\times \frac{1 \text{ mol } OH^-}{1 \text{ mol } H^+} \times \frac{1 \text{ liter soln}}{0.1 \text{ mol KOH}} \times \frac{1000 \text{ mL}}{1 \text{ liter soln}}$$

$$= 150 \text{ mL KOH soln}$$

Titration is a very useful laboratory technique. You will more than likely have to use it to determine the concentration of an unknown acid or base. You may also be required to calculate the formula weight or molar concentration of an acid or a base using titration data. With your understanding of moles and molarity, you should be able to calculate them if required to do so.

We now introduce a new concentration term that relates specifically to H^+ and OH^- concentrations in aqueous solutions. It is used to indicate the degree of acidity or alkalinity (or basicity) of an aqueous solution.

15 Acidity of an aqueous solution is measured by the hydrogen ion concentration of the solution. If you remember the ion product constant for pure water at 25°C, write it below. If not, use the following information to calculate it. Pure water has a H^+ concentration of $1 \times 10^{-7}\ M$ (at 25°C). The OH^- concentration of pure water is also $1 \times 10^{-7}\ M$ (at 25°C). The ion product constant for pure water at 25°C is _____.

Answer: $K_w = [H^+][OH^-] = 1 \times 10^{-14}\ M^2$

16 The K_w for water is usually listed as 1×10^{-14} with no reference to units or temperature. The temperature of 25°C and moles²/liter² (which is the same as M^2) are both assumed.

If the H^+ concentration is changed to $1 \times 10^{-3}\ M$ for an aqueous solution, what is the OH^- concentration? (Remember that K_w is a constant.)

_____

Answer: $[OH^-] = \dfrac{K_w}{[H^+]} = \dfrac{1 \times 10^{-14}\ M^2}{1 \times 10^{-3}\ M} = 1 \times 10^{-11}\ M$

17 If $[OH^-]$ is equal to $1 \times 10^{-9}\ M$, then the $[H^+]$ must be equal to _____ M.

Answer: $[OH^-] = \dfrac{K_w}{[H^+]} = \dfrac{1 \times 10^{-14}\ M^2}{1 \times 10^{-9}\ M} = 1 \times 10^{-5}\ M$

18 If the hydrogen ion concentration is known, then the hydroxide ion concentration is easily determined, and vice versa. A solution with a hydrogen ion concentration of *greater* than $1 \times 10^{-7}\ M$ is *acidic*. A solution with a hydrogen ion concentration of *less* than 1×10^{-7} M is *basic*.

Is a solution with a hydroxide ion concentration $[OH^-]$ of $1 \times 10^{-10}\ M$ acidic or basic? (Find the $[H^+]$ first.) _____

Answer: $[H^+] = \dfrac{K_w}{[OH^-]} = \dfrac{1 \times 10^{-14}\ M^2}{1 \times 10^{-10}\ M} = 1 \times 10^{-4}\ M$

Since $1 \times 10^{-4}\ M$ is greater than $1 \times 10^{-7}\ M$, the solution is definitely acidic.

pH AND pOH

19 Instead of writing the hydrogen ion concentration to indicate the acidity (or alkalinity) of a solution, a simple number indicating the **pH** of a solution is often used. The pH, sometimes called the "hydrogen ion exponent," is defined mathematically as the negative logarithm to the base 10 of the hydrogen ion concentration. For the following H^+ ion concentrations, compare the power of 10 exponent with the corresponding value of pH.

$$\text{For } [H^+] = 1 \times 10^{-2} \text{ the pH is 2}$$
$$\text{For } [H^+] = 1 \times 10^{-4} \text{ the pH is 4}$$
$$\text{For } [H^+] = 1 \times 10^{-9} \text{ the pH is 9}$$
$$\text{For } [H^+] = 1 \times 10^{-6} \text{ the pH is}\rule{1.5cm}{0.4pt}$$

Answer: 6

20 If the hydrogen ion concentration is $1 \times 10^{-(\text{exponent})}$, then the pH is equal to the exponent. Or, expressed another way:

$$pH = -\log[H^+] \quad (\text{log to the base 10})$$

If a solution has a pH of 3, what is the hydrogen ion concentration? _____

Answer: $[H^+] = 1 \times 10^{-3}$ *M*

21 What is the pH of pure neutral water? _____

Answer: $[H^+] = 1 \times 10^{-7}$ *M*
$$\therefore pH = 7$$

22 A solution is acidic if $[H^+]$ is greater than 1×10^{-7} *M*. A solution is alkaline if $[H^+]$ is less than 1×10^{-7} *M*. Is a solution with a pH of 6 acidic or alkaline? _____

Answer: acidic (The $[H^+] = 1 \times 10^{-6}$, which is greater than 1×10^{-7}.)

To determine the pH of a solution if the $[H^+]$ is something other than $1 \times 10^{-(\text{exponent})}$ requires use of common logarithms (log to the base 10). Some instructors may arrange problems so that all $[H^+]$ will be $1 \times 10^{-(\text{exponent})}$ to simplify calculations. More often than not, $[H^+]$ is something other than $1 \times 10^{-(\text{exponent})}$. We have chosen such examples and included a log table for your use in the Appendix. You may also use your scientific calculator or the handy calculators available on the Internet by way of a web browser and search engines. In this book, the answers to problems involving logarithms will assume you understand and know how to use logarithms.

23 To determine the pH of a hydrogen ion concentration of $n \times 10^{-(\text{exponent})}$, where n is a number other than 1, we must determine the logarithm of the number n.

$$pH = -\log[H^+]$$
$$pH = -\log[n \times 10^{-(\text{exponent})}]$$
$$pH = -\log n + \text{exponent}$$
$$pH = \text{exponent} - \log n$$

Determine the pH of a solution if $[H^+] = 3 \times 10^{-5}$ (to the nearest hundredth). Use the log table in the Appendix.
pH = _____

Answer: pH = exponent − log n
$[H^+] = 3 \times 10^{-5}$
pH = 5 − log 3 = 5 − 0.48 = 4.52

24 Determine the pH of a solution with $[H^+] = 6.4 \times 10^{-4}$ (to the nearest hundredth).
pH = _____

Answer: The log of 6.4 is 0.81 (see table).
If $[H^+] = n \times 10^{-(\text{exponent})}$, then
pH = exponent − log n = 4 − log 6.4 = 4 − 0.81 = 3.19.

25 Determine the pH of a 0.002 *M* aqueous solution of HCl (to the nearest hundredth). (Determine the $[H^+]$ first. The acid HCl is a strong electrolyte that dissociates completely.)
pH = _____

Answer: $[H^+] = 2 \times 10^{-3}$ *M*
pH = exponent − log n
pH = 3 − log 2 = 3 − 0.30 = 2.70

26 Conversely, we can determine $[H^+]$ if the pH is known. A pH of 8.58 indicates a hydrogen ion concentration of $1 \times 10^{-(8.58)}$ M. Such a number is not usable as is; it merely indicates an actual answer of between 1×10^{-9} and 1×10^{-8} M. We must convert it to a usable answer.

$$pH = -\log[H^+]$$
$$[H^+] = antilog(-pH)$$
$$[H^+] = antilog(-8.58)$$

The table of logs reads only positive logs. Therefore, we must change the negative antilog, antilog -8.58, to an expression of the next largest negative integer (-9) and find the antilog of the remainder ($-8.58 = -9 + 0.42$).

$$[H^+] = antilog(-8.58) = antilog(-9 + 0.42) = (antilog\ 0.42) \times (antilog\ -9)$$

The antilog of 0.42 is 2.6 (from the table). The antilog of -9 is 10^{-9}. Then $[H^+] = (antilog\ 0.42) \times (antilog\ -9) = 2.6 \times 10^{-9}$ M.

Determine the $[H^+]$ for a solution with pH of 4.72 (to the nearest tenth). $[H^+] = $ _____

Answer: $[H^+] = antilog(-pH) = antilog(-4.72) = antilog(-5 + 0.28)$
$$= (antilog\ 0.28) \times (antilog\ -5)$$
$$= 1.9 \times 10^{-5}\ M$$

27 Determine the $[H^+]$ of a solution if the pH is 10.8 (to the nearest tenth).
$[H^+] = $ _____

Answer: $[H^+] = antilog(-pH) = antilog(-10.8) = antilog(-11 + 0.2)$
$$= (antilog\ -11) \times (antilog\ 0.2) = 1.6 \times 10^{-11}\ M$$

28 Just as the pH is the negative log of $[H^+]$, we can evaluate pOH, which is the negative log of $[OH^-]$. Remember that for pure neutral water, $[H^+] = 1 \times 10^{-7}$ and $[OH^-] = 1 \times 10^{-7}$.

(a) What is the pH for pure water? _____

(b) What is the pOH for pure water? _____

(c) For pure water, pH + pOH = _____.

Answer: (a) 7; (b) 7; (c) 14

29 The pH and pOH of any aqueous solution add up to 14. If either the pH or the pOH is known for a solution, the other can quickly be determined. If the pOH is equal to 3, what is the pH? _____

Answer: pH + pOH = 14

pH = 14 − pOH = 14 − 3 = 11

Now that you can determine pH, what is its significance? Many chemical and physiological processes are pH dependent, that is, the pH of the reaction system determines whether or not the reaction will take place. The pH of your blood is normally around 7.6 (slightly basic). If it varies much, either higher or lower, you get sick. Your stomach contains HCl as part of your gastric juices and is normally acidic. What then is an "acid stomach"? The pH of the contents of your stomach has decreased (greater H^+ concentration), creating a greater than normal acidity. Electrolyte balance inside and outside the cells of our body is also pH dependent. To maintain the proper balance requires a carefully regulated system wherein the pH of the system remains virtually constant within very specific limits. We now discuss this system that is so prevalent in our bodies and that is extensively used in commercial processes to maintain a constant pH.

BUFFER SOLUTIONS

30 When chemists wish to keep the pH of a solution fairly constant even if some small amount of strong acid or base is added, they will use a buffer solution. A **buffer solution** involves a chemical equilibrium between either a weak acid and its salt or a weak base and its salt, and shows the common ion effect.

A typical buffer solution is one made up of acetic acid ($HC_2H_3O_2$), which dissociates to a small degree into H^+ and $C_2H_3O_2^-$ ions, and sodium acetate, a salt of acetic acid that dissociates completely into Na^+ and $C_2H_3O_2^-$ ions. Which ion is common to acetic acid and sodium acetate? _____

Answer: $C_2H_3O_2^-$, the acetate ion

31 A buffer solution can consist of a weak acid and its salt or a weak base and its salt, depending upon the desired pH of the buffer solution. A buffer solution with a pH in the acidic range $(1-7)$ can be made from a solution of a weak acid and its salt. A buffer solution with a pH in the basic range $(7-14)$ can be made from a solution of a weak base and its salt. $HC_2H_3O_2$ and its salt $NaC_2H_3O_2$ are useful for making a buffer solution with a pH in the _____ range.

Answer: acidic ($HC_2H_3O_2$ is a weak acid.)

32 The key to understanding the action of a buffer solution is to remember that a weak acid (or weak base) is only dissociated to a very small degree. Most of the $HC_2H_3O_2$ is still in molecular form when in aqueous solution. The salt, in contrast, is completely dissociated. All of the $NaC_2H_3O_2$ becomes Na^+ and $C_2H_3O_2^-$.

$$NaC_2H_3O_2 \rightarrow Na^+ + C_2H_3O_2^-$$

$$HC_2H_3O_2 \rightleftharpoons H^+ + C_2H_3O_2^-$$

If equal molar quantities of $NaC_2H_3O_2$ and $HC_2H_3O_2$ are mixed in an aqueous buffer solution, almost all of the acetate ion $(C_2H_3O_2^-)$ comes from $(HC_2H_3O_2$, $NaC_2H_3O_2)$ _____.

Answer: $NaC_2H_3O_2$

33 If just acetic acid is mixed in water, the result is a large amount of molecular $HC_2H_3O_2$ and only a little H^+ and a little $C_2H_3O_2^-$.

$$HC_2H_3O_2 \rightleftharpoons H^+ + C_2H_3O_2^-$$
a lot ⎯ a little ⎯ a little

If $NaC_2H_3O_2$ is mixed in water, the result is a large amount of Na^+ ion and a large amount of $C_2H_3O_2^-$ ion and no $NaC_2H_3O_2$.

$$NaC_2H_3O_2 \rightarrow Na^+ + C_2H_3O_2^-$$
none ⎯ a lot ⎯ a lot

If equal molar amounts of $HC_2H_3O_2$ and $NaC_2H_3O_2$ were added together in the same aqueous solution, the result would be:

(a) (a lot, a little) _____ $HC_2H_3O_2$

(b) (a lot, a little) _____ H^+

(c) (a lot, a little) _____ $C_2H_3O_2^-$

Answer: (a) a lot; (b) a little; (c) a lot

34 The resulting combination of both the weak acid and its fully dissociated salt results in a buffer solution with a large proportion of weak acid $(HC_2H_3O_2)$ in molecular form, a large proportion of the common ion from the salt $(C_2H_3O_2^-)$, and a small proportion of H^+ ion. We can ignore the large proportion of the positive ion from the salt (Na^+). The addition of extra $C_2H_3O_2^-$ from the salt shifts the equilibrium to increase the $HC_2H_3O_2$ and decrease the H^+, according to Le Chatelier's Principle.

$$HC_2H_3O_2 \rightleftharpoons H^+ + C_2H_3O_2^-$$

$$\text{a lot} \qquad \text{a little} \qquad \text{a lot}$$

According to Le Chatelier's Principle, what would happen to this new equilibrium if some additional H^+ were added to the solution? The concentration of $HC_2H_3O_2$ would (increase, decrease) _____. The concentration of $C_2H_3O_2^-$ would (increase, decrease) _____.

Answer: Adding some H^+ would shift the equilibrium to increase the concentration of $HC_2H_3O_2$ and, in the process, some of the $C_2H_3O_2^-$ would be used up so that its concentration would decrease. The equilibrium would shift to the left.

35 In this buffer system some additional hydrogen ion is added to the solution. The additional hydrogen ion, instead of remaining in the solution to increase the hydrogen ion concentration and decrease the pH, will combine with the acetate ion to form $HC_2H_3O_2$. In this buffer system, extra H^+ ion added to the solution shifts the equilibrium so that more ($HC_2H_3O_2$, $C_2H_3O_2^-$) _____ is produced and some of the ($HC_2H_3O_2$, $C_2H_3O_2^-$) _____ is used up.

Answer: $HC_2H_3O_2$; $C_2H_3O_2^-$

36 If some additional OH^- ion is added to the solution, a neutralization reaction would immediately occur.

$$H^+ + OH^- \rightarrow H_2O$$

Some of the H^+ ion from the equilibrium would be used up. The result is a shift in the equilibrium so that more H^+ is produced. The reaction would shift to the right temporarily ($HC_2H_3O_2 \rightarrow H^+ + C_2H_3O_2^-$) until the extra OH^- ion added had combined with the H^+ ion. At that point, the equilibrium is again restored.

Adding extra OH^- to the solution causes the equilibrium to shift in order to produce more H^+ ion. In order to produce more H^+ ion, the equilibrium shifts so that some $HC_2H_3O_2$ is (produced, used up) _____ and some $C_2H_3O_2^-$ is (produced, used up) _____.

Answer: used up; produced

37 This chemical equilibrium represents a buffer system.

$$HC_2H_3O_2 \rightleftharpoons H^+ + C_2H_3O_2^-$$

a lot   a little   a lot

(a) The addition of a little H^+ from a strong acid such as HCl would shift the equilibrium so that the concentration of $HC_2H_3O_2$ is (increased, decreased) _____ and the concentration of $C_2H_3O_2^-$ is (increased, decreased) _____.

(b) The addition of a little OH^- from a strong base such as NaOH has the opposite effect. It will remove some of the H^+ and shift the equilibrium so that the concentration of $HC_2H_3O_2$ is (increased, decreased) _____ and the concentration of the $C_2H_3O_2^-$ is (increased, decreased) _____.

Answer: (a) increased, decreased; (b) decreased, increased

38 In a buffer system such as acetic acid and the acetate ion from sodium acetate, the pH remains relatively constant because extra hydrogen ion (H^+) undergoes a chemical reaction rather than remaining in the solution. Extra hydroxide ion (OH^-) is neutralized by additional hydrogen ion produced by the acetic acid as needed. The extra OH^- ion also does not remain in solution since it is neutralized. An aqueous buffer system will continue to absorb most of the extra H^+ or OH^- until one of the equilibrants is depleted.

$$HC_2H_3O_2 \rightleftharpoons H^+ + C_2H_3O_2^-$$

a lot   a little   a lot

This buffer system will continue to absorb additional H^+ ion until the $(HC_2H_3O_2, C_2H_3O_2^-)$ _____ is used up.

Answer: $C_2H_3O_2^-$

39 A buffer solution useful for maintaining a pH of greater than 7 is made up of a weak base and a salt of the base. Such a buffer system involves an aqueous solution of ammonia ($NH_3 \cdot H_2O$) and its salt (NH_4Cl).

$$NH_4Cl \rightarrow NH_4^+ + Cl^-$$

none   a lot   a lot

$$NH_3 \cdot H_2O \rightleftharpoons NH_4^+ + OH^-$$

a lot   a little   a little

A buffer solution may be made from equal molar amounts of $NH_3 \cdot H_2O$ and NH_4Cl. Would the resulting solution contain a lot or a little of the following substances? (Ignore the Cl^- ion.)

(a) $NH_3 \cdot H_2O$ _____

(b) NH_4^+ _____

(c) OH^- _____

Answer: (a) a lot; (b) a lot; (c) a little

40 The resulting buffer solution equilibrium can be represented as follows if the Cl^- ion is ignored.

$$NH_3 \cdot H_2O \rightleftharpoons NH_4^+ + OH^- \ (K_b = 1.8 \times 10^{-5})$$

a lot   a lot   a little

According to Le Chatelier's Principle, the addition of a small quantity of extra OH^- ion from a strong base would shift the equilibrium so that the concentration of $NH_3 \cdot H_2O$ would (increase, decrease) _____ and the concentration of NH_4^+ would (increase, decrease) _____.

Answer: increase; decrease

41 The addition of more H^+ from a strong acid would cause an immediate neutralization reaction ($H^+ + OH^- \rightarrow H_2O$). The neutralization reaction would remove OH^- ions from the buffer equilibrium system. To compensate for a loss of OH^- ions (according to Le Chatelier's Principle), the equilibrium will shift so that the concentration of $NH_3 \cdot H_2O$ would (increase, decrease) _____ and the concentration of NH_4^+ would (increase, decrease) _____.

Answer: decrease; increase

42 Both the acidic and the basic buffer systems react with small additional quantities of H^+ and OH^- to shift a buffer equilibrium. Instead of remaining in solution, the extra H^+ or OH^- ions serve to shift the equilibrium and result in an increase (or decrease) of the weak acid or base and a decrease (or increase) in the salt of the acid or base. The final result is a relatively small change in the pH rather than a large change. Buffering action will continue until so much H^+ or OH^- is added that one of the equilibrants is mostly used up.

A buffer solution is useful for maintaining a relatively constant pH if small quantities of _____ or _____ are added to the solution.

Answer: H⁺; OH⁻

Here are other important equilibria involved in physiological processes.

$$H_2CO_3 \rightleftharpoons H^+ + HCO_3^- \text{ (digestive processes)}$$

$$HCO_3^- \rightleftharpoons H^+ + CO_3^{2-} \text{ (digestive processes)}$$

$$H_2PO_4^- \rightleftharpoons H^+ + HPO_4^{2-} \text{ (cell fluids)}$$

We now examine the situation where just adding a salt to water may change the pH of the water.

HYDROLYSIS OF SALTS

A salt is a product of the neutralization of an acid and a base. Common table salt (NaCl) is a product of the neutralization reaction involving HCl acid and Na OH base.

$$HCl + NaOH \rightarrow H_2O + NaCl$$

A salt can be a product of (a) a strong acid and a strong base, (b) a strong acid and a weak base, (c) a weak acid and a strong base, or (d) a weak acid and a weak base. HCl is a strong acid (completely dissociated). NaOH is a strong base (completely dissociated). NaCl is a product of a _____ acid and a _____ base.

Answer: strong; strong

By the Arrhenius definition of acids and bases, an acid and a base react to yield water and a salt.

NaOH is a strong (completely dissociated) base.
KOH is a strong base.
H_2SO_4 is a strong acid.
HCl is a strong acid.
$HC_2H_3O_2$ is a weak (partially dissociated) acid.
H_2CO_3 is a weak acid.
NH_4OH is a weak base. [Also written $NH_3 \cdot H_2O$ or NH_3(aq).]

NaCl is a product of a strong acid and a strong base.

(a) K_2CO_3 is a product of a _____ acid and a _____ base.

(b) NH_4Cl is a product of a _____ acid and a _____ base.

(c) Na_2SO_4 is a product of a _____ acid and a _____ base.

(d) $NH_4C_2H_3O_2$ is a product of a _____ acid and a _____ base.

Answer:

(a) weak; strong (H_2CO_3, KOH)

(b) strong; weak (HCl, NH_4OH)

(c) strong; strong (H_2SO_4, NaOH)

(d) weak; weak ($HC_2H_3O_2$, NH_4OH)

45   The general formula for the neutralization of Arrhenius acids and bases is:

$$acid + base \rightarrow H_2O + salt$$

The reverse reaction is possible, especially since weak acids or bases usually involve reversible equilibrium reactions. When a salt is mixed with water, it dissociates. The resulting ions may react with water in a reaction known as **hydrolysis**.

Pure water is ionized (dissociated) to a small degree. The ions of some salts may react with the ions of pure water. What ions result from the ionization of pure water? _____

Answer: H^+ and OH^-

46   The ions of pure water could possibly react with the ions of the various salts. Table salt (NaCl) is produced by reaction of a strong acid (HCl) and a strong base (NaOH). If NaCl is mixed with water, the equation could be as follows.

$$NaCl + H_2O \rightarrow HCl + NaOH$$

Both the acid (HCl) and the base (NaOH) are completely dissociated and can be shown as ions. NaCl is also completely dissociated.

$$Na^+ + Cl^- + H_2O \rightarrow H^+ + Cl^- + Na^+ + OH^-$$

Eliminate the spectator ions in the above reaction and write the net ionic equation. _____

Answer: $H_2O \rightarrow H^+ + OH^-$

47 The net ionic equation of the reaction of NaCl with water is simply $H_2O \rightarrow H^+ + OH^-$, the ionization of water. The addition of a salt from a strong acid and a strong base has no effect on the H^+ concentration or the OH^- concentration. The aqueous solution of such a salt remains neutral.

The salt Na_2SO_4 is the product of H_2SO_4 (a strong acid) and NaOH (a strong base). Would the addition of Na_2SO_4 to pure neutral water have any effect on the H^+ concentration or the OH^- concentration? _____

Answer: no (because it is a salt of a strong acid and a strong base)
$2Na^+ + SO_4^{2-} + 2H_2O \rightarrow 2Na^+ + SO_4^{2-} + 2H^+ + 2OH^-$, or $2H_2O \rightarrow 2H^+ + 2OH^-$

48 NH_4Cl is a salt of a strong acid and a weak base. The hydrolysis reaction is:

$$NH_4Cl + H_2O \rightarrow HCl + NH_4OH$$

The salt (NH_4Cl) dissociates completely and so does the strong acid (HCl). The H_2O and the weak base (NH_4OH) remain largely in molecular form. The complete ionic equation results in:

$$NH_4^+ + Cl^- + H_2O \rightarrow H^+ + Cl^- + NH_4OH$$

Eliminate the spectator ions in this equation. The net ionic equation is

_____.

Answer: $NH_4^+ + H_2O \rightarrow H^+ + NH_4OH$

49 The hydrolysis of a salt from a strong acid and a weak base produces a solution that is acidic. The net ionic equation of the hydrolysis of such a salt shows the production of additional H^+ ions. The concentration of H^+ ions is therefore increased and the solution becomes acidic.

The pH of neutral water is 7. The pH of a solution containing a salt from a strong acid and a weak base is (greater than, less than) _____ 7.

Answer: less than

50 $NaC_2H_3O_2$ is the salt of a weak acid and a strong base. The hydrolysis reaction is:

$$NaC_2H_3O_2 + H_2O \rightarrow HC_2H_3O_2 + NaOH$$

The salt ($NaC_2H_3O_2$) dissociates completely. The strong base (NaOH) also dissociates completely. The weak acid ($HC_2H_3O_2$) and H_2O remain largely in molecular form. The complete ionic equation becomes:

$$Na^+ + C_2H_3O_2^- + H_2O \rightarrow HC_2H_3O_2 + Na^+ + OH^-$$

The net ionic equation (eliminate spectator ions) is _____.

Answer: $C_2H_3O_2^- + H_2O \rightarrow HC_2H_3O_2 + OH^-$

51 The hydrolysis of a salt from a weak acid and a strong base results in a solution that is basic. The net ionic equation of the hydrolysis of such a salt shows the production of OH^- ions. The concentration of OH^- ions is increased, and therefore the solution becomes basic.

The pH of neutral water is 7. The pH of a solution containing a salt from a weak acid and a strong base is (greater than, less than) _____ 7.

Answer: greater than

52 The hydrolysis of a salt from:

a strong acid and a weak base yields an acidic solution.
a weak acid and a strong base yields a basic solution.

If the acid is stronger than the base, the salt solution will be acidic. If the base is stronger than the acid, the salt solution will be basic. If the acid and the base are both strong, the solution will be neutral.

$NaNO_3$ is the salt of HNO_3 (a strong acid) and $NaOH$ (a strong base). Which will an aqueous solution of $NaNO_3$ be? (acidic, basic, or neutral) _____

Answer: neutral

53 The only combination not yet covered is the salt of a weak acid and a weak base. In the case of a weak acid or a weak base, both may be weak but one may be slightly stronger than the other. Strength of a weak acid is measured by its K_a (ionization constant of a weak acid). The strength of a weak base is measured by its K_b (ionization constant of a weak base).

Imagine that a salt is produced from a weak acid and a weak base. The weak acid, however, is stronger than the weak base (K_a is larger than K_b). Based upon what you have already learned from the hydrolysis of a salt from a strong acid and a weak base, what would you expect from the hydrolysis of this salt?

Answer: It would be slightly acidic, because the weak acid is stronger than the weak base.

54 Another salt of a weak acid and a weak base undergoes hydrolysis.

$$K_a \text{ for the weak acid is } 2.0 \times 10^{-6}$$
$$K_b \text{ for the weak base is } 2.0 \times 10^{-4}$$

The weak base is slightly (weaker, stronger) _____ than the weak acid.

Answer: stronger (K_b is larger than K_a, therefore the weak base is somewhat stronger than the weak acid.)

55 A salt is produced from a weak acid and a weak base.

$$K_a \text{ for the weak acid is } 2.0 \times 10^{-6}$$
$$K_b \text{ for the weak base is } 2.0 \times 10^{-4}$$

Hydrolysis of the salt produces an aqueous solution that is slightly (acidic, basic) _____.

Answer: basic

56 $NH_4C_2H_3O_2$ is a salt of a weak acid ($HC_2H_3O_2$) and a weak base (NH_4OH).

$$K_a \text{ for } HC_2H_3O_2 \text{ is } 1.8 \times 10^{-5}$$
$$K_b \text{ for } NH_4OH \text{ is } 1.8 \times 10^{-5}$$

In this case, the K_a and the K_b are equal. The weak acid and the weak base are equal in strength. You learned earlier that a solution of a salt of a strong acid (completely dissociated) and a strong base (completely dissociated) would be neutral because the base and acid were equal in strength. If $NH_4C_2H_3O_2$ undergoes hydrolysis, would you expect the aqueous solution to be slightly acidic, slightly basic, or neutral? _____

Answer: neutral (Since neither the acid nor the base is stronger than the other, the salt solution will be neutral.)

57 NH_4CN is a salt of a weak acid (HCN) and a weak base ($NH_3(aq)$).

$$K_b \text{ for } NH_3 = \frac{[NH_4^+][OH^-]}{[NH_3]} = 1.8 \times 10^{-5}$$

$$K_a \text{ for } HCN = \frac{[H^+][CN^-]}{[HCN]} = 4.0 \times 10^{-10}$$

The aqueous solution resulting from the hydrolysis of NH_4CN will be (acidic, basic, neutral) _____.

Answer: basic

58 The aqueous solution of a salt derived from:

(a) a strong acid and strong base is _____.

(b) a strong acid and weak base is _____.

(c) a weak acid and strong base is _____.

(d) a weak acid and a weak base (K_a is larger than K_b) is _____.

(e) a weak acid and a weak base (K_b is larger than K_a) is _____.

(f) a weak acid and a weak base (K_a equals K_b) is _____.

Answer: (a) neutral; (b) acidic; (c) basic; (d) acidic; (e) basic; (f) neutral

The Arrhenius Theory is the most extensively used theory in beginning chemistry classes today. However, many chemists have used solvents other than water in their research. They noticed that strong Arrhenius acids and bases that were 100% ionized in water and therefore considered to be of the same strength did not all ionize to the same extent in other solvents. The strength of the acid or base depended upon the solvent used. The Arrhenius acid–base definitions limit acids to the presence of H^+ in solution and bases to the presence of OH^- in solution. A broader definition is required when using solvents other than water. We now discuss the theory that was developed for use in other solvents (nonaqueous solvents).

BRØNSTED–LOWRY ACIDS AND BASES: LEVELING EFFECT OF THE SOLVENT

59 HCl, HNO_3, and H_2SO_4 are all strong acids. They each dissociate completely in an aqueous solution. In aqueous solution, all of these acids are equal in strength because they all dissociate completely.

Perchloric acid ($HClO_4$) also dissociates completely in aqueous solution. Since acid strength is based upon the extent to which an acid dissociates, $HClO_4$ is (equal to, stronger than, weaker than) _____ HCl, HNO_3, and H_2SO_4 in aqueous solutions.

Answer: equal to

60 For simplicity, the dissociation of HCl in water is written as follows.

$$HCl(aq) \rightarrow H^+(aq) + Cl^-(aq)$$

All strong acids form a positive ion and a negative ion. In reality, however, the H^+ ion from an acid combines with H_2O to form H_3O^+, which is known as the **hydronium ion**.

(a) What is the positive ion from all strong acids? _____

(b) In aqueous solutions, this ion combines with H_2O to form what ion?

Answer: (a) H^+ or hydrogen ion; (b) H_3O^+ or hydronium ion

61 The polar nature of H_2O molecules causes the H^+ ion to combine with H_2O and form the H_3O^+ ion. An ion surrounded by H_2O molecules is said to be **hydrated**. The aqueous solution of any strong acid is actually composed of H_3O^+ ions and a negative ion.

$$HCl + H_2O \rightarrow H_3O^+ + Cl^-$$
$$HNO_3 + H_2O \rightarrow H_3O^+ + NO^{3-}$$
$$H_2SO_4 + 2H_2O \rightarrow 2\,H_3O^+ + SO_4^{2-}$$
$$HClO_4 + H_2O \rightarrow H_3O^+ + ClO_4^-$$

What ion is common to all of these strong acids? _____

Answer: H_3O^+ or hydronium ion

62 Because they dissociate completely, all strong acids form aqueous solutions that are composed of H_3O^+ ions and a negative ion. The negative ion has no effect on acid strength. The strongest acid that can exist in an aqueous solution is the hydronium ion (H_3O^+). All strong acids that dissociate completely in H_2O are of equal strength because all form the hydronium ion (H_3O^+). Even though $HClO_4$ may be stronger than HNO_3 in some other solvent, they are of equal strength in H_2O solution because both dissociate completely to form hydronium ions. In aqueous solution, no acid can be stronger than what ion? _____

Answer: hydronium ion, H_3O^+

63 In some solvent other than water, $HClO_4$ may be stronger than HNO_3. In aqueous solution, both are of equal strength because both form H_3O^+ equally well. All acids that dissociate completely in H_2O form H_3O^+ equally well. The strength of acid is limited by the solvent because H_3O^+ is the actual acid in aqueous solutions. This equalization of acid strength depending upon the solvent is known as the **leveling effect of the solvent**. Because the H$^+$ ion is hydrated when introduced in H$_2$O solvent, all acids are leveled to the strength of the H$_3$O$^+$ ion. The leveling effect of the solvent holds true also for bases. The strongest base that can exist in aqueous solution is the OH$^-$ ion.

The hydroxide ion (OH$^-$) can be described as a water molecule (H$_2$O) *without* a hydrogen ion (H$^+$). The hydronium ion (H$_3$O$^+$) can be described as a water molecule (H$_2$O) *with* a hydrogen ion (H$^+$).

(a) When the hydronium ion and the hydroxide ion combine, what is formed? (H$_3$O$^+$ + OH$^-$ →?) _____

(b) The OH$^-$ is the strongest base that can exist in what kind of solution? _____

(c) All strong bases are leveled to the strength of what ion in aqueous solutions? _____

Answer: (a) H$_2$O; (b) aqueous or H$_2$O; (c) OH$^-$ or hydroxide ion

64 Acids that dissociate completely in water may dissociate only partially in other solvents. A process similar to hydration occurs in such solvents. For solvents other than water this process is called **solvation**. Acids or bases in solvents other than water are no longer under the Arrhenius definition of acids and bases. Two scientists named Brønsted and Lowry formed a wider definition of acids and bases that includes solvents other than water.

The word "hydration" and the Arrhenius definition of an acid or base are both limited to what kind of solutions? _____

Answer: aqueous (H$_2$O or water)

65 Brønsted and Lowry defined an acid as a substance capable of donating a proton (**proton donor**) and a base as a substance capable of accepting a proton (**proton acceptor**). The proton is exactly the same as the H$^+$ ion in Arrhenius acid–base equations. (The H$^+$ ion is viewed as the hydrogen atom without the electron. The only subatomic particle left after removal of the electron is the proton.)

$$HCN \rightleftharpoons H^+ + CN^-$$

According to Brønsted–Lowry and the equation above, HCN is acid because it can donate _____.

Answer: H$^+$ (a proton)

66 Identifying the proton donor (acid) is not very difficult. Brønsted–Lowry bases (proton acceptors), however, include a variety of substances and ions in addition to the OH^- ion. In the following equation, the cyanide ion (CN^-) can act as a proton acceptor in the reverse reaction.

$$HCN \rightleftharpoons H^+ + CN^-$$

The CN^- ion is therefore the Brønsted–Lowry base in the equation. Using Brønsted–Lowry definitions, identify the acid, base, and proton in the equation below.

$$HCN \rightleftharpoons H^+ + CN^-$$

_____   _____   _____

Answer: $HCN \rightleftharpoons H^+ + CN^-$
   acid proton base

67 The Brønsted–Lowry acid (HCN) and the Brønsted–Lowry base (CN^-) are called a **conjugate acid–base pair** because the only difference between them is the presence or absence of a proton. HCN and CN^- are exactly the same with the exception of the proton (H^+).

The following substances are Brønsted–Lowry acids or bases. Only one conjugate pair is present, however.

$$NH_3, \quad H_2O, \quad NH_4^+, \quad Cl^-, \quad CO_4^{2-}, \quad HCN$$

The conjugate acid–base pair is _____ and _____.

Answer: NH_4^+; NH_3 (NH_4^+ is the same as NH_3 plus a proton.)

68 NH_4^+ and NH_3 are a conjugate acid–base pair.

(a) Which is the acid (proton donor)? _____

(b) Which is the base (proton acceptor)? _____

Answer: (a) NH_4^+; (b) NH_3
    $NH_4^+ \rightleftharpoons H^+ + NH_3$
    donor   acceptor

69 Since the proton **(H^+)** does not generally exist by itself, Brønsted–Lowry equations usually show two sets of conjugate acid–base pairs and no protons. A Brønsted–Lowry reaction generally shows an acid and a base as reactants and an acid and a base as products.

In the equation below, identify the acids and bases by writing the correct names in the blanks.

$$HCN + NH_3 \rightleftharpoons CN^- + NH_4^+$$

_____ _____ _____ _____

Answer: $HCN + NH_3 \rightleftharpoons CN^- + NH_4^+$
    acid   base   base   acid

70   In the equation in frame 69 identify the two conjugate acid–base pairs.
_____ and _____
_____ and _____

Answer:

HCN (acid); CN⁻ (base)

NH_4^+ (acid); NH_3 (base)

(The only difference between the acid and base of a conjugate pair is the proton.)

71   Identify the acids and bases in the following Brønsted–Lowry equation. Draw a line joining the acid and base of each conjugate pair. For example:

$$A + B \rightleftharpoons C + D$$
    acid   base   acid   base

$$HCl + NH_3 \rightleftharpoons Cl^- + NH_4^+$$

_____ _____ _____ _____

$$HCl + NH_3 \rightleftharpoons Cl^- + NH_4^+$$
    acid   base   base   acid
Answer:

72   Identify the acids, bases, and conjugate pairs in the following reactions as you did in frame 71.

(a) $HCl + H_2O \rightleftharpoons H_3O^+ + Cl^-$

_____ _____ _____ _____

(b) $NH_3 + H_2O \rightleftharpoons NH_4^+ + OH^-$

_____ _____ _____ _____

Answer:

(a) $HCl \; + \; H_2O \; \rightleftharpoons \; H_3O^+ \; + \; Cl^-$
     acid   base   acid   base

(b) $NH_3 \; + \; H_2O \; \rightleftharpoons \; NH_4^+ \; + \; OH^-$
     base   acid   acid   base

73 In one reaction above, H_2O accepted a proton to become H_3O^+. In another reaction, H_2O donated a proton to become OH^-.

$$HCl + H_2O \rightleftharpoons H_3O^+ + Cl^-$$

$$NH_3 + H_2O \rightleftharpoons NH_4^+ + OH^-$$

A substance that can act as *both* a Brønsted–Lowry acid and a base is said to be **amphiprotic**. In the above reactions, what substance is amphiprotic?

_____

Answer: H_2O (It can either donate or accept a proton. Donating a proton is the definition for a Brønsted–Lowry acid. Accepting a proton is the definition for a Brønsted–Lowry base.)

74 Some pure solvents ionize slightly. For example, we have shown that water ionizes ($H_2O \rightleftharpoons H^+ + OH^-$). The ionization is more correctly written to include the hydronium (H_3O^+) ion ($H_2O + H_2O \rightleftharpoons H_3O^+ + OH^-$). Whenever a pure solvent ionizes in this manner, the process is called **self-ionization** or **auto-ionization**. Identify the acids, bases, and conjugate pairs in the equation of H_2O ionization.

$$H_2O + H_2O \rightleftharpoons H_3O^+ + OH^-$$

Answer:

$H_2O + H_2O \rightleftharpoons H_3O^+ + OH^-$ or $H_2O + H_2O \rightleftharpoons H_3O^+ + OH^-$

acid   base   acid   base   base   acid   acid   base

(It doesn't matter which H_2O is labeled the acid or which is the base as long as the conjugate pairs are correct.)

75 We have seen that the strongest acid that can exist in an *aqueous* solution is the positive ion formed by the solvent plus a proton (H_2O + proton = H_3O^+). The strongest base that can exist is the negative ion formed by the solvent minus a proton (H_2O − proton = OH^-). Other solvents behave in a similar manner.

Liquid ammonia (NH_3) can be used as a solvent for Brønsted–Lowry acid base reactions. It self-ionizes to produce two ions.

$$NH_3(l) + NH_3(l) = NH_4^+ + NH_2^-$$

(a) Identify the strongest acid that can exist in a liquid ammonia solution.

—————————————

(b) Identify the strongest base that can exist in a liquid ammonia solution.

—————————————

Answer: (a) NH_4^+ (solvent + proton); (b) NH_2^- (solvent − proton)

76 In the self-ionization equation in frame 75, ammonia is acting as both an acid and a base. Since it acts as both a proton acceptor and a proton donor, ammonia can be called —————————.

Answer: amphiprotic

77 Pure acetic acid ($HC_2H_3O_2$) can also be used as a solvent for Brønsted–Lowry acid–base reactions. The self-ionization equation for acetic acid is as follows.

$$HC_2H_3O_2(l) + HC_2H_3O_2(l) \leftrightharpoons HC_2H_3O_2^+ + C_2H_3O_2^-$$

(a) What is the strongest acid that can exist when acetic "acid" is the solvent?

—————————

(b) What is the strongest base that can exist when acetic "acid" is the solvent?

—————————

Answer: (a) $H_2C_2H_3O_2^+$ (solvent + proton); (b) $C_2H_3O_2^-$ (solvent − proton)

78 In the Brønsted–Lowry concept of acids and bases, the stronger the acid, the weaker its conjugate base. Conversely, the stronger the base, the weaker its conjugate acid. In the following Brønsted–Lowry equation, the hydration of HCl produces the hydronium ion and the Cl^- ion. The reaction is virtually complete. A negligible amount of HCl remains in molecular form. Identify the acids, bases, and conjugate pairs in the reaction.

$$HCl + H_2O \rightarrow H_3O^+ + Cl^-$$

Answer:

$$HCl + H_2O \rightarrow H_3O^+ + Cl^-$$
acid   base   acid   base

79 In this reaction, HCl is a much stronger acid than H_3O^+ since HCl donates virtually all of its protons to H_2O to form H_3O^+. On the other hand, H_2O is a much stronger base than Cl^- since it accepts virtually all of the protons while leaving almost none for the Cl^- ion.

(a) Is the weaker base (Cl^-) the conjugate of the weaker acid or the stronger acid? _____

(b) Is the weaker acid (H_3O^+) the conjugate of the weaker base or the stronger base? _____

Answer: (a) stronger; (b) stronger

80 Every Brønsted–Lowry acid–base reaction favors the production of the weaker acid and the weaker base. A reaction with the stronger acid and base as reactants and the weaker acid and base as products will proceed favorably to produce a relatively large proportion of products.

stronger acid + stronger base → weaker acid + weaker base

The reverse reaction below is *not* favored and very little product will be formed.

weaker acid + weaker base $\not\rightarrow$ stronger acid + stronger base

The following reaction produced a large proportion of products with very little of the reactants left over. Identify the stronger acid and the stronger base.

$$HI + NH_3 \rightarrow NH_4^+ + I^-$$

(a) The stronger acid is _____.

(b) The stronger base is _____.

Answer: (a) HI; (b) NH_3 (The weaker acid and base are favored in a reaction. If a large proportion of products results from a reaction, then the reactants are stronger.)

81 Using solvents other than water, we can differentiate between strengths of those acids that are completely dissociated in water. The leveling effect of water is thus avoided. For example, HCl and HI are both strong acids (dissociate completely) in aqueous solution. However, in a solvent such as liquid ammonia, it has been determined that HI dissociates to a greater degree than HCl. HI is therefore stronger than HCl.

Strengths of Brønsted–Lowry Conjugate Acid–Base Pairs

Name	Formula	Formula	Name
Strongest acid			*Weakest base*
perchloric acid	$HClO_4$	ClO_4^-	perchlorate ion
hydrogen iodide	HI	I^-	iodide ion
sulfuric acid	H_2SO_4	HSO_4^-	hydrogen sulfate ion
hydrogen chloride	HCl	Cl^-	chloride ion
nitric acid	HNO_3	NO_3^-	nitrate ion
hydronium ion	H_3O^+	H_2O	water
hydrogen sulfate ion	HSO_4^-	SO_4^{2-}	sulfate ion
acetic acid	$HC_2H_3O_2$	$C_2H_3O_2^-$	acetate ion
ammonium ion	NH_4^+	NH_3	ammonia
hydrogen cyanide	HCN	CN^-	cyanide ion
water	H_2O	OH^-	hydroxide ion
ethyl alcohol	C_2H_5OH	$C_2H_5O^-$	ethoxide ion
ammonia	NH_3	NH_2^-	amide ion
Weakest acid			*Strongest base*

A table of some Brønsted–Lowry acids and bases appears above. The acids are listed in order of decreasing strength. Their conjugate bases are listed in order of increasing strength.

(a) What is the strongest Brønsted–Lowry acid listed? _____

(b) The conjugate base of the strongest acid is the weakest base. What is the weakest base? _____

Answer: (a) perchloric acid ($HClO_4$); (b) perchlorate ion (ClO_4^-)

82 The above table of Brønsted–Lowry acid–base strengths can be used to determine whether a reaction will proceed to favor the formation of products. If the product acid and base are weaker than the reactant acid and base, the forward reaction will proceed favorably to form products. If the product acid and base are stronger than the reactant acid and base, the reaction will not proceed to favor the products.

Will the following Brønsted–Lowry reaction proceed and favor the formation of products? (Use the table of acid and base strengths to determine the relative strengths of the acids and bases in the reaction.) _____

$$HI + C_2H_3O_2^- \rightarrow I^- + HC_2H_3O_2$$

acid   base   base   acid

Answer: Yes. HI is a stronger acid than $HC_2H_3O_2$. $C_2H_3O_2^-$ is a stronger base than I^-. Because the acid and base are the reactants, the reaction will proceed to favor the products.

83 Which does this Brønsted–Lowry reaction favor: the products or the reactants? _____

$$HSO_4^- + NH_2^- \rightarrow NH_3 + SO_4^{2-}$$

acid   base   acid   base

Answer: the products (HSO_4^- is stronger than NH_3. NH_2^- is stronger than SO_4^{2-})

84 Will the following Brønsted –Lowry reaction favor the reactants or the products? (Hint: First determine the acids and bases.) _____

$$C_2H_5OH + Cl^- \rightarrow C_2H_5O^- + HCl$$

Answer: $C_2H_5OH + Cl^- \rightarrow C_2H_5O^- + HCl$

       acid   base   base   acid

HCl is a stronger acid than C_2H_5OH. $C_2H_5O^-$ is a stronger base than Cl^-. Since the products are stronger than the reactants, the reactants are favored and only very slight amounts of products are formed.

85 In the following reaction, are the reactants or the products favored? _____

$$HCl + CN^- \rightarrow HCN + Cl^-$$

Answer: The products are favored. (HCl is a stronger acid than HCN. CN^- is a stronger base than Cl^-. The weaker acid and base of the products are favored.)

86 The general term that covers all Brønsted–Lowry acid–base reactions is **protolysis**. All protolytic reactions favor the production of the (weaker, stronger) _____ acid and base.

Answer: weaker

Arrhenius defined an acid–base neutralization as a reaction in which H_2O is always a product. Brønsted–Lowry defines an acid–base neutralization as a reaction in which the solvent is always a product, and an acid–base reaction is referred to as protolysis.

The Brønsted–Lowry definition broadens our concepts of acids and bases by permitting more compounds to be considered as acids or bases that are not under the Arrhenius definition. The Brønsted–Lowry definitions are particularly useful to organic chemists, who deal quite often with nonaqueous systems. However, the Brønsted–Lowry idea still requires the presence of a proton that may be transferred from an acid to a base. This is still somewhat limiting in scope, so a third concept involving something that is present in all substances, electrons, has been developed. We now discuss that third concept.

THE LEWIS ACID–BASE CONCEPT

87 The Arrhenius concept of acids and bases covers aqueous solutions of acids and bases. The Brønsted–Lowry concept of acids and bases covers acids and bases in other solvents as well as water. Some reactions have the characteristics of an acid–base reaction but do not undergo proton exchange.

An "acid" that does not have a proton to donate would not fit the Brønsted–Lowry concept. For all practical purposes, a proton is the same as what ion? _____

Answer: H^+

88 A scientist by the name of Lewis developed an acid–base theory to include those reactions that seem to behave like acid–base reactions but do not involve proton (H^+) exchange. According to Lewis, an acid is any substance that can accept a pair of electrons to form a coordinate covalent bond. A base, on the other hand, is any substance that is capable of donating a pair of electrons to form a coordinate covalent bond. To avoid confusion between the Brønsted–Lowry concept and the Lewis concept, remember that the electron and proton are oppositely charged.

(a) A Brønsted–Lowry acid is a proton (acceptor, donor) _____

(b) A Lewis acid is an electron pair (acceptor, donor) _____

(c) A proton has what charge? _____

(d) An electron has what charge? _____

Answer: (a) donor; (b) acceptor; (c) +; (d) –

89 Brønsted–Lowry protolysis normally deals with a single proton. Lewis acids and bases deal with a pair of electrons in coordinate covalent bonds. A Lewis base can be an electron-rich negative ion, such as OH^-, HSO_4^-, and O_2^-, or a molecule that has one or more pairs of electrons that are not already part of a covalent bond. Examples include H_2O and NH_3. Lewis acids include positive ions, such as H^+, Al^{3+}, and Ag^+, as well as molecules that have room for an additional pair of electrons. Examples include BF_3, SO_3, and CO_2.

HCO_3^- is an ion with a pair of available electrons. This ion is probably a Lewis (acid, base) _____.

Answer: base (A Lewis base is an electron pair donor.)

90 The reactions of Lewis acids and bases are best depicted by dot symbols that are appropriately known as Lewis symbols. Dot symbols were encountered in Chapter 3. Remember that dots are used to symbolize electrons.

The H^+ ion is a Lewis acid since it can accept a pair of electrons to share in a coordinate covalent bond. The NH_3 molecule is a Lewis base since it has an unshared pair of electrons.

$$H^+ + \overset{\displaystyle H}{\underset{\displaystyle H}{:\!N\!:\!H}} \longrightarrow \left[\overset{\displaystyle H}{\underset{\displaystyle H}{H\!:\!N\!:\!H}}\right]^+$$

Look at this Lewis reaction depicted by Lewis symbols. In a coordinate covalent bond, the electron pair forming the bond comes from one substance. In this reaction, both electrons come from what substance? _____

Answer: NH_3 (Since H^+ has no electrons at all, both bonding electrons must come from the NH_3 molecule.)

91 The hydrolysis of HCl to form the hydronium ion (H_3O^+) can also be interpreted through the Lewis concept.

$$H^+ : \overset{..}{\underset{..}{Cl}} :^- + : \overset{..}{\underset{\displaystyle H}{O}} : H \longrightarrow \left[H : \overset{..}{\underset{\displaystyle H}{O}} : H\right]^+ + [: \overset{..}{\underset{..}{Cl}} :]^-$$

Remember that HCl dissociates completely in aqueous solutions and that the Cl^- ion is a spectator ion.

(a) Which reactant is the Lewis acid? _____

(b) Which reactant is the Lewis base? _____

Answer: (a) HCl (Actually only the H^+ ion. It accepts a pair of electrons.); (b) H_2O (It shares a pair of its free electrons to form a coordinate covalent bond.)

92 One example of a Lewis acid–base reaction has been encountered previously as an example of a coordinate covalent bond. That is the reaction between BF_3 and NH_3 to form NH_3BF_3.

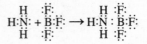

(a) The two electrons that form the coordinate covalent bond come from (NH_3, BF_3) _____.

(b) The Lewis acid in this example is _____.

(c) The Lewis base is _____.

Answer: (a) NH_3; (b) BF_3 (electron pair acceptor); (c) NH_3 (electron pair donor)

You have just learned three ways chemists use to describe acids, bases, and neutralization reactions while also learning about pH, buffer solutions, and titration. If you take more chemistry courses you will encounter these subjects again. This understanding will provide a firm foundation for further study. In any case, this general knowledge will help you understand the behavior of some common substances and their uses.

SELF-TEST

This self-test is designed to show how well you have mastered this chapter's objectives. Correct answers and review instructions follow the test.

1. One concept of acidity holds that hydrogen is not required to be present in an acid. This concept was proposed by:

   (a) Brønsted–Lowry

   (b) Arrhenius

   (c) Lewis

   (d) Lavoisier

2. According to Brønsted–Lowry concept, only one of the following statements is true. Which one is it?

   (a) A base is a neutron acceptor.

   (b) H_3O^+ and OH^- are a conjugate acid–base pair.

   (c) An acid is a proton donor.

   (d) The stronger the acid, the stronger is its conjugate base.

3. The strongest acid that can exist in substantial amounts in aqueous solution is:

   (a) $HClO_4$

   (b) H_2O

   (c) H_3O^+

   (d) OH^-

   (e) O^{2-}

   (f) NH_3

4. Which of the following is (are) true in the reaction $H_3N + BF_3 \rightarrow H_3NBF_3$?

   (a) H_3N is a Brønsted–Lowry acid.

   (b) H_3N is a Lewis acid.

   (c) BF_3 is a Brønsted–Lowry acid.

   (d) BF_3 is a Lewis acid.

5. What kind of a solution will the salt of a strong acid and a weak base give when dissolved in H_2O? (acidic, basic, neutral) _____

6. How many milliliters of 0.60 M HNO_3 are required to neutralize 40 milliliters of 1.20 M $Ca(OH)_2$? _____

7. Of the acids H_2SO_4, HCl, HNO_3, $HClO_4$, all have the same strength in aqueous solutions. This is an example of what is called the _____.

8. What is the pH of a solution that is 1.00×10^{-7} M HCl? _____

9. What is the pH of a solution that is 2.25×10^{-2} M HBr?

10. What is the H^+ concentration of a solution whose pH is 4.0? _____.

11. What is the H^+ concentration of a solution whose pH is 10.5?

12. According to Lewis theory, any negative ion is a _____.

13. Would a solution of $NaC_2H_3O_2$ be acidic, basic, or neutral? _____ Why?

   _____

14. Use the table on page 344 to predict if the formation of the products is favored. Explain why or why not. _____

    (a) $Ca(OH)_2 + H_2SO_4 \rightarrow CaSO_4 + 2H_2O$ _____

    (b) $C_2H_5OH + H_2O \rightarrow C_2H_5O^- + H_3O^+$ _____

    (c) $NH_4^+ + H_2O \rightarrow NH_3 + H_3O^+$ _____

15. Listed below are some statements. Indicate whether they are true or false.

    (a) A buffer solution is prepared by mixing solutions of a strong acid and a salt of that acid.

    (b) A solution whose H^+ concentration is 1×10^{-4} is acidic.

    (c) NH_4OH is a typical weak base.

    (d) $pH = -\log(OH^-)$.

    (e) The strongest base that can exist in substantial concentrations in a solvent that can act as an acid is the conjugate base of the solvent acid.

ANSWERS

Compare your answers to the self-test with those given below. If you answer all questions correctly, congratulations! If you miss any, review the frames indicated in parentheses following the answers. If you miss several questions, you should probably reread the chapter carefully.

1. (c) (frames 87, 88)

2. (c) (frame 65)

3. (c) (frame 62)

4. (d) (frame 88)

5. acidic (frame 58)

6. $Ca(OH)_2 + 2HNO_3 \rightarrow Ca(NO_3)_2 + 2H_2O$

$$\text{mL } HNO_3 \text{ soln} = 40 \text{ mL } Ca(OH)_2 \text{ soln} \times \frac{1.20 \text{ mol } Ca(OH)_2}{1 \text{ liter soln}} \times \frac{1 \text{ liter}}{1000 \text{ mL}} \times \frac{2 \text{ mol } OH^-}{1 \text{ mol } Ca(OH)_2}$$

$$\times \frac{1 \text{ mol } H^+}{1 \text{ mol } OH^-} \times \frac{1 \text{ liter soln}}{0.6 \text{ mol } HNO_3} \times \frac{1000 \text{ mL}}{1 \text{ liter}} = 160 \text{ mL } HNO_3 \text{ soln}$$

(frames 3, 9–14)

7. leveling effect (All strong acids are "leveled" to the strength of the H_3O^+ ion in aqueous solution.) (frame 63)

8. $[H^+] = 1 \times 10^{-7}$ M
   $pH = -\log 1 \times 10^{-7} = 7$ (frames 19–25)

9. $[H^+] = 2.25 \times 10^{-2}$ M
   $pH = -\log (2.25 \times 10^{-2}) = 1.65$ [frames 19 – 25]

10. $[H^+] = $ antilog($-pH$) = antilog(-4) = 1×10^{-4} M (frames 26, 27)

11. $[H^+] = $ antilog($-pH$) = antilog(-10.5) = 3.16×10^{-11} M H^+ [frames 26, 27]

12. base (frames 87–92)

13. basic, because it is the salt of a weak acid and a strong base (frames 46–58)

14. (a) Yes. A weaker conjugate acid and base formed.

    (b) No. A stronger conjugate acid and base formed.

    (c) No. A stronger conjugate acid and base formed so the reverse reaction is favored. (frames 80–86)

15. (a) F (frames 30–35)

    (b) T (frame 18)

    (c) T (frame 39, 40)

    (d) F (frames 19–23)

    (e) T (frame 63)

SUCROSE AND COCAINE - CRITICAL THINKING

The next chapter deals with organic compounds with carbon as the core atom. But based upon what you have learned thus far, we pose a question for you with a real-life situation involving two very different compounds whose molecular formulas are somewhat similar: sucrose and cocaine.

Sucrose is the chemical name for table sugar, and it has the molecular formula $C_{12}H_{22}O_{11}$. This compound can be purchased in any grocery store, and it is used to give a sweet taste to foods. People around the world consume sugar. However, there are warnings about consuming too much of it. Excessive sugar consumption is linked to diabetes, obesity, and coronary heart disease. Sugar can make up a large portion of a person's regular food intake.

In 2014 the National Institute for Health reported that Americans consume about 15% of their daily calories come from added sugars (~22 teaspoons of sugar per day). A number of published sources on nutrition have claimed that sucrose is a compound that should be consumed in moderation.

As odd as it may sound, sucrose has been compared to another compound that also contains carbon, hydrogen, and oxygen: cocaine ($C_{17}H_{21}O_4N$).

According to the National Institute on Drug Abuse, cocaine is a powerfully addictive stimulant. It can be used as a local anesthetic for some surgeries, but its recreational use is illegal. This compound is highly dangerous. The United States Drug Enforcement Agency classifies cocaine as a Schedule 1 drug, which means it has a high potential for abuse with little to no medical use in the United States.

Why has sucrose been compared to cocaine? That is a good question that you, the reader, should try to answer. An argument has come from the fact that both compounds contain carbon, hydrogen, and oxygen. But is this argument valid? Consider the following statement found on a number of websites and attributed to psychiatrist David Reuben, the author of *Everything You Always Wanted to Know About Nutrition*.

> [W]hite refined sugar is not a food. It is a pure chemical extracted from plant sources, purer in fact than cocaine, which it resembles in many ways. Its true name is sucrose and its chemical formula is $C_{12}H_{22}O_{11}$. It has 12 carbon atoms, 22 hydrogen atoms, 11 oxygen atoms, and absolutely nothing else to offer. ... The chemical formula for cocaine is $C_{17}H_{21}NO_4$. Sugar's formula again is $C_{12}H_{22}O_{11}$. For all practical purposes, the difference is that sugar is missing the "N", or nitrogen atom.

While researchers have noted similarities in the cravings for sugar and drugs such as cocaine in animal studies, does the argument that just because the chemical formulas seem to be similar, is that sufficient to equate the two? Is it enough to compare two compounds because they have similar elements in their molecular formula? Consider dimethyl ether and ethanol, two compounds you will learn about in the next chapter. Both compounds are made up of carbon, hydrogen, and oxygen.

A number of published sources on nutrition have claimed that we can make a chemical reactivity comparison between dimethyl ether and ethanol. However, these two compounds have very different chemical and physical properties, even though they are made up of the same types of elements.

Ethanol is a liquid at room temperature. It is also a common organic solvent and is used in making a variety of organic compounds. It is flammable and has a boiling point of 78.4 °C and a density of 0.789 g/cm^3. Dimethyl ether is a gas at room temperature and has a much lower boiling point of −24 °C. Its density is similar to that of ethanol, just a bit lower (0.735 g/cm^3). Dimethyl ether is used in aerosols as a propellant in a variety of fuel applications.

Dimethyl ether and ethanol not only possess carbon, hydrogen, and oxygen; both compounds also have identical molecular formulas, C_2H_6O. Since dimethyl ether and ethanol have the same molecular formula, both have the identical elemental compositions. Yet even with identical molecular formulas and the same elemental composition, dimethyl ether and ethanol do not have the same physical and chemical properties. Something else must account for these differences.

Dimethyl ether and ethanol (C_2H_6O; molecular mass = 46.07 g/mol)

$$\% \text{ carbon } \frac{(2 \times 12.011 \text{ grams})}{46.07 \text{ grams}} \times 100\% = 52.14\% \text{ C}$$

$$\% \text{ hydrogen } \frac{(6 \times 1.0079 \text{ grams})}{46.07 \text{ grams}} \times 100\% = 13.13\% \text{ H}$$

$$\% \text{ oxygen } \frac{(1 \times 16.00 \text{ grams})}{46.07 \text{ grams}} \times 100\% = 34.73\% \text{ O}$$

Dimethyl ether and ethanol, 52.14%C, 13.13% H, 34.73% O

Before we answer the question of why dimethyl ether and ethanol react so differently, let's return to sucrose and cocaine. Sucrose and cocaine also both are made from carbon, hydrogen, and oxygen, but their similarities stop there. Sucrose and cocaine's elemental compositions are not identical even though both compounds are composed of the same elements, with the exception of cocaine having nitrogen.

Sucrose ($C_{12}H_{22}O_{11}$, molecular mass = 342.29 g/mol)

$$\% \text{ carbon } \frac{(12 \times 12.011 \text{ grams})}{342.29 \text{ grams}} \times 100\% = 42.11\% \text{ C}$$

$$\% \text{ hydrogen } \frac{(22 \times 1.0079 \text{ grams})}{342.29 \text{ grams}} \times 100\% = 6.48\% \text{ H}$$

$$\% \text{ oxygen } \frac{(11 \times 16.00 \text{ grams})}{342.29 \text{ grams}} \times 100\% = 51.42\% \text{ O}$$

Sucrose: 42.11%C, 6.48% H, 51.42% O

Cocaine ($C_{17}H_{21}O_4N$, molecular mass = 303.35 g/mol)

$$\% \text{ carbon } \frac{(17 \times 12.011 \text{ grams})}{303.35 \text{ grams}} \times 100\% = 67.31\% \text{ C}$$

$$\% \text{ hydrogen } \frac{(21 \times 1.0079 \text{ grams})}{303.35 \text{ grams}} \times 100\% = 6.98\% \text{ H}$$

$$\% \text{ oxygen } \frac{(4 \times 16.00 \text{ grams})}{303.35 \text{ grams}} \times 100\% = 21.10\% \text{ O}$$

Cocaine: 67.31% C, 6.98% H, 21.10% O

As you can probably tell, merely having similar elements (C, H, and O) is not enough to validate the argument that two compounds *must* behave roughly the same way. What else is going on then? More important than possessing similar elements is the way in which these atoms are connected together. In other words, it is the structure of these compounds that makes all the difference.

In chemistry and biology, structure determines function. As you can see from the following figure, even though sucrose and cocaine have similar types of atoms, they have two very different structures. These two different structures will therefore have two different functions. Not only that, both will have different chemical and physical properties. Cocaine has a melting point of 98°C; sucrose does not even have a melting point. Instead it decomposes at 186°C. Other physical differences include different boiling points, different densities, and different solubilities in water.

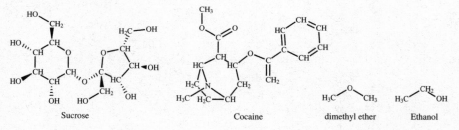

Sucrose   Cocaine   dimethyl ether   Ethanol

As for chemical reactivity, both obviously behave very differently. Even dimethyl ether and ethanol, which are more similar elementally than sucrose and cocaine (same elements, same molecular formula, same elemental composition), have different physical and chemical properties because these compounds have different structures.

As you can see, elemental similarity does not account for much more than that. Even compounds with similar elemental composition may not have similar physical and chemical properties. Rather, it is how the atoms are arranged in the structure of the compound that determines the chemical's function. So we ask you, is it good chemistry to compare sucrose and cocaine based on them both having carbon, hydrogen, and oxygen in their molecular formulas?

14 Organic Chemistry

The chemistry of carbon-containing compounds is broadly known as organic chemistry. This science is a subdiscipline of chemistry and it focuses on the chemistry within medicine, natural products, and the materials we come into contact with daily. In this chapter we will focus on hydrocarbons, nomenclature of hydrocarbons, oxygen-containing organic compounds, and nitrogen containing organic compounds. Additionally, we will focus on some basic properties that each class of organic compounds possess and look at their applications.

OBJECTIVES

After completing this chapter, you will be able to

- name alkanes and alkene hydrocarbons;

- identify and label constitutional isomers of hydrocarbons;

- identify and describe the structures and properties for alcohols, ethers, ketones, aldehydes, esters, and carboxylic acids;

- identify and describe the structures and properties for amines, imines, nitriles, and amides.

1 The simplest organic compounds are known as **hydrocarbons**. These compounds are composed of only two types of elements: carbon and hydrogen. Even when a compound is mostly carbon and hydrogen and has only one other element within its structure, then it is not a hydrocarbon. Consider the following **molecular formulas** for these common organic compounds. A molecular formula provides the number of atoms of each type of element that makes up a molecule.

Propane is used as a fuel. It has the molecular formula C_3H_8.

Acetone is an organic compound that is found in fingernail polish remover. It has the molecular formula C_3H_6O.

Ethanol is known as drinking alcohol and it also is an organic compound. It has the molecular formula C_2H_6O.

Octane is the main component found in gasoline and has the molecular formula C_8H_{18}.

Ethylamine is a common organic solvent used and has the molecular formula C_2H_7N.

Which of the previous organic compounds are hydrocarbons?

_____

Answer: propane and octane

2 Molecular formulas provide some information about organic compounds but they don't help in terms of the actual structure of the molecule. Structural information can help us understand the physical and chemical properties of a molecule. As such we will look at **structural formulas**. Structure formulas help us see how the atoms of a molecule are connected together.

Before we begin making hydrocarbon structures we must remember that the carbon atom has four valence electrons, since it is a Group IVA element. This means carbon will require four bonds in its structure and it will obtain this by the sharing of four additional electrons. By sharing the electrons the carbon atom will satisfy its octet (Chapter 3).

Let's draw the structure for the simplest hydrocarbon, CH_4, using a type of structural formula known as the **dash formula**. The dash formula is a Lewis structure that shows bonds as dashes and lone pairs of electrons where they exist on a molecule. The dash formula is not often accurate to the actual three-dimensional structure but it does serve well as a two-dimensional representation of a molecule.

The simplest hydrocarbon has one carbon atom and it is known as methane, CH_4.

There are four hydrogen atoms connected to the carbon atom. The dash formula for methane is

Methane

The structure shows us how the atoms are connected to one another.

How many valence electrons does the carbon atom possess? _____

Answer: Four

3   The next simplest hydrocarbon is known as ethane and has the molecular formula C_2H_6.

Notice that there are now two carbons with four single bonds each. Connecting the two carbon atoms together with three additional bonds each will give

Ethane

What is the molecular formula for ethane? _____

Answer: C_2H_6

4   Before we go any further let's discuss the classifications for hydrocarbons. Hydrocarbons can be separated into three main groups:

Saturated hydrocarbons: this comprises the **alkanes** and the **cycloalkanes**.
Unsaturated hydrocarbons: this comprises the **alkenes** and the **alkynes**.
Aromatic hydrocarbons: this comprises the **benzene** rings.

The following structures illustrate these groups.

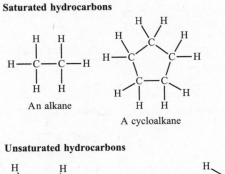

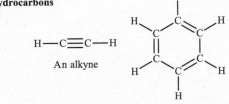

Saturated/Unsaturated hydrocarbons

What common feature can you find in unsaturated hydrocarbons?

_____

Answer: double or triple carbon bonds

BOND GEOMETRY

A hydrocarbon that is saturated contains only carbon and hydrogen single bonds. These are known as alkanes. Alkanes can be **cyclic** or **acyclic**. A cyclic compound does not have an end to it but resembles a ring. An acyclic is a molecule that has at least two ends to it. We also say that these carbons are sp^3 hybridized. In other words, the carbon atom contains four "bonding directions." You can think of this when considering the sp^3 hybridization ($1s + 3p = 4$ total orbitals). The four orbitals contribute to four bonding directions. The carbon in methane has four bonding directions as seen in its dash structure. Therefore, the carbon atom in methane is sp^3 hybridized. The language of hybridization will be seen as we cover various organic compounds in this chapter.

An unsaturated hydrocarbon is a compound that possesses a carbon–carbon double or triple bond. An **alkene** is an unsaturated hydrocarbon that has at least one carbon–carbon double bond. An **alkyne** is also an unsaturated hydrocarbon that has at least one carbon–carbon triple bond. An alkene carbon has a total of three bonding directions (two single bonds and one double bond). Since the alkene carbon has three bonding directions it is sp^2 hybridized ($1s + 2p = 3$ total orbitals). The alkyne has two bonding directions (one single bond and one triple bond) and is sp hybridized ($1s + 1p = 2$ total orbitals). Each hybridization has its own geometry and set of bond angles. See the following table.

Hybridization	Geometry	Bond angle (°)
sp	Linear	180
sp^2	Trigonal planar	120
sp^3	Tetrahedral	109.5

An **aromatic hydrocarbon** possesses a **benzene ring** or some similar structure. The benzene ring is a six-membered ring with three alternating carbon–carbon double bonds.

5 Below are a variety of hydrocarbons represented by dash formula structures. Dash formulas show all atoms and bonds within a structure.

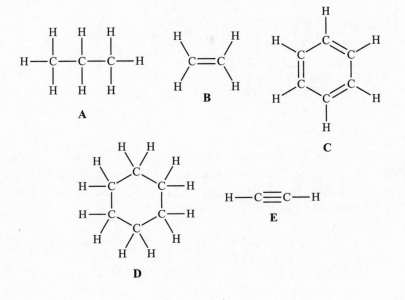

Which hydrocarbons are saturated and why? _____

Answer: Compounds A and D. Each comprises only carbon and hydrogen single bonds.

Which hydrocarbons are unsaturated and why? _____

Answer: Compound B, C, and E. These have either carbon–carbon double or triple bonds.

Which hydrocarbon is an alkyne? _____

Answer: E

Which hydrocarbon is an alkene? _____

Answer: B

Which hydrocarbon is an aromatic? _____

Answer: C

Which hydrocarbons are alkanes? _____

Answer: A and D

Which hydrocarbon is cyclic? _____

Answer: D

Which hydrocarbons are acyclic? _____

Answer: A, B, and E

6 Remember that methane is the simplest alkane. The number of hydrogens can be obtained quickly by knowing the number of carbons in an alkane. Alkanes have the general formula C_nH_{2n+2}. Methane has one carbon atom $n = 1$ and the number of hydrogen atoms is $2n + 2 = 2(1) + 2 = 4$. The molecular formula is therefore CH_4.

What is the molecular formula for butane, a fuel found in lighters? Butane has a total of four carbons. _____

Answer: C_4H_{10}

7 Following ethane, the name of the next molecule is propane. This gas is used as a fuel for heating and cooking, and it has a total of three carbons in its molecular formula. Using the formula C_nH_{2n+2} a three-carbon alkane will have eight hydrogen atoms.

What is the molecular formula of this alkane? _____

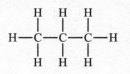

Answer: C_3H_8

8 What are the names of the first four hydrocarbons (containing 1, 2, 3, and 4 saturated carbon atoms)? _____

Answer: Methane, ethane, propane, and butane

9 After butane, the next alkanes incorporate Greek prefixes within their names. This makes it easier to know the number of carbons. The following table lists the Greek prefixes and their corresponding alkane names.

Greek prefix	Number of carbon atoms	Alkane name
Penta–	Five	Pentane
Hexa–	Six	Hexane
Hepta–	Seven	Heptane
Octa–	Eight	Octane
Nona–	Nine	Nonane
Deca–	Ten	Decane

What is the molecular formula and dash formula for heptane?

Answer: C_7H_{16}

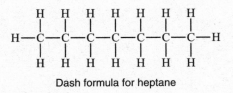

Dash formula for heptane

10   Another formula representation is known as the **condensed structural formula** or **condensed formula**. In this representation the bonds are not explicitly drawn but there is more information to their structure than the simpler molecular formula. For instance, the molecular formula for hexane is C_6H_{14}. Additionally, its condensed formula is $CH_3CH_2CH_2CH_2CH_2CH_3$. With the condensed formula we see how the atoms are grouped but we do not see how each is bonded, as in the case with the dashed formula structure.

    Write the condensed formula for nonane. _____

Answer: $CH_3CH_2CH_2CH_2CH_2CH_2CH_2CH_2CH_3$

11   As you can imagine there is a lot of drawing to do as the molecular formula increases in the number of carbon and hydrogen atoms. The most common type of structural formula that is fastest to draw is the **bond–line formula**. There are a few rules to know when drawing molecules using the bond-line formula.

- Bonds are represented by lines. Each bend in a line represents a carbon atom, unless another atom is explicitly shown.

- Elemental symbols for carbon atoms are not drawn. Elemental symbols for hydrogen atoms are not drawn on carbon atoms unless there is a need to represent a three-dimensional perspective.

- The number of hydrogen atoms are inferred by assuming carbon must complete its octet.

- Hydrogen atoms bonded to **heteroatoms** (atoms other than C and H) are drawn as well as the elemental symbol for the heteroatom.

12 Below is a bond-line formula. Notice how the carbons have been highlighted with an asterisk (⋆).

Carbon atoms = 5

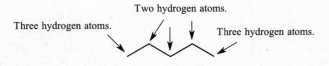

starred parent chain

Identify the alkane represented by this bond-line formula. _____

Answer: pentane

13 Although not drawn, hydrogen atoms are inferred on the above structure for pentane. The end carbons only have one bond drawn. Therefore the end carbons must have three hydrogen atoms in order for them to have four bonds in total. Each of the three middle carbons has two bonds drawn. Therefore, each middle carbon must have two hydrogens, providing a total of four bonds. The following highlights the number of hydrogen atoms bonded to each carbon atom in pentane. Again, these hydrogen atoms are inferred, not drawn onto the bond-line structure.

Two hydrogen atoms.

Three hydrogen atoms.   Three hydrogen atoms.

How many carbon atoms and hydrogen atoms are there for the following structure?

Answer: This structure has nine carbon atoms and 20 hydrogen atoms.

14 So far you have only been shown the first 10 alkanes, beginning with methane and ending in decane. These alkanes belong to a **homologous series**, which is a series of compounds differing from each other by a fixed group of atoms. In the case with alkanes the fixed group is an additional $-CH_2-$.

Are ethane, propane, butane, and pentane part of a homologous series? _____

What is the difference in the structure between ethane and propane? Butane and pentane? _____

Answer: yes; The difference between ethane and propane is a $-CH_2-$ group. The difference between butane and pentane is a $-CH_2-$ group.

15 You have also seen these alkanes as **straight-chain alkanes**. In other words, the carbon atoms are bonded to one another in a linear fashion. However, there are also **branched-chain alkanes** too. In a branched-chain alkane a hydrogen along the longest chain (known as the parent chain) is replaced with a CH group known as an *alkyl* group. The name of the branched portion is similar to one of the 10 straight-chain alkanes but the "*–ane*" ending is replaced with "*–yl.*" Below is a table that shows the first four alkyl groups and the parent alkane each alkyl is derived from.

Branched group	Alkyl name	Derived from the alkane
CH_3-	Methyl	Methane
CH_3CH_2-	Ethyl	Ethane
$CH_3CH_2CH_2-$	Propyl	Propane
$CH_3CH_2CH_2CH_2-$	Butyl	Butane

What is the difference between propane and a propyl group? _____

Answer: one less hydrogen atom, thus leaving a bond for attachment to another atom or group

16 The simplest branched-chain alkane has the following structure.

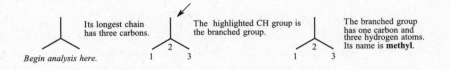

You'll notice the numbers are shown underneath the molecule. There is no standard placement for the numbers to go so long as these are in order. For

instance it would be just as accurate to label the molecule as follows. Again, the placement of the numbers does not matter in a molecule's numbering.

What is the parent chain and the branched alkyl group name for the following compound?

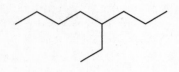

Answer:

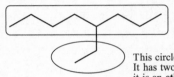

This top section is the longest carbon chain so its parent name is **octane**.

This circled section is the branched group. It has two carbons and five hydrogens, therefore it is an **ethyl** group.

17 In naming this branched alkane, we need to know to which carbon of the octane parent chain the ethyl group is attached. The ethyl group is located on which carbon? _____ (caution, tricky question)

Answer: Counting from the right, the fourth carbon; counting from the left, the fifth carbon. (We need some counting rules here.)

18 Although the parent chain is octane this is not its name. The actual name must include both its branched alkyl group and its parent name. This complicates naming so as a result we need to have rules for naming branched alkanes. Fortunately, the International Union for Pure and Applied Chemistry (IUPAC) has provided those rules.

   Rule 1 Count the longest carbon chain. This chain is not always straight. In fact it could be bent. However, before we go any further, you must count from the direction in which your **substituent** appears sooner. Remember, the substituent is the alkyl group(s) that is connected to the parent chain.

Rule 2 Identify the substituent by name and find its location along the parent chain.

Rule 3 Name the hydrocarbon by writing the substituent location first, then the substituent name followed by the parent name. There are no spaces in a hydrocarbon's name. However, numbers are written first followed by dashes with the alkyl name proceeding the parent name. If multiple substituents are present dashes are found between numbers.

Example 1: Number-alkyl group/parent name such as in 3-methyloctane

Example 2: Number-alkyl group–number–alkyl group by parent name such as in 4-ethyl-5-methyl-decane

19 Based on these rules, which is the correct name for the structure in frame 16?

(a) 5-methyloctane

(b) 4-octanemethyl

(c) 5-octanemethyl or

(d) 4-methyloctane

Answer: (d) (4-methyl octane follows the naming rules. The longest chain is octane, the smallest number for the carbon to which the substituent is attached is 4 on the parent chain from either side, and the substituent is named first.)

20 Identify this molecule's parent chain and substituent.

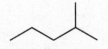

Answer:

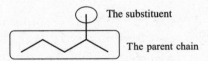

The substituent

The parent chain

21 Count the number of carbons in the parent chain. Make sure to number correctly.

Answer:

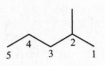

The parent chain is five carbons long so its parent name is *pentane*.

22 In the structure in frame 21, what is the substituent name and where is its location along the parent chain?

Answer: The substituent is a CH_3- group which is known as the methyl group. It is located at carbon-2.

23 What is the name for the organic compound in frame 21?

Answer: 2-Methylpentane

24 Simpler alkanes can be named quickly, but alkanes can have multiple substituents. When you have an alkane with multiple substituents you will need to use Greek prefixes prior to the substituent name. Additionally, for every substituent you must have a number written with it, even if it is on the same carbon!

Below are the first five Greek prefixes used in naming alkanes. Rarely does one exceed *tetra* or *penta* in a name.

Number of identical substituents	Greek prefix
Two	*Di*
Three	*Tri*
Four	*Tetra*
Five	*Penta*

What is the name for this alkane and why?

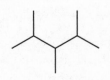

Answer: The name for this compound is 2,3,4-trimethylpentane. After counting the longest carbon chain you have identified the parent chain as pentane. The three substituents are all the same, methyl. They are located on carbons-2,3,4 along the chain. Since there are three identical substituents the Greek prefix *tri* is placed between the last number and the alkyl name.

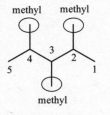

25 Remember the rule that if two substituents are on the same carbon, both numbers must be given. Try naming this alkane.

Answer: The name for this alkane is 2,2-dimethylpropane. The longest carbon chain has three carbons so its parent name is propane. It has two identical substituents, methyl. Both are located on carbon-2. Since there are two substituents the name must also have two numbers, even if they are the same number.

26 For your previous two alkanes, 2,3,4-trimethylpentane and 2,2-dimethyl-propane, did the direction for numbering matter when identifying the parent chain?

Answer: No. In both cases the substituents appeared at the same distance from either end so the direction for numbering didn't matter in naming this compound.

27 If a structure has two or more different alkyl groups, they must be named alphabetically. Let's look at a compound that offers a bit more challenge. What is the parent name, the substituent names (and their locations) for the following alkane? Finally, what is the overall name for this molecule?

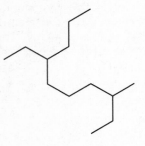

Answer: Since there are 10 carbons within the longest carbon chain the parent name is decane. Numbering from the bottom carbon will provide the first substituent at carbon-3 whereas numbering from the top carbon will provide the first substituent at carbon-4. Therefore, the numbering begins from the bottom carbon since it provides a substituent sooner.

The substituents and their locations are 3–methyl and 7–ethyl. Combining these substituents in alphabetical order with the parent name at the end provides 7–ethyl–3–methyldecane. The name 3–methyl–7–ethyldecane would be incorrect as the alkyl groups are not listed alphabetically. There is one more bit of information when it comes to listing alkyl groups alphabetically, which we will see in the following section.

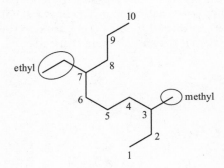

28   There is a rule for the order of naming multiple different substituents. The alkyl name is still listed by alphabetical order. For instance, butyl is written before ethyl. Ethyl will come before methyl, and methyl is listed before propyl. However, Greek prefixes do not play a role in the order. Consider the compound below. Its longest carbon chain is nine. Can you identify the longest carbon chain?

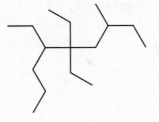

Answer:

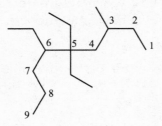

For instance, it would be incorrect to name this 3-methyl-5,5,6-triethylnonane thinking that the "*t*" for triethyl is part of the alphabetical order. Instead, triethyl's "*e*" for ethyl is used for alphabetizing. Therefore, the name of this compound is 5,5,6-triethyl-3-methylnonane.

 Here is a list of six alkane names. Which alkane name is written incorrectly? What is its correct name? Assume the carbon locations are correct (i.e., the numbering scheme) for each.

    3-Ethyl-4-methylheptane
    4-Methyl-4-propyloctane
    2,2-Dimethyl-4-ethylhexane
    3-Methyl-5,6-dipropyldecane
    2,2-Dimethylpropane

Answer: 2,2-Dimethyl-4-ethylhexane (the "*m*" for methyl was not alphabetized correctly). Its correct name (assuming the numbering is correct) is 4-ethyl-2,2-dimethylhexane.

 The variety of structures that arise from branched-chain alkanes allow for **constitutional isomers** (or **structural isomers**). These are compounds with the same molecular formula but different structural formulas.

Let's examine the two simplest alkane structural isomers, 2-methylpropane and butane.

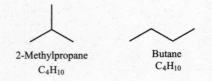

They both have the same molecular structure but their connectivity is different. In other words they have the same molecular formula but a different structure. As such, structural isomers will have different physical properties, such as different melting points and boiling points. When **heteroatoms** are involved (compounds other than hydrocarbons), chemical properties can drastically be different between structural isomers.

32 How many structural isomers exist for an alkane having the molecular formula C_4H_{10}?

Answer: Two

33 How many structural isomers exist for an alkane having the molecular formula C_5H_{12}? Make sure to draw and name all of these structural isomers.

Answer: Three

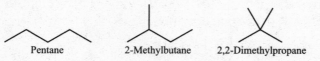

 Circle the alkane that is not a structural isomer for C_6H_{14}. Explain your choice.

Answer: The circled alkane has the formula C_5H_{12} while the other alkanes are isomers of C_6H_{14}.

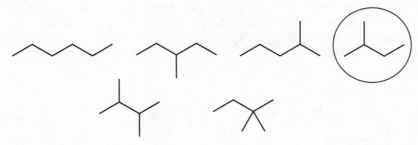

35 Cyclics are another group of alkanes but their structure does not have ends to them. Instead they are connected into a ring structure. The simplest cyclic is cyclopropane. Cyclopropane is a three-carbon ring shaped like a triangle. Notice that when naming cyclics the prefix *cyclo* is included. The following shows the first four cyclics' dash formula, bond-line formula, and name.

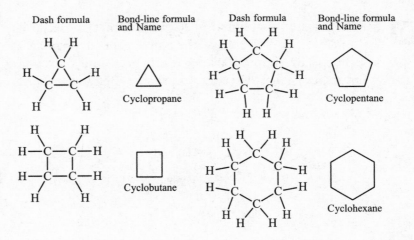

Cyclics have the formula C_nH_{2n}. Notice that it's different than the acyclic alkane formula C_nH_{2n+2}. Since the two ends connect to make a loop (aka, a ring) there are two fewer hydrogens. These compounds are still saturated since they are only composed of carbon and hydrogen single bonds.

Let's consider the four-carbon hydrocarbon chain butane, $CH_3CH_2CH_2CH_3$. Butane is fully saturated and as a result it cannot connect the two end carbons together in a loop. Remember, the maximum number of bonds to a carbon atom is four! The question is, how could we have a four-carbon chain loop up at the two end carbons?

Would the following structure work for looping the two end carbons? $CH_3CH_2CH_2CH_2-$

Answer: No. Notice that only one of the end carbons (the carbon on the far right) has a single hydrogen atom removed. The dash (–) reflects an available bond. However, if we were to remove two hydrogen atoms (one from each end carbon) we get $-CH_2CH_2CH_2CH_2-$. The two end carbons have only three bonds, leaving enough room to form a final bond to each other to give the cyclic known as cyclobutane.

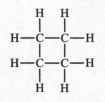

Cyclobutane (C_4H_8)

36 What are the molecular formulas for cycloheptane and cyclooctane?

Answer: cycloheptane: C_7H_{14}; cyclooctane: C_8H_{16}

37 **Alkenes** are unsaturated hydrocarbons that possess at least one carbon–carbon double bond. They have the same general formula as cyclic hydrocarbons, C_nH_{2n}. The simplest alkene is ethene. Note the ending is no longer "*–ane*" but rather "*–ene*" describing an alkene. Below are the various ways to represent ethene.

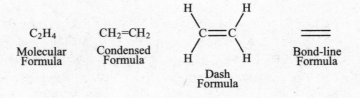

An alkene requires at least two carbon atoms. This is why the simplest alkene is ethene. The next larger alkene is three carbons long and is called propene. A four-carbon alkene is either 1-butene or 2-butene. The "1" in front of the name provides us the information as to where the C=C bond starts. If the double bond begins on the first carbon then it is 1-butene. If it begins at carbon-2, then it is 2-butene.

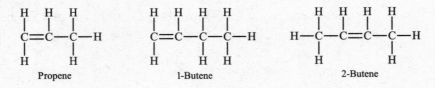

You can draw these compounds more quickly by using the condensed formula or bond-line formula.

$$CH_2{=}CH{-}CH_3 \quad CH_2{=}CH{-}CH_2{-}CH_3 \quad CH_3{-}CH{=}CH{-}CH_3$$

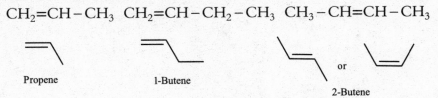

Propene   1-Butene   or

2-Butene

38 What are the condensed and bond line formulas for 1–heptene, 2–heptene, and 3–heptene?

Answer: $CH_2 = CH{-}CH_2{-}CH_2{-}CH_2{-}CH_2{-}CH_3$ (1-heptene); $CH_3{-}CH = CH{-}CH_2{-}CH_2{-}CH_2{-}CH_3$ (2-heptene); $CH_3{-}CH_2{-}CH = CH{-}CH_2{-}CH_2{-}CH_3$ (3-heptene)

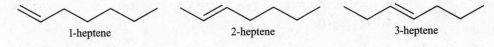

1-heptene   2-heptene   3-heptene

Answer:

39 A fact about alkenes is that the C=C bond is not allowed to rotate, unlike the carbon–carbon single bonds found within alkanes. Because of that rotational restriction alkenes can give rise to **geometric isomers**. These geometric isomers result when there are two different groups located on the alkene's double-bonded carbons. Geometric isomers are a specific type of isomer where the atoms are joined to one another in the same manner but differ in their orientation in space.

2-Butene has two geometric isomers. Notice that butene's C=C carbon atoms each have a (1) hydrogen atom and (2) a methyl group ($CH_3{-}$). When the larger groups are on the same side of the double bond, then the geometric isomer begins with the prefix "*cis*." If the two larger groups are on opposite sides of the double bond then the geometric isomer begins with the prefix "*trans*."

The double bond in each 2-Butene isomer is highlighted below with the larger groups (the methyl groups) circled. The first isomer shows the two methyl groups on opposite sides of the double bond while the second drawing shows the two methyl groups located on the same side of the double bond.

Based on what you have just learned, name each of the following structures as _____–2-butene.

Answer: ....

This is *trans*-2-butene   This is *cis*-2-butene

40 Given the information you have just learned about cis and trans isomers, circle the alkene that represents *cis*-2-pentene.

Answer:

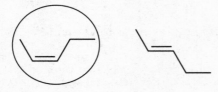

41 Can *cis–trans* isomers be possible if one of the carbons of a double bond has two identical groups?

Answer: No

Consider the following alkene to help illustrate this point. Notice the C=C bond and the carbon on the left. It is bonded to two other hydrogen atoms. Because of this we cannot say one hydrogen takes priority over the other. In other words, one group (in this case a hydrogen) does not have greater priority than the other group (also a hydrogen atom) since both are identical. Because of this we cannot label this as *cis* or *trans*.

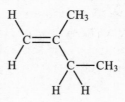

42 Name and draw the two geometric isomers for 3-hexene.

Answer:

cis-3-Hexene   trans-3-Hexene

43 Although *cis* and *trans* isomers have the same molecular formula they have different physical (melting points and boiling points) and chemical properties.
What is the name for the following alkene?

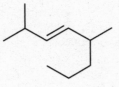

Answer:

1. First, let's identify if this alkene is *cis* or *trans*? Notice that the two circled groups are on opposite sides of the double bond. This is therefore a *trans* alkene.

2.

3. Next, count the parent chain from the end that produces the alkene double bond sooner and then identify the substituent groups. You'll notice on carbon-2 and carbon-5 methyl groups.

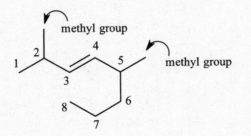

4. Now let us put all of this information together. The alkene's name is *trans*-2,5-dimethyl-3-octene.

44 **Alkynes** are unsaturated hydrocarbons that have at least one carbon–carbon triple bond. Alkynes have the formula C_nH_{2n-2}. The simplest alkyne is ethyne, which has two carbons, and according to the equation it will also have two hydrogen atoms. Below are the structures for ethyne, propyne, 1-butyne, and 2-butyne.

$$H—C \equiv C—H \qquad H—C \equiv C—CH_3 \qquad H—C \equiv C—CH_2—CH_3 \qquad H_3C—C \equiv C—CH_3$$

Ethyne   Propyne   1-Butyne   2-Butyne

What are the molecular formulas for an alkyne that has (a) five carbons, (b) six carbons, and (c) seven carbons in its structure?

Answer: Using the formula C_nH_{2n-2} and plugging in the number of carbon atoms for *n* we get (a) C_5H_8 (b) C_6H_{10} and (c) C_7H_{12}.

45 Naming alkynes is similar to alkenes. The triple bond is counted from the acyclic's closest end. Propyne's triple bond is not numbered since it gets a carbon-1 placement from either end of the molecule. However, butyne can have two possible locations for its triple bond. As such, both must be numbered to identify the two structural isomers.

Using bond-line formulas, draw all of the straight chain isomers for hexyne, C_6H_{10}.

Answer:

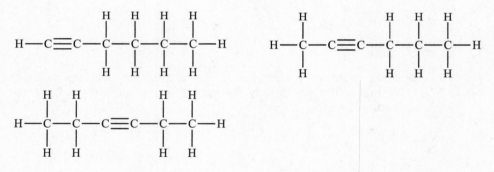

46 There is a great deal of stored chemical energy found within the double bond of the alkene and even more energy found within the triple bond of the alkyne. Acetylene (C_2H_2) is an alkyne that burns so hot that it is used in welding.

Which hydrocarbon should have the most stored chemical energy in it and why? (octane, 1-octene, or 1-octyne)

Answer: 1-Octyne should have the most stored chemical energy in it because it has a carbon–carbon triple bond.

47 Alkynes are either **internal** (found within the carbon chain of the molecule) or **terminal** (found at the end of the carbon chain). 3-Heptyne and 1-heptyne are alkyne isomers of one another. Which alkyne is the terminal alkyne and why?

Answer: 1-Heptyne is the terminal alkyne since the alkyne triple bond begins at the end carbon (or rather carbon-1) of the chain.

The triple bond begins at one end of the carbon chain, therefore this is a terminal alkyne.

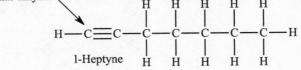

1-Heptyne

---

48 Draw the bond-line structure for 2-hexyne.

Answer:

---

49 Identify the structural isomers by circling them. Also, what are the isomers' molecular formula?

$H_2C = CH - CH = CH_2$ $H_3C - C \equiv C - CH_3$

Answer:

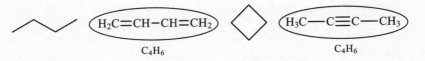

C_4H_6 C_4H_6

---

50 Aromatic hydrocarbons are compounds that have a six-membered ring with three alternating carbon–carbon single and double bonds. The simplest aromatic hydrocarbon is **benzene**, C_6H_6.

This is the structure for benzene.

Other aromatic hydrocarbons are found in fused-ring structures. A fused-ring structure has two or more aromatic rings fused together. Naphthalene, anthracene, phenanthrene, and pentacene are all aromatic hydrocarbons.

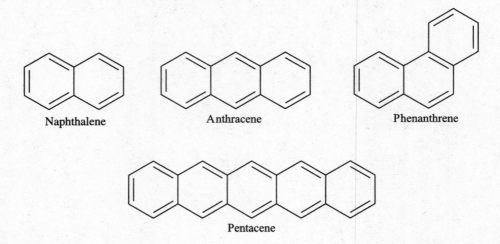

Naphthalene   Anthracene   Phenanthrene

Pentacene

All of these compounds are specifically known as **polycyclic aromatic hydrocarbons**. These are named polycyclic as they are made up of more than one aromatic ring. When looking at these compounds it is important to identify the aromatic ring structures found in each. Naphthalene is the main ingredient used in moth balls. This compound is used to make a variety of other compounds.

Anthracene is a compound found in coal tar and is used to make dyes. Phenanthrene is found in cigarette smoke and also a known skin irritant. Pentacene is a compound used in semiconductive materials.

Aromatic rings are also found in a variety of medications. Below is a list of compounds that possess the aromatic ring. The following are not considered aromatic hydrocarbons since there are atoms other than carbon and hydrogen within their structure. However, these compounds contain the aromatic ring structure.

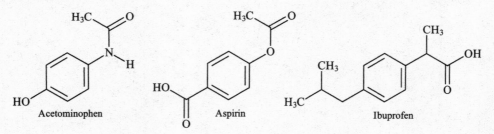

Acetominophen   Aspirin   Ibuprofen

Circle the aromatic ring found in the structure of ibuprofen.

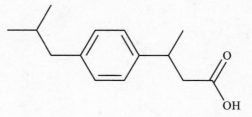

Answer:

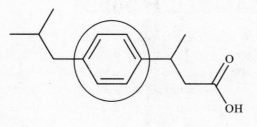

51   Aromatic rings are also found in a variety of natural products.

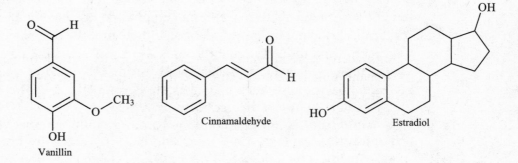

Vanillin   Cinnamaldehyde   Estradiol

Benzaldehyde (shown below) contributes to the smell of oyster mushrooms. Circle the aromatic ring found in benzaldehyde.

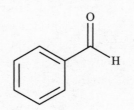

Answer:

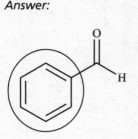

OXYGEN-CONTAINING ORGANIC COMPOUNDS

The next class of organic compounds are the ones that contain the oxygen atom as part of their functional group. A functional group is a particular connection of atoms that have unique chemical and physical properties. Functional groups are the reactive portion of a molecule. Alcohols, ethers, ketones, aldehydes, carboxylic acids, and esters all contain oxygen as part of their functional group.

Alcohols

Alcohols are an important class of organic compounds that are used as solvents, cleaners, and chemicals to make more complicated organic compounds. Common alcohols include isopropyl alcohol and ethanol. Isopropyl alcohol is used in antiseptics and disinfectants. Ethanol is also known as drinking alcohol.

When an sp^3 hybridized carbon atom is single-bonded to an −OH group that functional group is known as an alcohol group (aka, a hydroxyl group).

The basic structure of the alcohol functional group is C−O−H. However, the carbon atom requires its other connected atoms to fulfill its octet. The simplest alcohol is methanol, CH_3−OH.

Methanol can be represented the following ways.

$$CH_4O \qquad\qquad CH_3-OH$$
Molecular formula   Condensed formula

Naming alcohols is rather straightforward and is also similar to alkane nomenclature. The number of carbons is drawn from the parent alkane names. The suffix "*−ane*" is changed to "*−ol*" to reflect the hydroxyl functional group. Ethanol contains two carbons and the −OH functional group. Below are the different ways to represent ethanol.

$$C_2H_6O \quad H_3C\!-\!CH_2\!-\!OH$$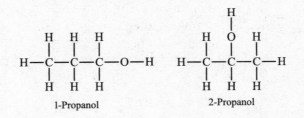

52

What is the molecular formula and condensed formula for propanol? Propanol is an alcohol with three carbons.

Answer: C_3H_8O, $CH_3CH_2CH_2OH$

53

As the number of carbon atoms in alcohols increase so does their complexity. The three carbon-containing alcohols have the molecular formula C_3H_8O and can come in the form of two structural isomers.

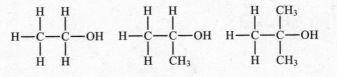

The numbering system shows where the −OH group is attached.

Alcohols are usually classified by the number of carbon atoms attached to the C−OH carbon. A primary (1°) alcohol has one carbon bonded to it, a secondary (2°) alcohol has two, and a tertiary (3°) alcohol has three.

For each of the following alcohols circle the carbon atom(s) directly attached to the C−OH carbon atom. Then label the C−OH carbon as primary (1°), secondary (2°), or tertiary (3°).

Answer:

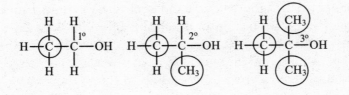

Draw the line structure for 3-hexanol and state whether it is a primary, secondary, or tertiary alcohol.

Answer:

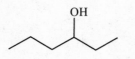

, this is a secondary alcohol.

Ethers

 Our next oxygen-containing functional group has the C–O–C connection. These are known as **ethers**. Ethers have been used in chemical reactions and in medicine. The simplest ether has two methyl groups connected by an oxygen atom. This compound is known as *dimethyl ether*. Below are the molecular formula, the condensed formula, the dashed formula, and the bond-line structure for dimethyl ether.

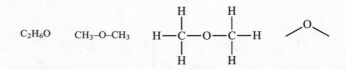

Are ethanol and dimethyl ether structural isomers? Why or why not?

Answer: Yes, because ethanol and dimethyl ether have the molecular formula C_2H_6O but different molecular structures.

CARBONYL GROUPS

Some carbon atoms are connected to oxygen atoms through double bonds. The C=O group is known as the *carbonyl* group and is common to several larger functional groups. In particular, the carbonyl group is part of the **aldehyde** and **ketone** functional groups. The simplest functional groups that contain these are known as the aldehyde and the ketone (shown below). The aldehyde and ketone are similar but there is a slight structural difference between the two. Both possess the C=O bond (the carbonyl group) but the aldehyde has a hydrogen atom bonded to the C=O carbon. The ketone has the basic structure C–C(=O)–C. The (=O) implies this bond is connected to the carbon to its left. A ketone is a C=O carbon connected to two other carbons through the C=O carbon.

Ethanal is the simplest aldehyde having the basic structure –C(=O)–H while propanone, the simplest ketone, has its C=O flanked by two alkyl groups. In this case, propanone's C=O carbon is bonded to two methyl groups.

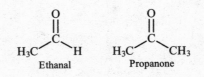

56 So far you have seen the alcohol, ether, ketone, and aldehyde functional groups. Circle and label the functional groups found in the following organic compounds.

Answer:

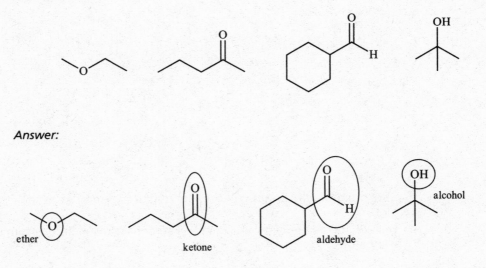

57 The next two oxygen–containing functional groups have the carbonyl group with its carbon also connected to one other oxygen atom through a single bond. This oxygen atom is an *sp*³ hybridized oxygen atom. The *R* in the formulas below represent any alkyl group. The oxygen atom in –OH or –OR has four total bonding directions (two bonds and two lone pairs). The two lone pairs on the oxygen atom count as two bonding directions. However, what is connected to the other end of the single-bonded oxygen atom determines the functional group's identity. Consider the basic structure of the **carboxylic acid** and ester, shown below. Notice that the carboxylic acid single-bonded oxygen has a bond to hydrogen. This is unique to the carboxylic acid, and this group

is often shortened to –COOH or –CO$_2$H. As for the **ester** its sp^3 hybridized oxygen is instead connected to an alkyl group.

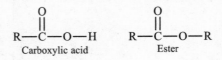

Circle and label the following oxygen-containing functional groups in vitamin C and aspirin.

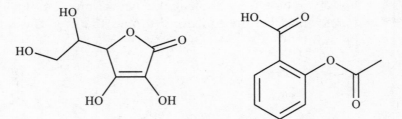

Answer:

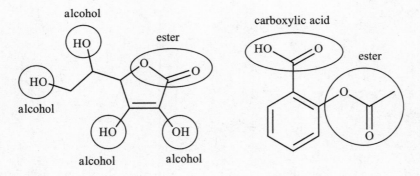

ORGANIC COMPOUNDS CONTAINING NITROGEN

58 Another type of organic compound has the nitrogen atom incorporated into its structure. As in the case with oxygen-containing organic compounds there are several types of nitrogen-containing functional groups. These include *amines*, *imines*, *nitriles*, and *amides*. These nitrogen-containing compounds have different chemical and physical properties than the oxygen-containing functional groups.

Most organic bases are *amines*, which are compounds derived from ammonia, NH_3. These derivatives are made by replacing one or more hydrogen atoms of ammonia with alkyl groups (CH groups generically written as R groups). The number of N–alkyl groups determines the substitution on the amine (1°, 2°, or 3°).

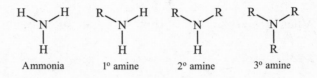

How many alkyl groups are directly attached to a secondary amine?

Answer: 2

59 How many lone pairs exist on the nitrogen atom for any nitrogen atom with three single bonds? Also, what is the hybridization on the amine's N atom?

Answer: 1, sp^3

60 As mentioned, amines are *bases*, since they have a lone pair available to accept a proton (H^+). After accepting a proton the amine becomes a substituted ammonium ion. In the example below methylamine (CH_3NH_2) deprotonates the water molecule to become the methylammonium ion ($CH_3NH_3^+$) and the hydroxide ion.

$$H_3C-\overset{..}{\underset{|}{N}}-H \;+\; H\overset{\overset{..}{\overset{..}{O}}}{\diagdown}H \;\rightleftharpoons\; H_3C-\overset{\overset{|}{H}}{\underset{\underset{H}{|}}{N^{\pm}}}-H \;+\; \overset{..}{\underset{..}{O}}{}^{-}\diagdown H$$

In this reaction, identify the acid, base, conjugate acid, and conjugate base.

Answer: acid = CH_3NH_2, base = H_2O, conjugate acid = $CH_3NH_3^+$, conjugate base = OH^-

Identify the acid, base, conjugate acid, and conjugate base for the reaction of ethylamine ($CH_3CH_2NH_2$) with water.

$$CH_3CH_2NH_2 + H_2 \rightleftarrows CH_3CH_2NH_3^+ + OH^-$$

Answer: acid = $CH_3CH_2NH_2$, base = H_2O, conjugate acid = $CH_3CH_2NH_3^+$, conjugate base = OH^-

61
Diethylamine, $(CH_3CH_2)_2NH$, has the dash line formula

What is the structure of diethylamine after it deprotonates a water molecules?

Answer:

62 Amines are one type of nitrogen-containing organic compounds. Remember, amines are nitrogen atoms with a lone pair on the central atom with three single bonds connected to alkyl groups. Just as hydrocarbons can have carbon–carbon single bonds, carbon–carbon double bonds, and carbon–carbon triple bonds so they can have multiple carbon–nitrogen bonds.

An *imine* functional group is composed of a nitrogen atom with a lone pair on the central atom having a single bond to one alkyl group and a double bond to a second alkyl group. The simplest amine (CH_2NH) is shown below.

$H_2C{=}\overset{..}{N}{-}H$

Consider the following nitrogen-containing compounds. Circle the one that is an imine.

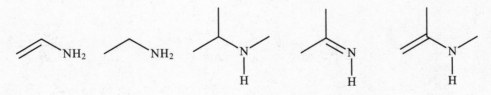

Answer:

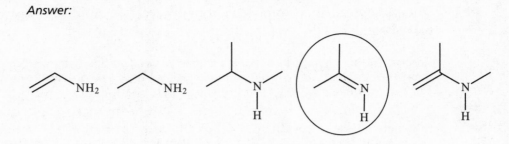

63 Is an imine basic? Why or why not? Also, what is the hybridization of the imine's nitrogen atoms?

Answer: The imine is basic since its nitrogen atom possesses a lone pair available to accept a proton. The N atom is *sp²* hybridized.

64 Imines can be made from reacting an aldehyde or ketone with a 1° amine. For instance, methanal, an aldehyde, reacts with methyl amine to produce an imine and water.

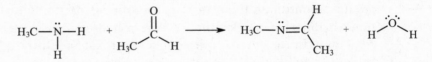

In the reaction with methanal and methyl amine circle the portion of the imine that came from the aldehyde.

Answer:

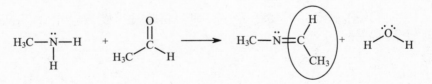

65 Can you make an imine from diethylamine, $(CH_3CH_2)_2NH$? Why or why not?

Answer: No, because diethylamine is a 2° amine.

66 Think about the imine produced from this reaction. Why must the amine be a primary amine? Hint: What other product must be made?

Answer: The amine must be a primary amine. Its two hydrogen atoms will combine with the ketone or aldehyde's oxygen atom to make water as a by-product.

67 The nitrile functional group contains a carbon–nitrogen triple bond with a lone pair on the nitrogen atom. The carbon atom of the nitrile connects to the alkyl group. The simplest nitrile is CH_3CN, known as acetonitrile, which is a common organic solvent. The hybridization of the nitrile's N atom is *sp*.

$$H_3C—C\equiv N\!:$$

Nitriles are often abbreviated in their structures with –CN. Acetonitrile can also be written as $^{H_3C—CN}$. The condensed formula provides CH_3CN and the bond-line formula is where only the nitrogen atom is written. All of these provide insight into the arrangement of atoms with the nitrile.

Nitriles undergo reactions that can make our final nitrogen-containing organic compound, the **amide**. The amide is a C=O directly connected to an sp^3-hybridized nitrogen atom. Amide bonds are the backbone that makes up proteins. The simplest amide is acetamide, CH_3CONH_2. Amides can also be made through a variety of other organic reactions.

$$\overset{\displaystyle O}{\underset{\displaystyle Acetamide}{H_3C—\overset{||}{C}—NH_2}}$$

Identify the nitrogen-containing functional groups for the following compounds.

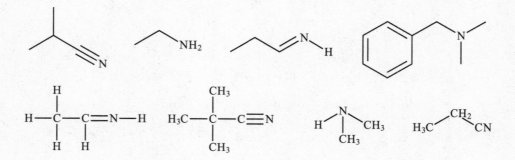

Answer:

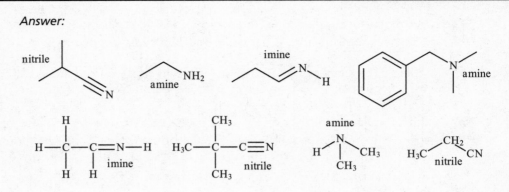

68 Amides can be made by reacting nitriles with acid. The simplest nitrile, ace-tonitrile, can be made by reacting acetonitrile with acid. Treating acetonitrile with acid can generate acetamide.

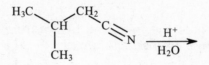

Draw the product that is made from the reaction.

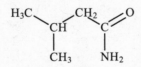

Answer:

$$H_3C \diagdown CH \diagup CH_2 \diagdown C \diagup O$$
$$CH_3 \quad NH_2$$

In the box provided draw the structure of the nitrile that makes the following amide.

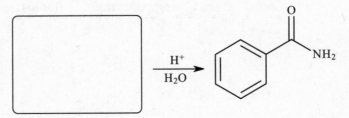

Answer:

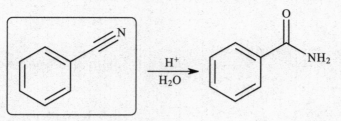

SUMMARY

You should now be able to name and draw the structures for a variety of alkanes and alkenes with a reasonable degree of proficiency. Also, you should be able to draw constitutional and geometric isomers for a variety of hydrocarbons. Finally, you should be able to identify several of the common organic chemistry functional groups.

SELF-TEST

1. Which of the following hydrocarbons is saturated: alkanes, cycloalkanes, alkenes, alkynes, or aromatics?

2. Fill in the following table for the labeled carbon atoms.

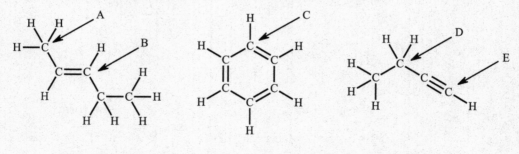

Carbon atom	Hybridization	Geometry	Bond angle (°)
A			
B			
C			
D			
E			

3. Fill in the missing content in the following table concerning straight-chain alkane molecular formulas, condensed formulas, bond-line formulas, and their names.

Molecular formula	Condensed formula	Bond-line formula	Name
	$CH_3CH_2CH_2CH_3$		
C_6H_{14}			
		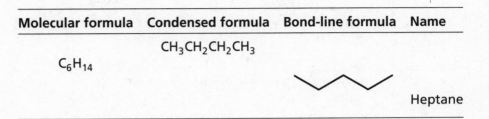	Heptane

4. Provide the names for alkanes #1 and #2.

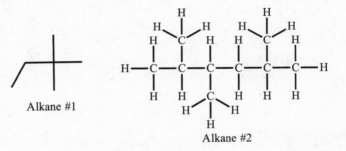

Alkane #1

Alkane #2

5. Draw the bond-line formula for 4,5-diethyl-2,9-dimethyldecane.

6. Provide the molecular formula for each of the structures below and circle the compounds that are isomers.

7. Draw the following compounds using bond-line formulas and name them.

   (a) A seven-membered cycloalkane.

   (b) An alkene that has eight carbons with the double bond beginning on carbon-1.

8. Consider an alkene that has six carbon atoms. Using bond-line formulas draw (a) all of the straight chain isomers and (b) at least four branched chain isomers where the parent chain is pentene.

9. Draw and name the two geometric isomers for 3-methyl-3-hexene.

10. What are the formulas for alkanes, alkenes, cycloalkanes, and alkynes?

11. Name the following alkynes.

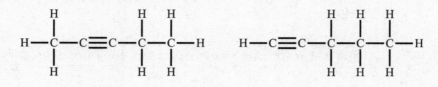

12. Identify the following organic compounds by their functional group.

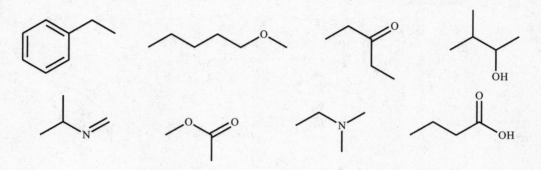

13. Identify the following organic compounds by their functional group.

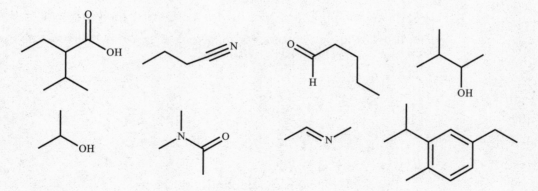

14. Label the following alcohols and amines as primary (1°), secondary (2°), or tertiary (3°).

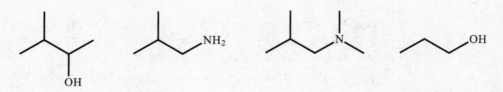

15. Draw the product of the following reaction.

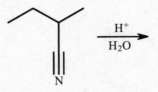

$$\xrightarrow[\text{H}_2\text{O}]{\text{H}^+}$$

ANSWERS

1. Alkanes and cycloalkanes [frame 4]

2.

Carbon atom	Hybridization	Geometry	Bond angle (°)
A	sp^3	Tetrahedral	109.5
B	sp^2	Trigonal planar	120
C	sp^2	Trigonal planar	120
D	sp^3	Tetrahedral	109.5
E	sp	Linear	180

[frame 4]

3.

Molecular formula	Condensed formula	Bond-line formula	Name
C_4H_{10}	$CH_3CH_2CH_2CH_3$		Butane
C_6H_{14}	$CH_3CH_2CH_2CH_2CH_2CH_3$		Hexane
C_5H_{12}	$CH_3CH_2CH_2CH_2CH_3$		Pentane
C_7H_{16}	$CH_3CH_2CH_2CH_2CH_2CH_2CH_3$		Heptane

[frames 6–11]

4. Alkane #1 is 2,2-dimethylbutane; alkane #2 is 2,3,5-trimethylhexane. [frames 12–29]

5.

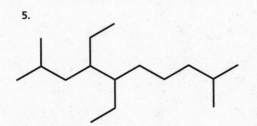

[frames 12–28]

6.

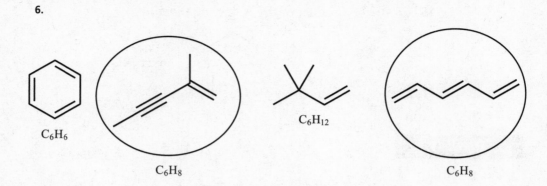

C_6H_6

C_6H_8

C_6H_{12}

C_6H_8

[frames 1, 29–34]

7. (a)

(b)

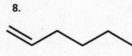

[frames 34–38]

8.

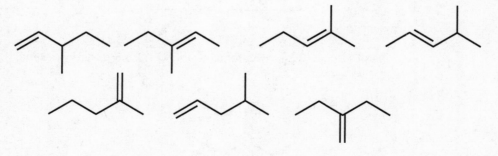

(a) Although you are asked for four, all are provided. [frames 37–41]

9.

trans-3-methyl-3-hexene   *cis*-3-methyl-3-hexene

[frames 37–43]

10. Alkanes (C_nH_{2n+2}); alkenes (C_nH_{2n}); cycloalkanes (C_nH_{2n}); alkynes (C_nH_{2n-2}) [frames 6, 35, 37, 45]

11. 2-pentyne and 1-pentyne [frames 44–48]

12.

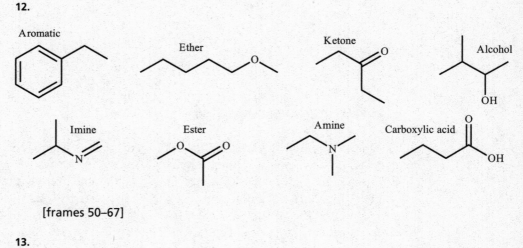

[frames 50–67]

13.

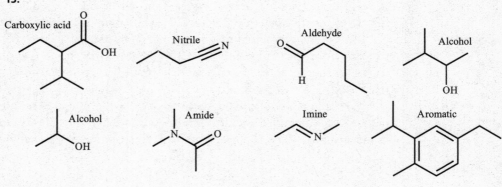

[frames 50–67]

14.

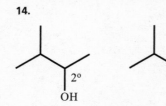

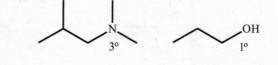

[frame 53, 58]

15.

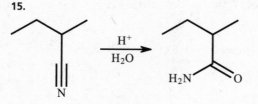

[frame 68]

EVERYDAY ORGANIC CHEMISTRY

Organic compounds are found in food and in medicines. There is no way around it, we consume chemical compounds that possess a wide variety of organic chemistry functional groups that serve biological and physiological functions.

Let's look at a common go-to drink for early risers. In the morning, many people around the world begin their day with a cup (or more!) of coffee. If that coffee is caffeinated (i.e., not decaf) then it has the caffeine molecule dissolved in it. Caffeine (shown below) comes from the coffee bean and this compound is known for stimulating the central nervous system, helping us feel more alert. Your coffee maker performs a hot water extraction that removes much of the caffeine from the coffee beans, as well as the many flavors, and they are collected in the decanter. From there the coffee is poured into a cup, ready for drinking.

The caffeine molecule is rather small but as you can see it is composed of several organic functional groups. Within the caffeine molecule you can find amide functional groups, an alkene, an amine, and an imine functional group.

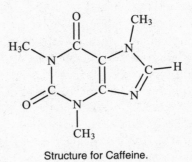

Structure for Caffeine.

Caffeine is one of the world's most popular drugs. It is considered a drug since it produces a physiological effect on the body once ingested. However, caffeine should be taken with caution. The Mayo Clinic states that "up to 400 mg of caffeine a day appears to be safe for most healthy adults." Any more than 400 mg can lead to health problems, which can include nervousness, irregular heartbeat, irritability, muscle tremors, anxiety, and insomnia. As you can see moderation is key!

There are other organic compounds taken regularly across the world. These include some popular over-the-counter medications such as aspirin, acetaminophen, and Aleve®. All of these organic compounds are taken as pain relievers and fever reducers. Aspirin is also used as a blood thinner as well as to treat inflammation. Aleve is taken for minor muscle aches, inflammation, and headaches. The active ingredient in Aleve is naproxen. This compound is responsible for the drug's effects.

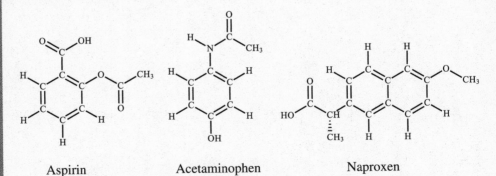

Aspirin   Acetaminophen   Naproxen

Medications are helpful in the dosages determined by your doctor and pharmacist. The dosage amounts promote the proper effect of the medication while minimizing their side effects. Doctors have advised low-dosage aspirin to many patients who are at risk of heart attack or stroke. Continued aspirin use does have side effects, ranging from upset stomach all the way up to stomach ulcers. Acetaminophen is relatively safe if kept at dosages no greater than 3 grams per day for a healthy adult. It is interesting to note that aspirin and naproxen are in a class of medications called NSAIDs (nonsteroidal anti-inflammatory drugs).

People who recently have had a heart attack should avoid naproxen unless directed to take it by their physician. NSAID side effects include adverse gastrointestinal issues. Since acetaminophen is not an NSAID it has fewer gastrointestinal side effects, which can include diarrhea, vomiting, and abdominal pain. This is one additional reason why pharmacies provide Children's TYLENOL® to treat children having discomfort from fever and/or headache. Tylenol is a trade name for acetaminophen.

Notice that all three pain relievers have one functional group in common, the aromatic ring. However, these organic compounds are different. Aspirin has the carboxylic acid functional group and the ester functional group. Acetaminophen has the amide functional group and the alcohol functional group. Finally, naproxen has the carboxylic acid functional group and the ether functional group. Naproxen also has two aromatic rings. You'll notice these two aromatic rings are fused together.

The structures of organic compounds are the result of the presence or absence of functional groups. Functional groups play a key role in the organic compound, not only in structure but also in function. These structures help determine the function of the molecule.

Appendix

TABLE OF ATOMIC WEIGHTS

Element Name	Symbol	Atomic Number	Standard Atomic Weight	Element Name	Symbol	Atomic Number	Standard Atomic Weight
Actinium	Ac	89	[227]	Mendelevium	Md	101	[258]
Aluminum*	Al	13	26.98	Mercury	Hg	80	200.59
Americium	Am	95	[243]	Molybdenum	Mo	42	95.95
Antimony	Sb	51	121.76	Moscovium	Mc	115	[289]
Argon	Ar	18	39.95	Neodymium	Nd	60	144.24
Arsenic	As	33	74.92	Neon	Ne	10	20.180
Astatine	At	85	[210]	Neptunium	Np	93	[237]
Barium	Ba	56	137.33	Nickel	Ni	28	58.693
Berkelium	Bk	97	[247]	Nihonium	Nh	113	[286]
Beryllium	Be	4	9.01	Niobium	Nb	41	92.906
Bismuth	Bi	83	208.98	Nitrogen	N	7	14.007
Bohrium	Bh	107	[270]	Nobelium	No	102	[259]
Boron	B	5	10.81	Oganesson	Og	118	[294]
Bromine	Br	35	79.904	Osmium	Os	76	190.23
Cadmium	Cd	48	112.41	Oxygen	O	8	15.999
Cesium*	Cs	55	132.91	Palladium	Pd	46	106.42
Calcium	Ca	20	40.08	Phosphorus	P	15	30.974
Californium	Cf	98	[251]	Platinum	Pt	78	195.08
Carbon	C	6	12.011	Plutonium	Pu	94	[244]
Cerium	Ce	58	140.12	Polonium	Po	84	[209]
Chlorine	Cl	17	35.45	Potassium	K	19	39.098
Chromium	Cr	24	51.996	Praseodymium	Pr	59	140.901
Cobalt	Co	27	58.933	Promethium	Pm	61	[145]
Copernicium	Cn	112	[285]	Protactinium	Pa	91	231.04
Copper	Cu	29	63.546	Radium	Ra	88	[226]
Curium	Cm	96	[247]	Radon	Rn	86	[222]
Darmstadtium	Ds	110	[281]	Rhenium	Re	75	186.21
Dubnium	Db	105	[270]	Rhodium	Rh	45	102.91
Dysprosium	Dy	66	162.50	Roentgenium	Rg	111	[281]
Einsteinium	Es	99	[252]	Rubidium	Rb	37	85.468
Erbium	Er	68	167.26	Ruthenium	Ru	44	101.07
Europium	Eu	63	151.96	Rutherfordium	Rf	104	[267]
Fermium	Fm	100	[257]	Samarium	Sm	62	150.36
Flerovium	Fl	114	[289]	Scandium	Sc	21	44.956
Fluorine	F	9	18.998	Seaborgium	Sg	106	[269]
Francium	Fr	87	[223]	Selenium	Se	34	78.971
Gadolinium	Gd	64	157.25	Silicon	Si	14	28.085
Gallium	Ga	31	69.723	Silver	Ag	47	107.87
Germanium	Ge	32	72.630	Sodium	Na	11	22.990
Gold	Au	79	196.97	Strontium	Sr	38	87.62
Hafnium	Hf	72	178.49	Sulfur	S	16	32.06
Hassium	Hs	108	[270]	Tantalum	Ta	73	180.95
Helium	He	2	4.0026	Technetium	Tc	43	[97]
Holmium	Ho	67	164.93	Tellurium	Te	52	127.60
Hydrogen	H	1	1.008	Tennessine	Ts	117	[293]
Indium	In	49	114.82	Terbium	Tb	65	158.93
Iodine	I	53	126.90	Thallium	Tl	81	204.38
Iridium	Ir	77	192.22	Thorium	Th	90	232.04
Iron	Fe	26	55.845	Thulium	Tm	69	168.93
Krypton	Kr	36	83.798	Tin	Sn	50	118.71
Lanthanum	La	57	138.91	Titanium	Ti	22	47.867
Lawrencium	Lr	103	[262]	Tungsten	W	74	183.84
Lead	Pb	82	207.2	Uranium	U	92	238.03
Lithium	Li	3	6.94	Vanadium	V	23	50.942
Livermorium	Lv	116	[293]	Xenon	Xe	54	131.29
Lutetium	Lu	71	174.97	Ytterbium	Yb	70	173.05
Magnesium	Mg	12	24.305	Yttrium	Y	39	88.906
Manganese	Mn	25	54.938	Zinc	Zn	30	65.38
Meitnerium	Mt	109	[278]	Zirconium	Zr	40	91.224

Values in square brackets are the mass number of the most stable or best-known isotope.
*also spelled as Aluminium and Caesium (IUPAC).

Periodic Table of the Elements

Hydrogen
1
H
1.008

→ atomic number
→ atomic weight

s orbitals

p orbitals

d orbitals

f orbitals

Research on the characteristics of newer laboratory elements in Period 7 is ongoing.

Period	IA	IIA	IIIB	IVB	VB	VIB	VIIB	VIII			IB	IIB	IIIA	IVA	VA	VIA	VIIA	Noble Gases VIIIA
1	Hydrogen 1 **H** 1.008																	Helium 2 **He** 4.0026
2	Lithium 3 **Li** 6.94	Beryllium 4 **Be** 9.0122											Boron 5 **B** 10.81	Carbon 6 **C** 12.011	Nitrogen 7 **N** 14.007	Oxygen 8 **O** 15.999	Fluorine 9 **F** 18.998	Neon 10 **Ne** 20.180
3	Sodium 11 **Na** 22.990	Magnesium 12 **Mg** 24.305											Aluminium 13 **Al** 26.982	Silicon 14 **Si** 28.085	Phosphorus 15 **P** 30.974	Sulfur 16 **S** 32.06	Chlorine 17 **Cl** 35.45	Argon 18 **Ar** 39.95
4	Potassium 19 **K** 39.098	Calcium 20 **Ca** 40.078	Scandium 21 **Sc** 44.956	Titanium 22 **Ti** 47.867	Vanadium 23 **V** 50.942	Chromium 24 **Cr** 51.996	Manganese 25 **Mn** 54.938	Iron 26 **Fe** 55.845	Cobalt 27 **Co** 58.933	Nickel 28 **Ni** 58.693	Copper 29 **Cu** 63.546	Zinc 30 **Zn** 65.38	Gallium 31 **Ga** 69.723	Germanium 32 **Ge** 72.630	Arsenic 33 **As** 74.922	Selenium 34 **Se** 78.971	Bromine 35 **Br** 79.904	Krypton 36 **Kr** 83.798
5	Rubidium 37 **Rb** 85.468	Strontium 38 **Sr** 87.62	Yttrium 39 **Y** 88.906	Zirconium 40 **Zr** 91.224	Niobium 41 **Nb** 92.906	Molybdenum 42 **Mo** 95.95	Technetium 43 **Tc** [97]	Ruthenium 44 **Ru** 101.07	Rhodium 45 **Rh** 102.91	Palladium 46 **Pd** 106.42	Silver 47 **Ag** 107.87	Cadmium 48 **Cd** 112.41	Indium 49 **In** 114.82	Tin 50 **Sn** 118.71	Antimony 51 **Sb** 121.76	Tellurium 52 **Te** 127.60	Iodine 53 **I** 126.90	Xenon 54 **Xe** 131.29
6	Cesium 55 **Cs** 132.91	Barium 56 **Ba** 137.33	Lanthanum 57 **La** 138.91 *	Hafnium 72 **Hf** 178.49	Tantalum 73 **Ta** 180.95	Tungsten 74 **W** 183.84	Rhenium 75 **Re** 186.21	Osmium 76 **Os** 190.23	Iridium 77 **Ir** 192.22	Platinum 78 **Pt** 195.08	Gold 79 **Au** 196.97	Mercury 80 **Hg** 200.59	Thallium 81 **Tl** 204.38	Lead 82 **Pb** 207.2	Bismuth 83 **Bi** 208.98	Polonium 84 **Po** [209]	Astatine 85 **At** [210]	Radon 86 **Rn** [222]
7	Francium 87 **Fr** [223]	Radium 88 **Ra** [226]	Actinium 89 **Ac** [227] **	Rutherfordium 104 **Rf** [267]	Dubnium 105 **Db** [268]	Seaborgium 106 **Sg** [269]	Bohrium 107 **Bh** [270]	Hassium 108 **Hs** [269]	Meitnerium 109 **Mt** [278]	Darmstadtium 110 **Ds** [281]	Roentgenium 111 **Rg** [282]	Copernicium 112 **Cn** [285]	Nihonium 113 **Nh** [286]	Flerovium 114 **Fl** [289]	Moscovium 115 **Mc** [290]	Livermorium 116 **Lv** [293]	Tennessine 117 **Ts** [294]	Oganesson 118 **Og** [294]

*Lanthanide series

Cerium 58 **Ce** 140.12	Praseodymium 59 **Pr** 140.91	Neodymium 60 **Nd** 144.24	Promethium 61 **Pm** [145]	Samarium 62 **Sm** 150.36	Europium 63 **Eu** 151.96	Gadolinium 64 **Gd** 157.25	Terbium 65 **Tb** 158.93	Dysprosium 66 **Dy** 162.50	Holmium 67 **Ho** 164.93	Erbium 68 **Er** 167.26	Thulium 69 **Tm** 168.93	Ytterbium 70 **Yb** 173.05	Lutetium 71 **Lu** 174.97

**Actinide series

Thorium 90 **Th** 232.04	Protactinium 91 **Pa** 231.04	Uranium 92 **U** 238.03	Neptunium 93 **Np** [237]	Plutonium 94 **Pu** [244]	Americium 95 **Am** [243]	Curium 96 **Cm** [247]	Berkelium 97 **Bk** [247]	Californium 98 **Cf** [251]	Einsteinium 99 **Es** [252]	Fermium 100 **Fm** [257]	Mendelevium 101 **Md** [258]	Nobelium 102 **No** [259]	Lawrencium 103 **Lr** [266]

☐ Metal
☐ Metalloid
☐ Nonmetal

TABLE OF FOUR-PLACE LOGARITHMS

No.	0	1	2	3	4	5	6	7	8	9
10	0000	0043	0086	0128	0170	0212	0253	0294	0334	0374
11	0414	0453	0492	0531	0569	0607	0645	0682	0719	0755
12	0792	0828	0864	0899	0934	0969	1004	1038	1072	1106
13	1139	1173	1206	1239	1271	1303	1335	1367	1399	1430
14	1461	1492	1523	1553	1584	1614	1644	1673	1703	1732
15	1761	1790	1818	1847	1875	1903	1931	1959	1987	2014
16	2041	2068	2095	2122	2148	2175	2201	2227	2253	2279
17	2304	2330	2355	2380	2405	2430	2455	2480	2504	2529
18	2553	2577	2601	2625	2648	2672	2695	2718	2742	2765
19	2788	2810	2833	2856	2878	2900	2923	2945	2967	2989
20	3010	3032	3054	3075	3096	3118	3139	3160	3181	3201
21	3222	3243	3263	3284	3304	3324	3345	3365	3385	3404
22	3424	3444	3464	3483	3502	3522	3541	3560	3579	3598
23	3617	3636	3655	3674	3692	3711	3729	3747	3766	3784
24	3802	3820	3838	3856	3874	3892	3909	3927	3945	3962
25	3979	3997	4014	4031	4048	4065	4082	4099	4116	4133
26	4150	4166	4183	4200	4216	4232	4249	4265	4281	4298
27	4314	4330	4346	4362	4378	4393	4409	4425	4440	4456
28	4472	4487	4502	4518	4533	4548	4564	4579	4594	4609
29	4624	4639	4654	4669	4683	4698	4713	4728	4742	4757
30	4771	4786	4800	4814	4829	4843	4857	4871	4886	4900
31	4914	4928	4942	4955	4969	4983	4997	5011	5024	5038
32	5051	5065	5079	5092	5105	5119	5132	5145	5159	5172
33	5185	5198	5211	5224	5237	5250	5263	5276	5289	5302
34	5315	5328	5340	5353	5366	5378	5391	5403	5416	5428
35	5441	5453	5465	5478	5490	5502	5514	5527	5539	5551
36	5563	5575	5587	5599	5611	5623	5635	5647	5658	5670
37	5682	5694	5705	5717	5729	5740	5752	5763	5775	5786
38	5798	5809	5821	5832	5843	5855	5866	5877	5888	5899
39	5911	5922	5933	5944	5955	5966	5977	5988	5999	6010
40	6021	6031	6042	6053	6064	6075	6085	6096	6107	6117
41	6128	6138	6149	6160	6170	6180	6191	6201	6212	6222
42	6232	6243	6253	6263	6274	6284	6294	6304	6314	6325
43	6335	6345	6355	6365	6375	6386	6395	6405	6415	6425
44	6435	6444	6454	6464	6474	6484	6493	6503	6513	6522
45	6532	6542	6551	6561	6571	6580	6590	6599	6609	6618
46	6628	6637	6646	6656	6665	6675	6684	6693	6702	6712
47	6721	6730	6739	6749	6758	6767	6776	6785	6794	6803
48	6812	6821	6830	6839	6848	6857	6866	6875	6884	6893
49	6902	6911	6920	6928	6937	6946	6955	6964	6972	6981
50	6990	6998	7007	7016	7024	7033	7042	7050	7059	7067
51	7076	7084	7093	7101	7110	7118	7126	7135	7143	7152
52	7160	7168	7177	7185	7193	7202	7210	7218	7226	7235
53	7243	7251	7259	7267	7275	7284	7292	7300	7308	7316
54	7324	7332	7340	7348	7356	7364	7372	7380	7388	7396
	0	1	2	3	4	5	6	7	8	9

TABLE OF FOUR-PLACE LOGARITHMS (continued)

No.	0	1	2	3	4	5	6	7	8	9
55	7404	7412	7419	7427	7435	7443	7451	7459	7466	7474
56	7482	7490	7497	7505	7513	7520	7528	7536	7543	7551
57	7559	7566	7574	7582	7589	7597	7604	7612	7619	7627
58	7634	7642	7649	7657	7664	7672	7679	7686	7694	7701
59	7709	7716	7723	7731	7738	7745	7752	7760	7767	7774
60	7782	7789	7796	7803	7810	7818	7825	7832	7839	7846
61	7853	7860	7868	7875	7882	7889	7896	7903	7910	7917
62	7924	7931	7938	7945	7952	7959	7966	7973	7980	7987
63	7993	8000	8007	8014	8021	8028	8035	8041	8048	8055
64	8062	8069	8075	8082	8089	8096	8102	8109	8116	8122
65	8129	8136	8142	8149	8156	8162	8169	8176	8182	8189
66	8195	8202	8209	8215	8222	8228	8235	8241	8248	8254
67	8261	8267	8274	8280	8287	8293	8299	8306	8312	8319
68	8325	8331	8338	8344	8351	8357	8363	8370	8376	8382
69	8388	8395	8401	8407	8414	8420	8426	8432	8439	8445
70	8451	8457	8463	8470	8476	8482	8488	8494	8500	8506
71	8513	8519	8525	8531	8537	8543	8549	8555	8561	8567
72	8573	8579	8585	8591	8597	8603	8609	8615	8621	8627
73	8633	8639	8645	8651	8657	8663	8669	8675	8681	8686
74	8692	8698	8704	8710	8716	8722	8727	8733	8739	8745
75	8751	8756	8762	8768	8774	8779	8785	8791	8797	8802
76	8808	8814	8820	8825	8831	8837	8842	8848	8854	8859
77	8865	8871	8876	8882	8887	8893	8899	8904	8910	8915
78	8921	8927	8932	8938	8943	8949	8954	8960	8965	8971
79	8976	8982	8987	8993	8998	9004	9009	9015	9020	9025
80	9031	9036	9042	9047	9053	9058	9063	9069	9074	9079
81	9085	9090	9096	9101	9106	9112	9117	9122	9128	9133
82	9138	9143	9149	9154	9159	9165	9170	9175	9180	9186
83	9191	9196	9201	9206	9212	9217	9222	9227	9232	9238
84	9243	9248	9253	9258	9263	9269	9274	9279	9284	9289
85	9294	9299	9304	9309	9315	9320	9325	9330	9335	9340
86	9345	9350	9355	9360	9365	9370	9375	9380	9385	9390
87	9395	9400	9405	9410	9415	9420	9425	9430	9435	9440
88	9445	9450	9455	9460	9465	9469	9474	9479	9484	9489
89	9494	9499	9504	9509	9513	9518	9523	9528	9533	9538
90	9542	9547	9552	9557	9562	9566	9571	9576	9581	9586
91	9590	9595	9600	9605	9609	9614	9619	9624	9628	9633
92	9638	9643	9647	9652	9657	9661	9666	9671	9675	9680
93	9685	9689	9694	9699	9703	9708	9713	9717	9722	9727
94	9731	9736	9741	9745	9750	9754	9759	9763	9768	9773
95	9777	9782	9786	9791	9795	9800	9805	9809	9814	9818
96	9823	9827	9832	9836	9841	9845	9850	9854	9859	9863
97	9868	9872	9877	9881	9886	9890	9894	9899	9903	9908
98	9912	9917	9921	9926	9930	9934	9939	9943	9948	9952
99	9956	9961	9965	9969	9974	9978	9983	9987	9991	9996
	0	1	2	3	4	5	6	7	8	9

Index